Antimicrobial Resistance
in Environmental Waters

Antimicrobial Resistance in Environmental Waters

Special Issue Editors

Karina Yew-Hoong Gin
Charmaine Ng

MDPI • Basel • Beijing • Wuhan • Barcelona • Belgrade

MDPI

Special Issue Editors

Karina Yew-Hoong Gin
National University of Singapore
Singapore

Charmaine Ng
National University of Singapore
Singapore

Editorial Office
MDPI
St. Alban-Anlage 66
4052 Basel, Switzerland

This is a reprint of articles from the Special Issue published online in the open access journal *Water* (ISSN 2073-4441) from 2017 to 2019 (available at: https://www.mdpi.com/si/water/Antimicrobial-Resistance-Environmental-Waters).

For citation purposes, cite each article independently as indicated on the article page online and as indicated below:

LastName, A.A.; LastName, B.B.; LastName, C.C. Article Title. *Journal Name* **Year**, *Article Number*, Page Range.

ISBN 978-3-03897-608-0 (Pbk)
ISBN 978-3-03897-609-7 (PDF)

Cover image courtesy of Charmaine Ng.

Contents

About the Special Issue Editors

Karina Yew-Hoong Gin is an Associate Professor with the Department of Civil and Environmental Engineering, National University of Singapore. She received her Bachelor degree in Civil Engineering from the University of Melbourne in 1988 and her M.Eng Degree from the National University of Singapore in 1991. She obtained her Doctor of Science (ScD) Degree jointly from the Massachusetts Institute of Technology and the Woods Hole Oceanographic Institution in 1996. Her research specialization is in the area of water quality, fate and transport of emerging contaminants, and ecosystem processes.

Charmaine Ng is a Research Fellow with the Department of Surgery, National University of Singapore. She graduated with a PhD in microbiology at the University of New South Wales, Australia in 2011. Her research interests include marine and freshwater microbial ecology, antimicrobial resistance, and the study of the human microbiome in relation to different disease states.

Preface to "Antimicrobial Resistance in Environmental Waters"

In recent years, the emergence of antimicrobial resistance has drawn heightened global concern because of its severe ramifications on the treatment of microbial infections. In particular, the issue of antibiotic resistance arises due to the overuse and misuse of antibiotics in both developed and developing countries. Bacteria develop antibiotic resistance in the presence of residual levels of antibiotics, and these antibiotic-resistant bacteria are in turn able to spread their resistance to other bacteria through mechanisms such as horizontal gene transfer, mediated by mobile genetic elements (e.g., plasmids, integrons) or co-selecting agents such as biocides and toxic metals. There is a worrying trend that pathogens are developing antibiotic resistance to a degree where last-resort antibiotics are no longer effective. This, in turn, has severe implications for public health and healthcare costs.

In an effort to better understand the rising levels of antimicrobial resistance, surveillance studies have been undertaken across countries in a common effort to explore the occurrence of antimicrobial resistance in both clinical and natural environments. Implementing such initiatives by assessing the types of antibiotics used, antibiotic-resistant bacteria (ARB) present, and associated antibiotic resistant genes (ARGs) in microbiomes enables a better understanding of the impact of antibiotics in the medicine, agriculture, and aquaculture industries. Aquatic environments harbor diverse freshwater bacterial communities which may be subjected to anthropogenic pressures, while domestic wastewaters receive direct loads of antibiotics and pathogenic bacteria from human excretion. The nature of these environments allows them to function as hotspots for resistance through the selection of ARB and the circulation of ARGs through the stimulation of horizontal gene transfer between members of the microbiome.

The aims of this Special Issue are to present current trends in antimicrobial/antibiotic resistance in diverse environmental waters, ranging from the detection and occurrence of antimicrobial factors (e.g., antimicrobials, antibiotics, ARB, ARGs) to their fate and transformations in different environments such as surface waters, groundwaters, biofilms, and water and wastewater treatment processes. This knowledge is needed to assist in the management and control of antimicrobial/antibiotic resistance and, ultimately, the protection of public health.

Karina Yew-Hoong Gin, Charmaine Ng
Special Issue Editors

water   MDPI

Editorial

Monitoring Antimicrobial Resistance Dissemination in Aquatic Systems

Charmaine Ng [1] and Karina Yew-Hoong Gin [2,3,*]

[1]   Department of Surgery, Yong Loo Lin School of Medicine, National University of Singapore,
      Singapore 117411, Singapore; ng.charmainemarie@gmail.com
[2]   Department of Civil and Environmental Engineering, National University of Singapore,
      Singapore 138602, Singapore
[3]   NUS Environmental Research Institute (NERI), Singapore 138602, Singapore
[*]   Correspondence: ceeginyh@nus.edu.sg; Tel.: +65-65168104

Received: 13 December 2018; Accepted: 28 December 2018; Published: 3 January 2019

check for updates

Abstract: This special issue on Antimicrobial Resistance in Environmental Waters features 11 articles on monitoring and surveillance of antimicrobial resistance (AMR) in natural aquatic systems (reservoirs, rivers), and effluent discharge from water treatment plants to assess the effectiveness of AMR removal and resulting loads in treated waters. The occurrence and distribution of antimicrobials, antibiotic resistant bacteria (ARB), antibiotic resistance genes (ARGs) and mobile genetic elements (MGEs) was determined by utilizing a variety of techniques including liquid chromatography—mass spectrometry in tandem (LC-MS/MS), traditional culturing, antibiotic susceptibility testing (AST), molecular and OMIC approaches. Some of the key elements of AMR studies presented in this special issue highlight the underlying drivers of AMR contamination in the environment and evaluation of the hazard imposed on aquatic organisms in receiving environments through ecological risk assessments. As described in this issue, screening antimicrobial peptide (AMP) libraries for biofilm disruption and antimicrobial candidates are promising avenues for the development of new treatment options to eradicate resistance. This editorial puts into perspective the current AMR problem in the environment and potential new methods which could be applied to surveillance and monitoring efforts.

Keywords: Antimicrobial Resistance; Environmental Waters; water treatment plants; water reuse; ecological risk assessment

1. Introduction

The release of antimicrobials, antibiotic-resistant bacteria (ARB) and antibiotic resistance genes (ARGs) originating from human and animal waste to the environment is a global problem which has serious ramifications on public health. In response to this growing health threat, the World Health Organization (WHO) launched a global action plan on Antimicrobial Resistance (AMR) in 2015 with 5 strategic objectives, one of which was to strengthen knowledge of the spread of AMR through surveillance and research [1]. As a guide, the WHO has drawn up a priority list of AMR pathogens based on the threat they pose on human infections, response to antibiotic treatment, transmissibility between humans and animals, and whether there are antibiotics in current research and development pipelines to treat infections caused by these pathogens. Those of highest priority are carbapenem-resistant *Acinetobacter baumannii*, carbapenem-resistant *Pseudomonas aeruginosa*, and carbapenem-resistant, extended spectrum beta-lactamase (ESBL)-producing *Enterobacteriaceae* [2]. The WHO's Global Antimicrobial Surveillance System (GLASS) report for 2018 revealed widespread occurrence of antibiotic resistance among half a million people with suspected bacterial infections across 22 countries [3]. An AMR risk assessment of the South East Asian Region by Chereau et al. [4] concluded a high likelihood of emergence and dissemination among humans. Low stewardship on antibiotic prescription in treatment of human infections and the absence of legal frameworks for antibiotic use in animal husbandry and aquaculture are the main drivers for the selection of ARB in South East Asia [5]. The direct release or insufficient treatment of wastewater effluents from healthcare, livestock, aquaculture, and agriculture sites into receiving environments also poses a significant risk. In 2013, China alone produced 92,700 tonnes of antibiotics, 48% of which were consumed by humans and 52% by animals [6]. It was reported that almost half of all antibiotics were released in rivers through wastewater effluents and the practice of manure and sludge land spreading [6].

In South East Asia, AMR risk ranking across humans, animals, and environmental compartments show that human to human transmission in community and hospital settings represent the highest risk of the emergence and selection of AMR vectors (ARB, ARGs), followed by food- and waterborne transmission to humans through ingestion of contaminated sources [4]. Although transmission via contact with contaminated environments (through soil, water, and air), and livestock/animals is regarded as low risk in comparison to human to human transmission, it is still considered a valid route of exposure, particularly in countries that are water-scarce and reliant on water reuse. Hence, it is crucial to address the impact of antimicrobials, ARB and ARGs on AMR prevalence in receiving environments, specifically in countries where data availability is scant and guidelines for AMR stewardship frameworks have yet to be established. It is only through identifying and tracking sources and sinks of AMR in the environment, where intervention strategies can be devised to prevent and control the spread of the problem.

2. Measuring Vectors of AMR

2.1. Antimicrobials

Currently, there are various practices and methods of monitoring AMR dissemination and the fate of ARB and ARGs in aquatic environmental systems. Highly sensitive analytical protocols have been developed to detect antimicrobials using liquid chromatography-tandem mass spectrometry in environmental water samples and effluents from wastewater treatment plants (WWTPs) [7,8], while culture-based techniques are used to enumerate ARB.

2.2. ARB

Most environmental and wastewater treatment AMR surveys use methods applied in clinical settings, where media is supplemented with antibiotics at concentrations above the recommended minimum inhibitory concentration (MIC) breakpoints implemented by the Clinical and Laboratory Standards Institute [9] or the European Committee on antimicrobial Susceptibility Testing [10–17]. For non-clinically relevant bacteria (such as environmental bacteria), epidemiological cut-off values (ECOFF) are an alternative method used to gauge non-wild-type bacteria that display reduced susceptibility to certain antimicrobials or biocides [18]. Further testing of multidrug resistance (MDR) of ARB isolates are performed using broth dilution assays, Kirby-Bauer disk diffusion test on Muller-Hinton agar, or high-throughput platforms such as the VITEK system by bioMerieux (France).

2.3. ARGs, MGEs

To detect the prevalence of ARGs and vectors such as mobile genetic elements (MGE) that may facilitate horizontal gene transfer of ARGs, traditional quantitative polymerase chain reaction (qPCR), and the more recent high-throughput qPCR (HT-qPCR) platform with capabilities of detection of ~200 different ARGs and mobile genetic elements (MGE), has been used to compare relative concentrations of AMR contamination across a variety of aquatic environments including water treatment plants [19–24]. OMIC approaches such as metagenomics are able to provide a holistic picture of the diversity of ARGs, MGE and vectors (e.g., integrons, plasmids) that assist horizontal gene transfer, and the overall microbial community structure (bacteria, viruses) in environmental systems and wastewaters [25–29]. ResCap, a targeted capture platform (TCP) designed to analyze ~78,000 ARGs, metal resistance and plasmid markers is a targeted metagenomics approach for qualitative and quantitative resistome analysis [30]. Other OMIC approaches, such as metatranscriptomics, enable the identification of active microbial members within a community and, in the context of AMR, enables the measurement of transcription activity of ARB through ARG expression [31].

3. Key Outcomes of this Special Issue

This special issue comprises of 11 research articles that fall within the scope of AMR. Broadly, topics extend from AMR monitoring and surveillance in environmental resources and effluents from water/wastewater treatment plants [32–39], antimicrobial ecological risk assessments of two river reservoir systems in China that are sources of drinking water supply [40,41], and exploring novel strategies of using engineered antimicrobial peptides (AMP) to target specific bacteria to disrupt biofilms that are major causes of chronic and persistent infections [42].

AMR Monitoring and Surveillance

Knowledge of removal efficiencies of technology employed at WWTPs is essential in AMR surveillance, whether effluents are intended for discharge into the environment or reuse for irrigation purposes.

To determine the presence of antibiotic resistant *E. coli* in a conventional WWTP in Georgia, Aslan et al. [36] isolated *E. coli* from post-secondary, post-UV and post-chlorination effluent and performed antibiotic susceptibility tests on the isolates. They reported that ~5.2 log removal of *E. coli* and an additional 1.1 log reduction post chlorination was obtained. However, the MICs of *E. coli* isolated in the finished water were higher than those at the other treatment stages. The selection of more resistant organisms in the finished water underscores the urgent need to evaluate the health risks of using reclaimed water for downstream irrigation. In a qPCR assessment of ARGs of a full-scale tertiary water reclamation plant, Quach-Cu et al. [39] showed that tertiary-stage WWTPs with disinfection had superior removal of ARGs (*sul1*, *bla*$_{SHV/TEM}$), ~3–4 logs compared to reliance of secondary treatment alone where the removal was only ~1–3 logs. To assess the impact of treated effluent on receiving environments, Lambirth et al. [35] measured the removal efficiency of ten antibiotics and assayed resistomes upstream, downstream and within various treatment steps of two urban WWTPs (secondary and disinfection treatments). The authors found elevated concentration of all 10 antibiotics surveyed in downstream receiving waters compared to waters upstream of the WWTP. The relative abundance of ARG signatures encoding for resistance to carbapenems and ESBL antibiotics were much lower than those detected upstream and sampling points within the WWTP, which debunks the notion that the wastewater treatment process selects for ARG resistance. Instead, the authors hypothesized that antibiotics discharged from treated wastewater effluent into the downstream environment may have an effect on natural microbial communities.

Jumat et al. [37] conducted an ESBL study of diversity and transcriptional activity of bacteria in a WWTP in Saudi Arabia using metagenomics, metatranscriptomics and real-time qPCR. They found an increase in the relative abundance of *Acinetobacter junii* in MBR- and chlorinated treated effluent. Survival and predominance of *A. junii* was explained by metatranscriptomics data that showed an upregulation of gene associated with active cell repair, resistance, virulence (efflux transporters involved in metal and antibiotic resistance) and cell signaling. These adaptive cellular mechanisms enable *A. junii* to withstand depletion of nutrients and counter the effects of chlorination. However, the authors indicate that the low concentrations of viable *A. junii* isolated from MBR effluents may not present that huge a risk. The varying results from WWTP studies covered in this issue makes it challenging to establish whether WWTPs are indeed hotspots of AMR dissemination. Rather, it is likely that differences in global and plant operating process contribute to variation in antibiotic resistance elements detection.

In countries with high agricultural productivity, water reuse is commonly practiced to meet the high water demands of the industry [43]. On a global scale, there are no clear guidelines implemented on assessment of water quality for reuse purposes, although a few countries have drawn up recommended microbiological parameters to monitor the quality of recycled water [44]. A recent review by Hong et al. [44] highlights the urgency of understanding the risks of microbiological and ARG contamination linked to water reuse. In the Philippines, surface waters contaminated with fecal coliform are frequently used for irrigating urban farms in densely populated cities [45]. Vital et al. [33] evaluated the antibiotic resistance profiles of 212 *E. coli* strains isolated from irrigation water, soil, and vegetables from six urban agricultural farms. Of the total isolates, 36.5% were resistant to more than three antibiotics tested, with the most multidrug-resistant (MDR) isolates being detected in irrigation water, followed by soil and vegetables. Of the MDR *E. coli* isolated from irrigation water, 7 of them were ESBL producers that carried either bla_{TEM} or bla_{CTX-M} genes, which raises public health concerns in primary production environments such as agricultural soils and fresh produce grown in these areas.

The use of antimicrobials in food animals is widespread, and runoffs originating from animal waste may carry unmetabolized antibiotics or ARB and ARGs that have direct impact on surrounding water bodies [46]. Tsai et al. [32] linked significantly higher concentrations of *A. baumannii* along the Puzi River in China to sampling sites of livestock wastewater channels and tributaries adjacent to livestock farms. Further testing of 20 *A. baumannii* isolates against 7 antibiotics (ciprofloxacin, cefepime, gentamicin, imipenem, ampicillin-sulbactam, sulfamethoxazole/trimethoprim, tetracycline) by the Kirby-Bauer disk diffusion test showed that only 10% had resistance to sulfamethoxazole/trimethoprim and 5% had resistance to tetracycline. Although the *A. baumannii* isolated by Tsai et al. [32] did not display MDR patterns which are regarded a serious AMR threat by WHO, their epidemiological potential warrants further studies on prevalence and AMR developmental trends in the environment.

In China, the Yellow River in the North serves as an important source of drinking water and is flanked by cities along its banks and watersheds. Previous studies have reported high turbidity and concentrations of antimicrobials in the Yellow River Catchment [47]. This prompted Lu et al. [38] to investigate the distribution and abundance of ARGs in Sand Settling Reservoirs (SSR) and Drinking Water Treatment Plants (DWTP) along the Yellow River. By targeting 17 ARGs as a proxy for AMR removal through the treatment process, the total concentrations of ARGs decreased from 10^4 copies/mL in influent river waters to 10^3 copies/mL in SSR effluent to 10^2 copies/mL in finished water. The 2 MGE targets decreased by at least an order of magnitude, from 10^6 copies/mL in influent river waters to 10^5 copies/mL in finished waters.

Horizontal gene transfer (HGT) of ARGs in the mammalian gut through the ingestion of contaminated food or water generally poses a low risk due to harsh conditions in the gut. However, this may present a greater risk for propagation in the environment through transformation and transduction [44]. In another study by Xu et al. [34], the authors studied the relationship of 16 antibiotics with environmental water quality parameters and the impact of antibiotic concentrations on the microbial community structure along Qingcaosha Reservoir, the largest estuary reservoir in China. This reservoir has a similar function to the Yellow River, in that it compensates for drinking water shortages in Shanghai. From the study, the authors concluded that upstream runoffs and anthropogenic activity along the river contributed to the concentrations of antibiotics measured within the reservoir, and that tylosin, penicillin G and erythromycin-H_2O showed significant correlations with variations in bacterial community structure. Further to this study, Jiang et al. [40] studied seasonal variations of antibiotics in surface waters of Qingcaosha Reservoir. By using risk quotients (RQs) based on the European technical guidance documentation (TGD) on risk assessment, they showed that out of the 17 antibiotics monitored, four antibiotics (doxycycline, penicillinV, norfloxacin, ciprofloxacin) posed a high risk to relevant sensitive aquatic organisms, as well as imposed selective stress on microbial communities. In another study, of a subtropical river-reservoir system located in the Headwater Region of the Dongjiang River which supplies drinking water to three major cities in China, Chen et al. [41] conducted an ecological risk assessment which showed that concentrations of ciprofloxacin and norfloxacin posed a moderate risk, while tetracycline posed a higher risk to the aquatic ecosystem.

Finally, through screening a local antimicrobial peptide (AMP) library, Chin et al. [42] identified LG21, an AMP that specifically binds to exopolysaccharide PsI of *P. aeruginosa* that has a functional role of biofilm formation, which provides a protective environment for tolerance and resistance towards antibiotic treatment. This strategy of exploring AMP to target specific biofilm matrix components to disrupt formation and development of biofilms is a promising line of treatment to eradicate antibiotic resistant biofilms in both environmental and clinical settings.

Acknowledgments: The guest editors are grateful to the authors that contributed their work to this special issue. We would also like to thank the reviewers and journal editors who dedicated their time and expertise towards evaluating the articles.

Conflicts of Interest: The authors declare no conflict of interest.

References

1. World Health Organization. Global Action Plan on Antimicrobial Resistance. Available online: https://www.who.int/antimicrobial-resistance/global-action-plan/en/ (accessed on 29 December 2018).
2. World Health Organization. WHO Publishes List of Bacteria for Which New Antibiotics are Urgently Needed. Available online: http://www.who.int/mediacentre/news/releases/2017/bacteria-antibiotics-needed/en/ (accessed on 29 December 2018).
3. World Health Organization. Global Action Plan on Antimicrobial Resistance (GLASS) Report. Available online: https://www.who.int/glass/resources/publications/early-implementation-report/en/ (accessed on 29 December 2018).
4. Chereau, F.; Opatowski, L.; Tourdjman, M.; Vong, S. Risk assessment for antibiotic resistance in South East Asia. *BMJ* **2017**, *358*. [CrossRef] [PubMed]
5. Holloway, K.A.; Kotwani, A.; Batmanabane, G.; Puri, M.; Tisocki, K. Antibiotic use in South East Asia and policies to promote appropriate use: Reports from county situational analyses. *BMJ* **2017**, *358*. [CrossRef] [PubMed]
6. Zhang, Q.Q.; Ying, G.G.; Pan, C.G.; Liu, Y.S.; Zhao, J.L. Comprehensive evaluation of antibiotics emission and fate in the river basins of China: Source analysis, multimedia modelling, and linkage to bacterial resistance. *Environ. Sci. Technol.* **2015**, *49*, 6772–6782. [CrossRef] [PubMed]
7. Tran, N.H.; Chen, H.; Do, T.V.; Van Do, T.; Reinhard, M.; Ngo, H.H.; He, Y.; Gin, K.Y.H. Simultaneous analysis of multiple classes of antimicrobials in environmental water samples using SPE coupled with UHPLC-ESI-MS/MS and isotope dilution. *Talanta* **2016**, *159*, 163–173. [CrossRef] [PubMed]
8. Tran, N.H.; Chen, H.; Reinhard, M.; Mao, F.; Gin, K.Y.H. Occurrence and removal of multiple classes of antibiotic snad antimicrobial agents in biological wastewater treatment processes. *Water Rese.* **2016**, *104*, 461–472. [CrossRef] [PubMed]
9. CLSI. Clinical and Laboratory Standards Institute. Available online: https://clsi.org/meetings/microbiology/ (accessed on 29 December 2018).
10. EUCAST. European Committee on Antimicrobial Susceptibility Testing. Available online: http://www.eucast.org/clinical_breakpoints/ (accessed on 29 December 2018).
11. Le, T.H.; Ng, C.; Tran, N.H.; Chen, H.; Gin, K.Y.H. Removal of antibiotic residues, antibiotic resistant bacteria and antibiotic resistance genes in municipal wastewater by membrane bioreactor systems. *Water Res.* **2018**, *145*, 498–508. [CrossRef] [PubMed]
12. Ng, C.; Chen, H.; Goh, S.G.; Haller, L.; Charles, F.R.; Wu, Z.; Trottet, A.; Gin, K.Y.H. Microbial water quality and the detection of multidrug resistant *E. coli* and antibiotic resistance genes in aquaculture sites of Singapore. *Mar. Pollut. Bull.* **2018**, *135*, 475–480. [CrossRef] [PubMed]
13. Ng, C.; Goh, S.G.; Saeidi, N.; Gerhard, W.A.; Gunsch, C.K.; Gin, K.Y.H. Occurrence of *Vibrio* species, beta-lactam resistant *Vibrio* species, and indicator bacteria in ballast and port waters of a tropical harbour. *Sci. Total Environ.* **2018**, *610*, 651–656. [CrossRef] [PubMed]

14. Haller, L.; Chen, H.; Ng, C.; Le, T.H.; Koh, T.H.; Barkham, T.; Sobsey, M.; Gin, K.Y.H. Occurrence and characteristics of extended-spectrum ß-lactamase-and carbapenemase-producing bacteria from hospital effluents in Singapore. *Sci. Total Environ.* **2018**, *615*, 1119–1125. [CrossRef] [PubMed]

15. Low, A.; Ng, C.; He, J. Identification of antibiotic resistant bacteria community and a GeoChip based study of resistome in urban watersheds. *Water Res.* **2016**, *106*, 330–338. [CrossRef] [PubMed]

16. Le, T.H.; Ng, C.; Chen, H.; Yi, X.Z.; Koh, T.H.; Barkham, T.M.S.; Zhou, Z.; Gin, K.Y.H. Occurrences and Characterization of Antibiotic Resistant Bacteria and Genetic Determinants of Hospital Wastewaters in a Tropical Country. *Antimicrob. Agents Chemother.* **2016**, *21*, 7449–7456. [CrossRef] [PubMed]

17. Huang, Y.; Zhang, L.; Tiu, L.; Wang, H.H. Characterization of antibiotic resistance in commensal bacteria from an aquaculture ecosystem. *Front. Microbiol.* **2015**. [CrossRef] [PubMed]

18. Morrissey, I.; Oggioni, M.R.; Knight, D.; Curiao, T.; Coque, T.; Kalkanci, A.; Martinez, J.L. Evaluation of epidemiological cut-off values indicates that biocide resistant subpopulations are uncommon in nature isolates of clinically-relevant microoriganisms. *PLoS ONE* **2014**. [CrossRef] [PubMed]

19. Liu, L.; Su, J.Q.; Guo, Y.; Wilkinson, D.M.; Liu, Z.; Zhu, Y.G.; Yang, J. Large-scale biogeographical patterns of bacterial antibiotic resistome in the waterbodies of China. *Environ. Int.* **2018**, *117*, 292–299. [CrossRef] [PubMed]

20. Zhu, Y.G.; Zhao, Y.; Li, B.; Huang, C.L.; Zhang, S.Y.; Yu, S.; Chen, Y.S.; Zhang, T.; Gillings, M.R.; Su, J.Q. Continental-scale pollution of estuaries with antibiotic resistance genes. *Nat. Microbiol.* **2017**, *2*, 16270. [CrossRef]

21. Muziasari, W.I.; Pitkanen, L.K.; Sorum, H.; Stedtfeld, R.D.; Tiedje, J.M.; Virta, M. The Resistome of Farmed Fish Feces Contributes to the Enrichment of Antibiotic Resistance Genes in Sediments below Baltic Sea Fish Farms. *Front. Microbiol.* **2017**. [CrossRef]

22. Xu, L.; Ouyang, W.; Qian, Y.; Su, C.; Su, J.; Chen, H. High-throughput profiling of antibiotic resistance genes in drinking water treatment plants and distribution systems. *Environ. Pollut.* **2016**, *213*, 119–126. [CrossRef]

23. An, X.L.; Su, J.Q.; Li, B.; Ouyang, W.Y.; Zhao, Y.; Chen, Q.L.; Cui, L.; Chen, H.; Gillings, M.R.; Zhang, T.; et al. Tracking antibiotic reistome during wastewater treatment using high throughput quantitative PCR. *Environ. Int.* **2018**, *117*, 146–153. [CrossRef]

24. Karkman, A.; Johnson, T.A.; Lyra, C.; Stedtfeld, R.D.; Tamminen, M.; Tiedje, J.M.; Virta, M. High-throughput quantification of antibiotic resistance genes from an urban wastewater treatment plant. *FEMS Microbiol. Ecol.* **2016**, *92*. [CrossRef]

25. Bondarczuk, K.; Piotrowska-Segat, Z. Microbial diversity and antibiotic resistance in a final effluent-receiving lake. *Sci. Total Environ.* **2018**, *650*, 2951–2961. [CrossRef]

26. Chu, B.T.T.; Petrovich, M.L.; Chaudhary, A.; Wright, D.; Murphy, B.; Wells, G.; Poretsky, R. Metagenomics Reveals the Impact of Wastewater Treatment Plants on the Dispersal of Microorganisms and Genes in Aquatic Sediments. *Appl. Environ. Microbiol.* **2018**. [CrossRef] [PubMed]

27. Gupta, S.K.; Shin, H.; Han, D.; Hur, H.G.; Unno, T. Metagenomic analysis reveals the prevalence and persistence of antibiotic- and heavy metal-resistance genes in wastewater treatment plant. *J. Microbiol.* **2018**, *56*, 408–415. [CrossRef] [PubMed]

28. Ng, C.; Tay, M.; Tan, B.; Le, T.H.; Haller, L.; Chen, H.; Koh, T.H.; Barkham, T.; Thompson, J.R.; Gin, K.Y.H. Characterization of Metagenomes in Urban Aquatic Compartments Reveals High Prevalence of Clinically Relevant Antibiotic Resistance Genes in Wastewaters. *Front. Microbiol.* **2017**. [CrossRef] [PubMed]

29. Guo, J.; Li, J.; Chen, H.; Bond, P.L.; Yuan, Z. Metagenomic analysis reveals wastewater treatment plants as hotspots of antibiotic resistance genes and mobile genetic elements. *Water Res.* **2017**, *123*, 468–478. [CrossRef] [PubMed]

30. Lanza, V.F.; Baquero, F.; Martinez, J.L.; Ramos-Ruiz, R.; Gonzalez-Zorn, B.; Andremont, A.; Sanchez-Valenzuela, A.; Ehrlich, S.D.; Kennedy, S.; Ruppe, E.; et al. In-depth resistome analysis by targeted metagenomics. *Microbiome* **2018**. [CrossRef] [PubMed]

31. Rowe, W.P.M.; Baker-Austin, C.; Verner-Jeffreys, D.W.; Ryan, J.J.; Micallef, C.; Maskell, D.J.; Pearce, G.P. Overexpression of antibiotic resistance genes in hospital effluents over time. *J. Antimicrob. Chemother.* **2017**, *72*, 1617–1623. [CrossRef]

32. Tsai, H.-C.; Chou, M.-Y.; Shih, Y.-J.; Huang, T.-Y.; Yang, P.-Y.; Chiu, Y.-C.; Chen, J.-S.; Hsu, B.-M. Distribution and Genotyping of Aquatic Acinetobacter baumannii Strains Isolated from the Puzi River and Its Tributaries Near Areas of Livestock Farming. *Water* **2018**, *10*, 1374. [CrossRef]

33. Vital, P.G.; Zara, E.S.; Paraoan, C.E.M.; Dimasupil, M.A.Z.; Abello, J.J.M.; Santos, I.T.G.; Rivera, W.L. Antibiotic Resistance and Extended-Spectrum Beta-Lactamase Production of Escherichia coli Isolated from Irrigation Waters in Selected Urban Farms in Metro Manila, Philippines. *Water* **2018**, *10*, 548. [CrossRef]

34. Xu, Z.; Jiang, Y.; Te, S.H.; He, Y.; Gin, K.Y.H. The Effects of Antibiotics on Microbial Community Composition in an Estuary Reservoir during Spring and Summer Sessions. *Water* **2018**, *10*, 154. [CrossRef]

35. Lambirth, K.; Tsilimigrae, M.; Lulla, A.; Johnson, J.; Al-Shaer Am Wynblatt, O.; Sypolt, S.; Brouwer, C.; Clinton, S.; Keen, O.; Redmond, M.; et al. Microbial Community Composition and Antibiotic Resistance Genes within a North Carolina Urban Water System. *Water* **2018**, *10*, 1539. [CrossRef]

36. Aslan, A.; Cole, Z.; Bhattacharya, A.; Oyibo, O. Presence of Antibiotic-Resistant *Escherichia coli* in Wastewater Treatment Plant Effluents Utilized as Water Reuse for Irrigation. *Water* **2018**, *10*, 805. [CrossRef]

37. Jumat, M.R.; Haroon, M.F.; Al-Jassim, N.; Cheng, H.; Hong, P.-Y. An Increase of Abundance and Transcriptional Activity for Acinetobacter junii Post Wastewater Treatment. *Water* **2018**, *10*, 436. [CrossRef]

38. Lu, J.; Tian, Z.; Yu, J.; Yang, M.; Zhang, Y. Distribution and Abundance of Antibiotic Resistance Genes in Sand Settling Reservoirs and Drinking Water Treatment Plants across the Yellow River, China. *Water* **2018**, *10*, 246. [CrossRef]

39. Quach-Cu, J.; Herrera-Lynch, B.; Marciniak, C.; Adams, S.; Simmerman, A.; Reinke, R.A. The Effect of Primary, Secondary, and Tertiary Wastewater Treatment Processes on Antibiotic Resistance Gene (ARG) Concentrations in Solid and Dissolved Wastewater Fractions. *Water* **2018**, *10*, 37. [CrossRef]

40. Jiang, Y.; Xu, C.; Wu, X.; Chen, Y.; Han, W.; Gin, K.Y.H.; He, Y. Occurrence, Seasonal Variation and Risk Assessment of Antibiotics in Qingcaosha Reservoir. *Water* **2018**, *10*, 115. [CrossRef]

41. Chen, Y.; Chen, H.; Zhang, L.; Jiang, Y.; Gin, K.Y.H.; He, Y. Occurrence, Distribution, and Risk Assessment of Antibiotics in a Subtropical River-Reservoir System. *Water* **2018**, *10*, 104. [CrossRef]

42. Chin, J.S.F.; Sinha, S.; Nalaparaju, A.; Yam, J.K.H.; Qin, Z.; Ma, L.; Liang, Z.-X.; Lu, L.; Bhattacharjya, S.; Yang, L. Pseudomonas aeruginosa PsI Exopolysaccharide Interacts with the Antimicrobial Peptide LG21. *Water* **2018**, *9*, 681. [CrossRef]

43. Hong, P.-Y.; Julian, T.R.; Pype, M.-L.; Jiang, S.C.; Nelson, K.L.; Graham, D.; Pruden, A.; Manaia, C.M. Resing Treated Wastewater: Consideration of the Safety Aspects Associated with Antibiotic-Resistant Bacteria and Antibiotic Resistance Genes. *Water* **2018**, *10*, 244. [CrossRef]

44. Jaramillo, M.F.; Restrepo, I. Wastewater Reuse in Agriculture: A Review about Its Limitations and Benefits. *Sustainability* **2017**, *9*, 1734. [CrossRef]

45. Garcia, B.C.B.; Dimasupil, M.A.Z.; Vital, P.G.; Widmer, K.W.; Rivera, W.L. Fecal contamination in irrigation water and microbial quality of vegetable primary production in urban farms of Metro Manila, Philippines. *J. Environ. Sci. Health* **2015**, *50*, 734–743. [CrossRef]

46. Landers, T.F.; Cohen, B.; Wittum, T.E.; Larson, E.L. A Review of Antibiotic Use in Food Animals: Perspective, Policy, and Potential. *Public Health Rep.* **2012**, *127*, 4–22. [CrossRef] [PubMed]

47. Xu, W.; Zhang, G.; Zou, S.; Ling, Z.; Wang, G.; Yan, W. A preliminary investigation on the occurrence and distribution of antibiotics in the Yellow River and its tributaries, China. *Water Environ. Res.* **2009**, *81*, 248–254. [CrossRef] [PubMed]

Article

Microbial Community Composition and Antibiotic Resistance Genes within a North Carolina Urban Water System

Kevin Lambirth [1,2], Matthew Tsilimigras [1], Anju Lulla [1], James Johnson [1], Abrar Al-Shaer [3], Orion Wynblatt [2], Shannon Sypolt [4], Cory Brouwer [1,2], Sandra Clinton [5], Olya Keen [6], Molly Redmond [7], Anthony Fodor [1] and Cynthia Gibas [1,*]

[1] Department of Bioinformatics and Genomics, The University of North Carolina at Charlotte, 9201 University City Blvd, Charlotte, NC 28223, USA; kclambirth@uncc.edu (K.L.); mbrown67@uncc.edu (M.T.); alulla@uncc.edu (A.L.); jjohn443@uncc.edu (J.J.); cbrouwer@uncc.edu (C.B.); afodor@uncc.edu (A.F.)

[2] Bioinformatics Services Division, The University of North Carolina at Charlotte, 150 N Research Campus Dr, Kannapolis, NC 28081, USA; orion.s.wynblatt@gmail.com

[3] Department of Nutrition, The University of North Carolina at Chapel Hill, 135 Dauer Drive, Chapel Hill, NC 27599, USA; abrar_alshaer@unc.edu

[4] Charlotte Water, Charlotte-Mecklenburg Utility Department, 5100 Brookshire Blvd, Charlotte, NC 28216, USA; ssypolt@ci.charlotte.nc.us

[5] Department of Geography and Earth Sciences, The University of North Carolina at Charlotte, 9201 University City Blvd, Charlotte, NC 28223, USA; sclinto1@uncc.edu

[6] Department of Civil and Environmental Engineering, The University of North Carolina at Charlotte, 9201 University City Blvd, Charlotte, NC 28223, USA; okeen@uncc.edu

[7] Department of Biological Sciences, The University of North Carolina at Charlotte, 9201 University City Blvd, Charlotte, NC 28223, USA; mcredmond@uncc.edu

* Correspondence: cgibas@uncc.edu; Tel.: +1-704-687-8378

Received: 13 September 2018; Accepted: 24 October 2018; Published: 29 October 2018

check for updates

Abstract: Wastewater treatment plants (WWTPs) are thought to be potential incubators of antibiotic resistance. Persistence of commonly used antibiotics in wastewater may increase the potential for selection of resistance genes transferred between bacterial populations, some of which might pose a threat to human health. In this study, we measured the concentrations of ten antibiotics in wastewater plant influents and effluents, and in surface waters up- and downstream from two Charlotte area treatment facilities. We performed Illumina shotgun sequencing to assay the microbial community and resistome compositions at each site across four time points from late winter to mid-summer of 2016. Antibiotics are present throughout wastewater treatment, and elevated concentrations of multiple antibiotics are maintained in moving stream water downstream of effluent release. While some human gut and activated sludge associated taxa are detectable downstream, these seem to attenuate with distance while the core microbial community of the stream remains fairly consistent. We observe the slight suppression of functional pathways in the downstream microbial communities, including amino acid, carbohydrate, and nucleic acid metabolism, as well as nucleotide and amino acid scavenging. Nearly all antibiotic resistance genes (ARGs) and potentially pathogenic taxa are removed in the treatment process, though a few ARG markers are elevated downstream of effluent release. Taken together, these results represent baseline measurements that future studies can utilize to help to determine which factors control the movement of antibiotics and resistance genes through aquatic urban ecosystems before, during, and after wastewater treatment.

Keywords: metagenomics; antibiotic resistance; wastewater; environmental ecology

1. Introduction

Urbanization has the potential to affect surface water quality and alter microbial community composition [1–4]. One mechanism by which human activity directly affects surface waters is through the wastewater treatment process, in which human waste is collected, treated, and the residual water eventually released back into surface waterways [3,5,6]. Pharmaceuticals and antimicrobial compounds that are not fully metabolized or that are disposed of improperly make treated wastewater a significant source of pharmaceuticals in surface waters [7]. Significant effects of pharmaceuticals, including metformin, estrogens, and illegal drugs on native flora and fauna have recently been reported as well [8–11].

With the growth of antibiotic resistance as a public health threat, there has been increased interest in the prevalence of antibiotics and associated resistance elements released to the environment, as well as their removal from wastewater systems [12]. Both antivirals [13] and antibiotics [14] have been found in treated effluent waters in recent studies, exposing the native microbial flora to sub-lethal levels of antibiotics, and contributing to selective pressures potentially resulting in the emergence of resistant strains [15,16]. Agricultural runoff from antibiotic-administered livestock is also of concern [17,18], as is the use of reclaimed water in public locations, such as water, amusement, and grassy parks [19]. It is hypothesized that any antibiotic resistance genes (ARGs) persisting in reclaimed water may become long-term environmental contaminants, potentially creating hot spots for breeding resistant microbe populations [20]. Further exploration of reclaimed water systems and efforts to reduce the presence of antibiotics and the associated resistance factors are described in a recent review [21].

Results from prior studies of treated effluent impacts on surface waters are highly varied. In areas where water treatment infrastructure is sparse, human microbial resistomes have been shown to pass into the environment through mobile genetic elements [22]. Even in nations with state-of-the-art water treatment systems, the outcome of water treatment is highly dependent on the specific technology used and on operating parameters. For example, a 2011 study showed that treated wastewater was a significant source of antibiotic resistance markers in Minnesota's Duluth Harbor [23]. In contrast, a recent study from Denmark suggests that the dissemination of the resistome is fairly limited [24]. A concurrent study demonstrated that aerobic treatment procedures may significantly reduce antibiotic resistance elements that are present in wastewater processing locations [25].

The potential for human impact on surface water composition and microbiomes in Charlotte, NC is significant as a city of approximately 800,000 people within a larger metropolitan area of 2.3 million. Charlotte has no large body of water within the city itself, but the surrounding Mecklenburg County does have a network of over 3000 miles of small creeks and streams, many of which are integral features of popular public greenway and park facilities. The regional water utility, Charlotte Water, operates five major wastewater treatment and two minor package plant facilities that release treated wastewater into local creeks. The purpose of this study is to establish a baseline understanding of the impact of treated wastewater release on the urban stream microbiome, and to quantify the impact of released effluent on antibiotic concentrations and resistance elements that were observed in the streams.

2. Materials and Methods

2.1. Selection of Sampling Sites

The design of this study encompasses two systems: the urban stream upstream and downstream of treated wastewater release points, and the wastewater treatment plant and its various processing stages. Our primary interest was in the impact of treated wastewater release on the stream environment, which has the most obvious consequences for the population of Charlotte that interacts with these streams in public areas, but we also sought insight into the impact of sewage input sources and various stages in the water treatment process on the final released product.

The sites selected for this study, Mallard Creek (MC) and Sugar Creek (SC) reclamation facilities, are activated sludge plants that release treated wastewater into Mallard Creek and Little Sugar Creek.

These streams are in two different watersheds; Mallard Creek feeds into the Yadkin Pee-Dee watershed, while Little Sugar Creek feeds into Sugar Creek, and ultimately into the Catawba River basin. Sampling sites chosen in the urban streams include two points upstream of treated wastewater release and two points downstream, at distances of approximately 3300 and 2700 m from effluent release for Little Sugar and Mallard Creek, respectively. Two remote locations were also selected in the Appalachian Mountains (MNTA/MNTB) and Uwharrie forest preserve (UWHA/UWHB) to give insight into the microbial community and antibiotic load at sites more removed from urban activities. While these sites differ in their elevation and therefore in their underlying geology, they were the best option for a remote background due to the lack of Piedmont stream sites that are relatively untouched by human activity.

Several sites were sampled inside each WWTP, representing processing stages. Five sites were selected inside the Mallard Creek plant, including the raw influent (INF), the primary clarifier influent (PCI), the primary clarifier effluent (PCE), the aeration tank effluent (ATE), and the final clarifier effluent (FCE). Samples from both hospital-adjacent (HOSP) and residential (RES) sewage trunklines "upstream" of the plant were collected in each case, to allow the investigation of potential microbial community differences between residential and hospital waste. Corresponding locations were used at the Sugar Creek plant, with sites, including INF, PCE, ATE, and FCE. Location of sampling points at the Sugar Creek plant did not allow for collection of a PCI sample, but it was possible to directly sample ultraviolet disinfected effluent (UV), which had not been accessible at the Mallard Creek plant.

The locations of the plants and sampling sites are illustrated in Figure 1 with full details of sample collection and handling provided in Supplementary Materials.

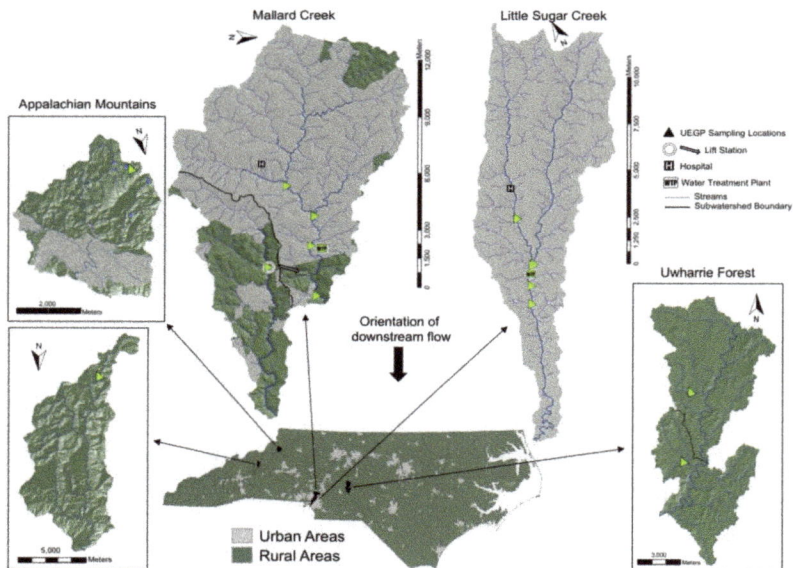

Figure 1. Geographical illustration of sampling sites and locations. Mountain samples were collected in Caldwell (Mountain A) and Yancey (Mountain B) counties, while both Uwharrie forest samples were collected in Stanly county. Mallard and Sugar Creek facilities and trunklines were all located in Mecklenburg County. Urban locations are marked in grey, while more rural locations are shaded green. Hospital and wastewater treatment plant locations are indicated in relation to sample collection sites, which are denoted with green triangles.

2.2. Treatment Methods and Conditions in Charlotte Water Facilities

Both plants studied use similar activated sludge treatment processes, beginning with physical bar screens, grit removal stations, and primary clarifiers, followed by activated sludge processing consisting of anoxic and aerobic zones and secondary clarifier tanks, and ending with ultraviolet disinfection at an intensity of 16,000 μW-s/cm^2. The Sugar Creek facility employs an additional filtration step with anthracite bed filters, followed by wet and dry odor scrubber units, prior to creek release. Anaerobic digestion is used for solids treatment with digestate that was returned to the activated sludge tank, and biosolids reclamation is conducted at both facilities. An overview of this process is shown in Figure 2. Final effluent is monitored for carbonaceous biochemical oxygen demand (CBOD), total suspended solids (TSS), ammonia, chronic toxicity (*Ceriodaphnia dubia*), fecal coliform, flow rates, dissolved oxygen, pH, and phosphorus. Sugar Creek also monitors nickel and copper concentrations. The Mallard Creek facility is rated for 12 million gallons daily (MGD) being released directly into Mallard Creek, while the Sugar Creek facility is rated for 20 MGD and discharges into Little Sugar Creek.

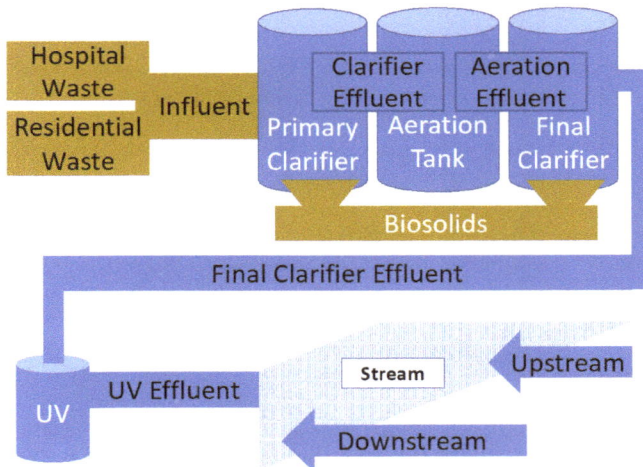

Figure 2. Summary of the wastewater treatment process in Charlotte facilities. Untreated wastewater from trunkline locations, including hospital and residential wastes, collects into the main influent trunkline before entry into the treatment plant. The primary clarification tank is the first stage in treatment, removing physical debris and solids. The aeration tank assists in removing oils and hydrocarbons before a final physical precipitation in the final clarifier. Solids are collected in both the primary and final clarifier tanks for reuse. Effluent from the final clarifier is sterilized in an ultraviolet treatment tank before discharge into environmental streams.

2.3. Sample Collection and Handling

Stream samples were collected by submerging sterile 1 L Nalgene screw cap bottles approximately six inches beneath the water surface until full. The bottle mouth was oriented against the direction of flow to prevent any disturbed upstream sediment from being received and was capped while submerged. A total of 6 L of creek water was collected from each creek sampling site and stored in an insulated cooler for transport. Sewage samples were collected using an ISCO 6712 auto-sampler (Teledyne, Lincoln, NE, USA), pulling 150 mL of sewage every 30 min peristaltically to generate a 24-h composite volume in sterile 2.5 gallon carboys. The carboys were kept in an ice water bath within the sampler during the collection time. Composite collections at the reclamation facilities also used the same sampling strategy with refrigerated ISCO 6712FR autosamplers (Teledyne, Lincoln, NE, USA) and carboy storage at 4 degrees Celsius for 24 h. Following collection, 1 L of volume was transferred via a peristaltic pump from the collection carboys after thorough mixing into 1 L sterile Nalgene bottles in a sterile hood for further processing. An additional 500 mL was pumped into sterile amber glass bottles containing 200 mg of EDTA to inhibit bacterial growth for mass spectrometry analysis of antibiotic residues. In total, four time points were collected beginning with late Winter (time point 1), early and late Spring (time points 2 and 3, respectively), and mid-Summer (time point 4). The pump tubing was replaced between each sample collection and the bottles were stored at 4 degrees Celsius until DNA extraction. At the time of sample retrieval, environmental metadata measurements were collected along with the sample material. These data were integrated in downstream correlation analyses and models. Metadata gathered at the time of collection included dates and times, latitude and longitude, ambient, sample, and storage temperatures, conductivity (pHmv), pH, humidity, and autosampler composite collection start and end times (if applicable).

2.4. Detection of Antibiotic Compounds by Mass Spectrometry

Ten antibiotic compounds representing a broad range of commonly prescribed classes were chosen for the study. These included sulfamethoxazole, trimethoprim, ciprofloxacin, cephalexin, levofloxacin, amoxicillin, clindamycin, doxycycline, ertapenem (used exclusively for multi-drug resistant infections in hospitals), and azithromycin. Standardized compounds were used to generate calibration curves for the detection of each antibiotic in the wastewater composites, treated wastewater, and stream collections.

Antibiotic standards for each were spiked into various solvents, according to their preparation instructions. Ciprofloxacin, doxycycline, ertapenem, and amoxicillin were spiked into sterile water in varying concentrations from 1.95 ng/mL to 1000 ng/mL. Azithromycin, clindamycin, and sulfamethoxazole were spiked into ethanol, and levofloxacin, trimethoprim, and cephalexin were spiked into methanol at the same concentrations. An additional curve was generated for amoxicillin at diluted concentrations from 0.12 ng/mL to 0.98 ng/mL. Ciprofloxacin required an additional standard curve, as the initial data points were not linear and could not be used for concentration detection. Five-hundred mL composites of previously described wastewater and creek samples were passed over Whatman microfiber and 0.47 micron membrane filters (GE Healthcare, Pittsburgh, PA, USA) prior to cartridge loading. Growth-inhibiting EDTA was added before and after filtering to inhibit bacterial growth and to chelate any metal ions. Each sample was adjusted to a pH of 3.5 with 10% formic acid, and 200 mL of each sample were loaded onto preconditioned Oasis HLB cartridges. Each cartridge was washed with 12 mL of sterile water and 12 mL of methanol and formic acid (99%/1% v/v) to elute the bound material, which was subsequently dried overnight in nitrogen. Following desiccation, 190 uL of 99%/1% methanol/formic acid with 10 uL of 200 ng/mL Cl-phenylalanine was added to the dried material. Samples were vortexed for 10 min and centrifuged at 10,000× g for 5 min before the supernatant was transferred to vials for ultra performance liquid chromatography tandem mass spectrometer (UPLC-MS) analysis. In total, 26 samples were loaded into an Acquity UPLC-Quattro Premier XE MS system (Waters Corp, Milford, MA, USA) and ran in positive electrospray ionization mode. Raw files were processed using the TargetLynx Application Manager (Waters Corp, Milford, MA, USA), obtaining values of peak area and retention time. Concentrations of antibiotic compounds from each sample were calculated from the spray spectra based on the standard antibiotic curves. A total of 92 samples were processed in four batches, so that the preparation of the samples and the UPLC-MS analysis could be conducted on the same day. Antibiotic standard dilutions were also run in series with each batch, along with an internal phenylalanine standard to ensure consistency in instrument performance.

2.5. DNA Extraction

All samples were passed in triplicate over 0.45 micron vacuum water filters (MOBIO, Carlsbad, CA, USA) in 100 mL volumes until the flow had stopped. Flow-through volumes varied from ~150 mL to

~1 L depending on the turbidity and origin of the sample. Filters were removed from the filtration unit with sterilized forceps and were halved with sterile scissors, where one half was sliced into ~5 mm wide strips and the other was stored at −80 Celsius for future redundancy. The filter material strips were then placed in a bead homogenizer tube of the FastDNA SPIN kit for soil (MP Biomedicals, Santa Ana, CA, USA) and homogenized for 60 s using a benchtop FastPrep-24 homogenizer (MP Biomedicals, Santa Ana, CA, USA), as recommended in the manufacturer's manual. DNA extraction and elution were conducted according to the manufacturer's protocol for increased yield and quantified using a Qubit fluorometer (Thermo-Fisher, Waltham, MA, USA) and Nanodrop spectrophotometer (Thermo-Fisher). Following DNA quality control, samples were frozen at −20 Celsius before delivery to the David H. Murdock Research Institute (DHMRI, Kannapolis, NC, USA) Genomics laboratory for sequencing library preparation. Primer, adapter, and barcode sequences are listed in the Supplementary Materials.

2.6. Shotgun Metagenomic Library Preparation and Sequencing

All sequencing was performed by the core laboratory at the David H. Murdock Research Institute (DHMRI, Kannapolis, NC, USA). Amplicon libraries were generated from collected DNA templates and validated using qPCR. Each was uniquely indexed and all the samples were pooled together in equimolar proportions and sequenced with 125 bp paired-end reads on an Illumina HiSeq 2500 flow cell (Illumina, San Diego, CA, USA). Time points 1–3 each comprised 66 total samples, while time point 4 consisted of 78 total samples, including the Uwharrie and Appalachian locations. Sequences were demultiplexed and debarcoded by the DHMRI team prior to delivery. Libraries were pooled nine per HiSeq 2500 lane, resulting in a sequencing depth of ~5 Gb per sample with 125 bp paired-end reads.

2.7. DNA Sequence Trimming and Quality Control

Raw DNA sequences were filtered and demultiplexed by the DHMRI using the Illumina HiSeq HCS software (Illumina, San Diego, CA, USA). Greater than 80% of the bases must have a quality score greater than 30, or an accuracy rate of 99.9%. Barcode, primer, and adapter sequences were also removed and verified in-house before proceeding with further analysis. Trimmomatic parameters for read clipping were ILLUMINACLIP:TruSeq3-PE:2:30:10 to ensure the complete removal of any remaining Illumina adapters. Leading and trailing base calls below a PHRED score of 3 were removed, and a sliding window approach was implemented with a required average PHRED score of 20 across a three-base section. Lastly, a minimum read length of 50 bases was specified, retaining ~80% of the total input reads, which were merged using PEAR with a minimum specified overlap of 10 bases, minimum assembled length of 50 bases, and a minimum alignment *p*-value of 0.01, resulting in an average assembly efficiency of 99%. In total, 1,344,546,211 paired reads for each deep-sequenced sample were obtained and used to ensure the multiplexing strategy did not affect or dilute the results returned for pooled samples. For the pooled metagenome shotgun sequences, approximately 88% of the total reads per sample contained surviving paired mates following quality control, yielding an average of 147,163,056 reads for each multiplexed sample.

2.8. Metagenomic Classification Analysis Using MetaPhlAn2

To determine species-level relative abundance, we analyzed the merged shotgun sequence datasets with the Metagenomic Phylogenetic Analysis for Metagenomic Taxonomic Profiling (MetaPhlAn) package, version 2.5.0 [26]. Sequences were aligned to the default MetaPhlAn2 marker database (v.20) for relative abundance measurements. Heat maps of the top 100 species-level taxa were generated in R [27] with ggplot2 [28], using hclust and Bray-Curtis distance calculations, along with alpha and beta diversity metrics (links provided in Supplementary Materials). Bar plots were generated using ggplot2.

2.9. Identification and Quantitation of Resistance Elements Using ShortBRED

Antibiotic resistant markers were identified and quantified using Short, Better Representative Extract Data (ShortBRED) [29]. To ensure the broadest available reference database for alignments, we created a custom database with ShortBRED-identify, consisting of the Comprehensive Antibiotic Resistance Database (CARD) [30] version 1.1.0 (August 2016) that was merged with the Lahey Clinic beta-lactamase database. ShortBRED-quantify was applied to the merged forward and reverse paired shotgun sequencing reads. The highest depth replicate was used to calculate the representative normalized reads per kilobase per million mapped reads (RPKM) counts of hits to each marker in the database for each sample. All the markers with zero RPKM hits were then removed. Heat maps were generated from the top 40 markers with the highest number of normalized counts and barplots were created to show overall ARG load at each sampling location.

2.10. Functional Classification Analysis Using HUMAnN2

To determine pathway-level relative abundance, we analyzed the merged shotgun sequencing datasets with the HMP Unified Metabolic Analysis Network (HuMAnN2) package, version V0.11.1. Sequences were aligned to the ChocoPhlAn nucleotide database and a translated search was performed using the Uniprot UniRef90 [31] database. Pathway abundance files that were generated by HuMAnN2 were used for downstream statistical analysis.

2.11. Statistical Methods

The significance of differences in microbial relative abundances, pathway relative abundances, and ShortBRED-derived resistance elements between sites were each assessed using linear regression models, via the lm function in R [27]. The relative abundance or RPKM of ShortBRED genes served as the response variable and the final explanatory variables consisted of the stream source (MC or SC), the sampling site (upstream, treatment plant influent, downstream, etc.), and a representation of the time at which the sample was taken (late winter, early spring, late spring, mid-summer). As the triplicates were clustered closely together, the most deeply sequenced sample from each set of replicates was taken as the representative for that measurement. A threshold frequency of a non-zero presence in at least 25% of all samples in the comparison was used to avoid wasting hypotheses on stochastic differences in rare taxa or genes. The Benjamini-Hochberg false discovery rate was used to perform multiple hypothesis correction [32]. All the p-values are Benjamini-Hochberg corrected, and complete statistical comparison results are available at the links provided in Supplementary Materials.

3. Results

Our study of antibiotics, taxa and resistance genes throughout the Charlotte, NC urban watershed surveyed several processing stages inside two wastewater treatment plants, as well as upstream and downstream sites across four time points in the 2016 calendar year. We found that the release of treated water maintains elevated concentrations of multiple antibiotics in downstream waters, and that some pathogens of interest are present in small quantities in the streams, both upstream and downstream of treatment sites. However, unlike antibiotic compounds, pathogenic taxa and antibiotic resistance gene markers are generally not significantly increased in concentration in moving waters downstream of treated water release.

3.1. Antibiotic Concentrations Are Elevated Downstream of Wastewater Treatment Plants

Ten compounds were chosen to represent a diversity of frequently used classes of antibiotics with different chemical properties and mechanisms of action. Eight of these (ciprofloxacin, doxycycline, azithromycin, clindamycin, sulfamethoxazole, cephalexin, trimethoprim, and levofloxacin) were detected in most of the samples that were collected (Supplementary Materials Figure S1). By contrast, ertapenem and amoxicillin were only seen in a limited number of samples at concentrations below the limit of quantification (1.95 and 0.24 ng/L, respectively).

Our survey included two urban upstream sites (Mallard and Little Sugar Creek) as well as two rural sites that were chosen for low human impact, which we anticipated would have low background antibiotic concentrations. Remote rural sites and upstream sites were substantially equivalent, with antibiotic concentrations changing by an average of less than 5 ng/L when compared to each other, with no significant differences with regards to the presence of specific antibiotics, and no significant seasonal changes ($p < 0.05$, Supplementary Materials File S2). Consistent with previous literature [14,33,34], concentrations of the eight consistently detected antibiotics were present in measurable concentrations throughout the treatment process (Figure 3). For both Little Sugar and Mallard Creek locations, downstream concentrations of all antibiotic compounds were significantly increased relative to their upstream concentrations, by an average factor of eight-fold, except for clindamycin (Figure S3). This demonstrates whole-compound persistence throughout the treatment plants and into the discharged treated wastewater. We also compared two downstream sites at distances of 3300 and 2700 m for both Little Sugar and Mallard Creek, respectively, and found that overall antibiotic concentrations decreased significantly by an average of 6 ng/L from the more proximal downstream site A to the more distal downstream B site (Figure 3), with the largest concentration reduction seen in ciprofloxacin. Overall, these data demonstrate that wastewater treatment plant effluents can act a source of antibiotics, sufficient to maintain elevated concentrations of antibiotics in moving water for a considerable distance downstream of the plant.

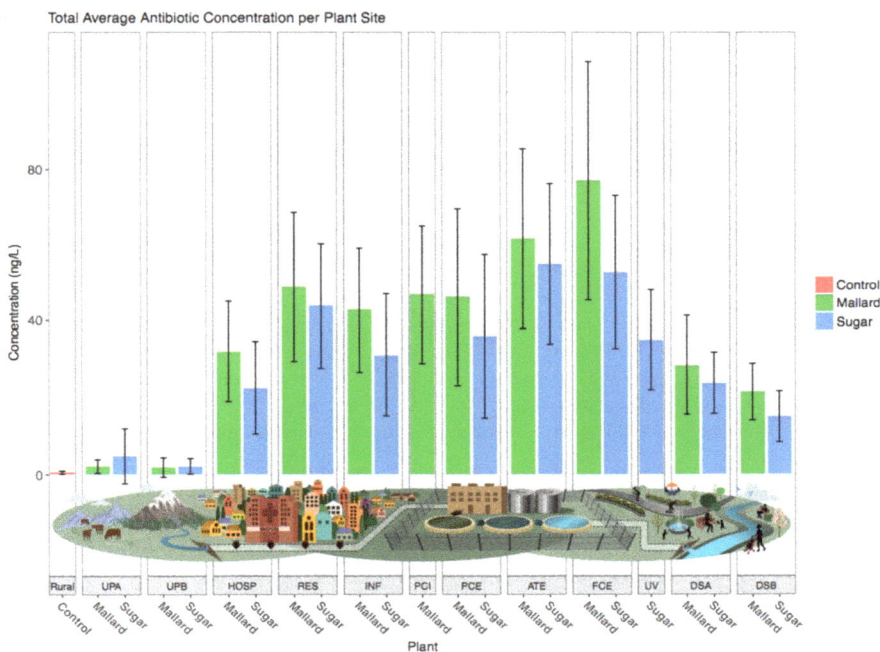

Figure 3. Antibiotic concentrations for each sampling site. Concentrations for all ten antibiotic compounds, reported in ng/L, were averaged across all four time points for each sampling location and treatment plant. Standard deviations are reported within each concentration bar. Rural sites are shown in red for comparison to urban wastewater, with Mallard and Little Sugar Creek sites in green and blue colors, respectively. Primary clarifier influent (PCI) is reported for Mallard Creek only, and UV is only reported for the Sugar Creek facility, as described in Methods.

3.2. Treated Wastewater Microbial Communities Become More like Fresh Water Communities as Waste Progresses through the Treatment Process

Relative abundances of microbial species were computed from shotgun sequencing data using MetaPhlan2 [26] (Figure 4). Influent samples and sewage prior to the ATE stage were the richest in species diversity, while stream samples and treated wastewater were the least diverse (Figure 5).

Beta-diversity measures showed that treated water samples clustered together, separately from stream and rural sites (Figure 6). Technical replicates conducted in triplicate with independent sample DNA extraction and sequencing displaying little variance (Figure S4).

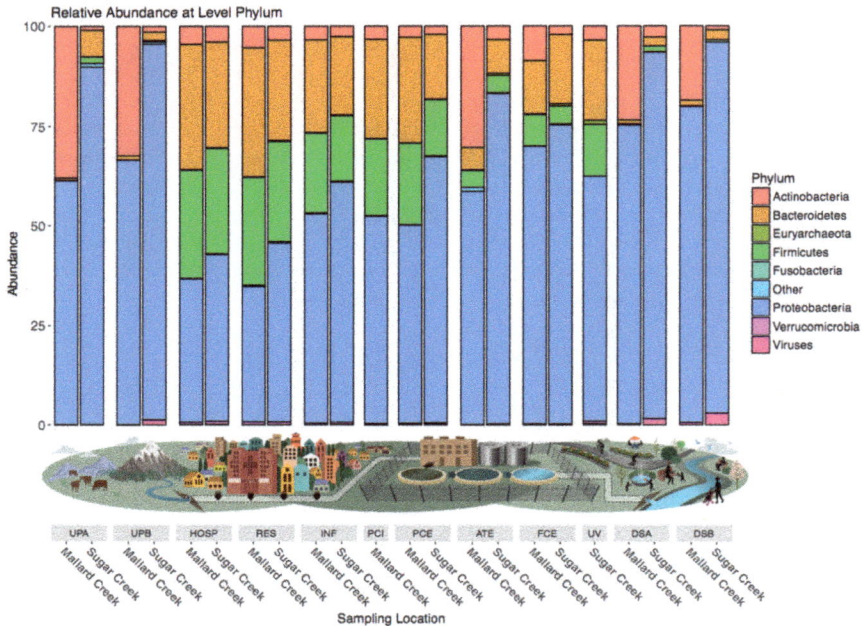

Figure 4. Relative abundance of phyla in collection sites for shotgun sequence data. Abundance values calculated from normalized reads per kilobase per million mapped reads (RPKM) counts at the phylum level are shown for all sampling locations in time point 1, including detected viral load. Phyla that are <1% of the total abundance are combined into the "Other" category. These abundances are representative of the remaining three timepoints, as variability was limited with respect to season.

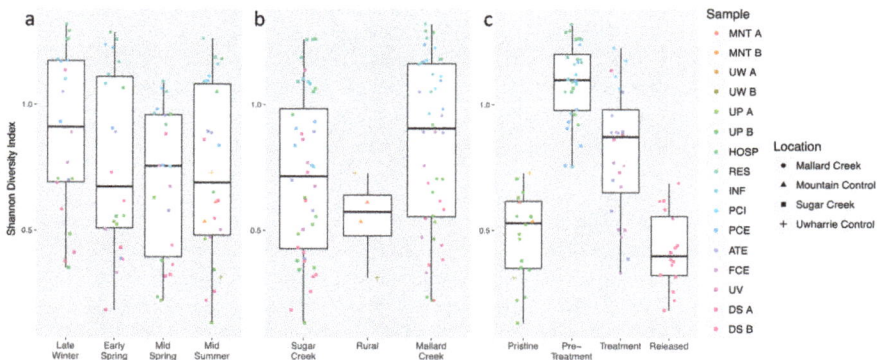

Figure 5. Phylum Shannon diversity for all four collection time points, sample types, and sampling locations is shown, with the mean and standard deviation for each (**a**–**c**). A color gradient denotes different samples, while shapes indicate location.

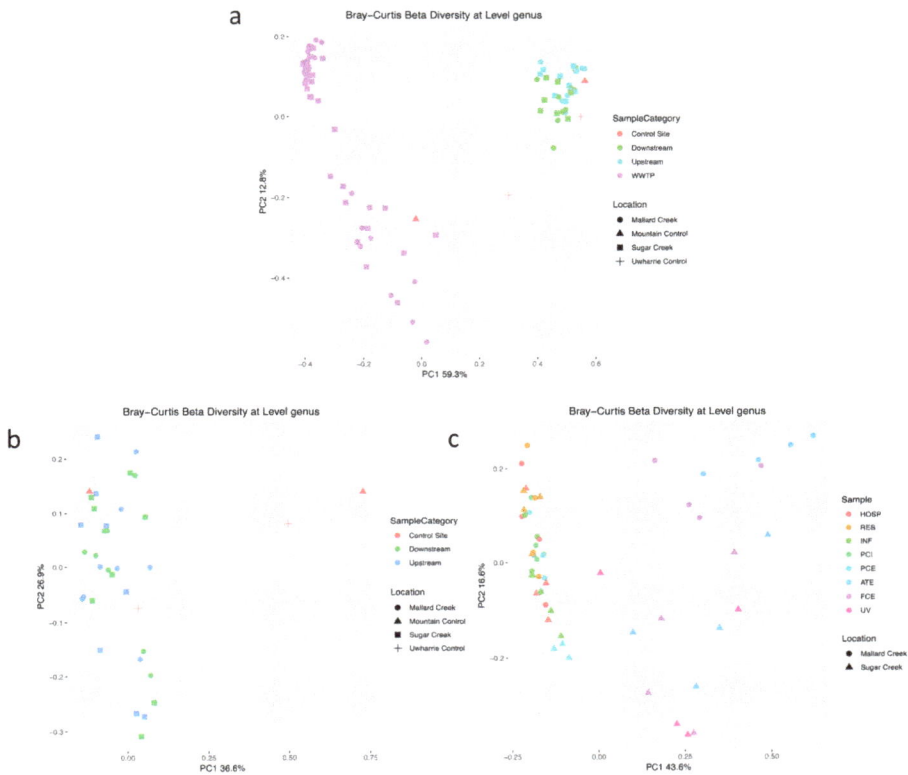

Figure 6. Alpha and beta diversities from shotgun sequencing. PCoA ordination at the genus level for all time points with PC1 and PC2 components. Data are clustered and colored by stream vs. wastewater samples (**a**), stream samples only (**b**), and wastewater samples only (**c**). Raw, unprocessed wastewater samples cluster together, while streams and processed wastewater (ATE+) have similar clustering patterns. Rural locations cluster with background urban streams, and the greatest separation between samples appears to be driven most by treatment type.

When we examine upstream and downstream stream samples at the phylum level Actinobacteria and Proteobacteria predominate (Figure 4), as is typical for freshwater communities [35]. At the genus level, 366 total taxa were detected, 16 of which were significantly different between upstream and downstream locations. Of these, 15 were higher in relative abundance downstream, and they are also present in at least two wastewater treatment plant locations at a relative abundance of greater than 0.1% (Table S1). These results are consistent with these taxa being introduced into the downstream ecosystem from the wastewater treatment plant. However, when we compared the closest (DSA) and most distant (DSB) downstream locations from the plant effluent discharge sites to see whether any of these species decrease in relative abundance with increased distance from the effluent sources we found no significant differences in enriched taxa with respect to distance from the discharge site (data not shown).

Within the influents and the treatment plant environment, Bacteroides and Firmicutes predominate (Figure 4), as we would expect from previous literature [36,37] for material mainly originating in the human gut. Collectively, raw sewage influents between the Mallard and Sugar Creek treatment facilities were comparable, with no significant population deviations between the two basins at the phylum level ($p < 0.05$). Minimal differences in microbial relative abundance were detected between mixed hospital waste and waste that was exclusively residential, with the soil dwelling bacterium *Kocuria rhizophila* ($p = 0.0318$) as the only significantly different taxon (Table 1). When influents pass through the aeration tank following the point at which activated sludge is introduced, there is a noticeable shift in dominant phyla with 23 taxa being significantly different (Table 1). *Actinobacteria* species are reintroduced, and the trending relative abundances of Bacteroides and Firmicutes drop, although those phyla are still more abundant than they are in the stream. When upstream locations and rural sites were compared, the soil-associated genus *Sphingobium* was the only significant differing taxon ($p = 0.01$) that was detected at elevated levels in the rural sites when compared to Charlotte upstream locations (Table 1).

Table 1. Taxa of significant differential abundance between wastewater treatment stages and stream sites. The sampling location containing a higher percent abundance of the listed taxa is shown in the "Higher Abundance" column, and the Bonferroni-corrected *p* value resulting from the mixed linear models is also shown.

Taxa	*p* Value	Higher Abundance
Peptostreptococcaceae	0.0039	Downstream to Upstream
Afipia	0.0093	Downstream to Upstream
Holospora	0.0039	Downstream to Upstream
Azoarcus	0.0114	Downstream to Upstream
Acinetobacter	0.013	Downstream to Upstream
Bppunalikevirus	0.0093	Downstream to Upstream
Yualikevirus	0.0096	Downstream to Upstream
Sphingobium	0.01	Rural to Upstream
Kocuria rhizophila	0.0318	Residential to Hospital
Nitrospira defluvii	0.0216	ATE to PCI
Caulobacter sp.	0.0058	ATE to PCI
Afipia clevelandensis	0.0048	ATE to PCI
Rhodopseudomonas paulustris	0.012	ATE to PCI
Hyphomicrobium denitrificans	0.0114	ATE to PCI
Mesorhizobium sp.	0.0183	ATE to PCI
Paracoccus sp.	0.0439	ATE to PCI
Reyranella massiliensis	0.0111	ATE to PCI
Sphingobium xenophagum	0.0184	ATE to PCI
Sphingopyxis sp.	0.0003	ATE to PCI
Alicycliphilus sp.	0.0004	ATE to PCI
Limnohabitans sp.	0.0005	ATE to PCI
Polaromonas sp.	0.0003	ATE to PCI
Variovorax sp.	0.0014	ATE to PCI
Azoarcus sp.	0.0006	ATE to PCI
Dechloromonas sp.	0.011	ATE to PCI
Methyloversatilis sp.	0.0008	ATE to PCI
Thauera aminoaromatica	0.0212	ATE to PCI
Actinobacter parvas	0.025	ATE to PCI
Turneriella parva	0.0058	ATE to PCI
Methanobrevibacter sp.	0.0357	ATE to PCI
Gordonia amarae	0.0476	ATE to PCI
Tetrasphera elongata	0.0218	ATE to PCI
Rhodococcus	0.0409	Downstream to FCE
Actinobacterium sp.	0.0116	Downstream to FCE
Polynucleobacter necessarius	0.00000007	Downstream to FCE
Limnohabitans	0.00000007	Downstream to FCE
Methylotenera	0.0404	Downstream to FCE
Bppunalikevirus	0.0132	Downstream to FCE
Yualikevirus	0.0266	Downstream to FCE

3.3. Microbial Community Shifts during Wastewater Processing

Comparisons between wastewater collection and treatment stages were constrained to processing stages that were identical between the two plants. No significant differences in taxa were observed between the PCI and PCE stages where there is no active processing. However, between the PCE and the aeration tank effluent (ATE), 221 unique taxa were significantly altered in relative abundance, with 192 of these being reduced overall. Of the taxa that increased in relative abundance in the aeration tank effluent, all have been previously characterized as digestors and denitrifiers within wastewater and/or activated sludge [36,38–55].

In the subsequent treatment stage from the ATE to the final clarifier effluent (FCE), we also observed very minor changes in relative abundance, although several differences were marginally significant (Supplementary Materials File S3). The composition of the FCE prior to ultraviolet disinfection in both plants differed significantly from downstream samples across 66 different taxa. Most of the differential taxa appeared to be reduced from FCE levels during the subsequent UV disinfection step (Table 1)

The internal population profiles of both Mallard and Sugar Creek, plants were generally quite similar. At both the late winter (Figure 4) and the other collection timepoints (Figure S4), the FCE of Sugar Creek had significantly lower relative abundances of the soil- and freshwater- associated genus *Thauera*, when compared to Mallard Creek FCE. All comparisons between plant stages can be accessed in their entirety from the Figshare links provided in the Supplementary Materials.

3.4. Shifts in Stream Microbial Community Function Are Observed Downstream of Treated Wastewater Release Points

In order to understand the potential functional significance of observed changes in the microbial communities among the locations sampled, we analyzed the data using HUMAnN [56] to determine the relative abundance of orthologous gene families making up known MetaCyc microbial pathways [57]. Between upstream and downstream sites, there are 24 significant pathway relative abundance changes ($p < 0.05$) (Table 2). Twenty-two of the twenty-four affected pathways are present in higher relative abundance upstream of treated water release. Pathways involved include nucleotide, amino acid, and carbohydrate biosynthesis, as well as nucleotide and peptide degradation and salvage pathways.

Table 2. Metabolic pathways significantly enriched in up or downstream locations. Significant MetaCyc functional pathways from HUMAnN are shown, along with their corresponding *p* values and whether they were enriched upstream or downstream of treatment plant effluent release.

MetaCyc Pathway	*p*-Value	Highest Abundance
PWY-5747 2-methylcitrate cycle II	0.009	Upstream
PWY-5659 GDP-mannose biosynthesis	0.010	Upstream
PWY0-42 2-methylcitrate cycle I	0.010	Upstream
GLYCOGENSYNTH-PWY glycogen biosynthesis I from ADP-D-Glucose	0.042	Upstream
ILEUSYN-PWY L-isoleucine biosynthesis I from threonine	0.042	Upstream
PWY-5109 2-methylbutanoate biosynthesis	0.042	Downstream
PWY-5973 cis-vaccenate biosynthesis	0.042	Upstream
PWY-6606 guanosine nucleotides degradation II	0.042	Upstream
PWY-6609 adenine and adenosine salvage III	0.042	Upstream
PWY-7111 pyruvate fermentation to isobutanol engineered	0.042	Upstream
PWY-7198 pyrimidine deoxyribonucleotides de novo biosynthesis IV	0.042	Upstream
PWY-7199 pyrimidine deoxyribonucleosides salvage	0.042	Upstream
PWY-7208 superpathway of pyrimidine nucleobases salvage	0.042	Upstream
PWY-7210 pyrimidine deoxyribonucleotides biosynthesis from CTP	0.042	Upstream
PWY-7211 superpathway of pyrimidine deoxyribonucleotides de novo biosynthesis	0.042	Upstream
PWY-7663 gondoate biosynthesis anaerobic	0.042	Upstream
SALVADEHYPOX-PWY adenosine nucleotides degradation II	0.042	Upstream
UNINTEGRATED	0.042	Upstream
UNMAPPED	0.042	Downstream
VALSYN-PWY L-valine biosynthesis	0.042	Upstream
PWY-3781 aerobic respiration I cytochrome c	0.042	Upstream
PWY0-1261 anhydromuropeptides recycling	0.042	Upstream
GLUTORN-PWY L-ornithine biosynthesis	0.045	Upstream
PWY-6608 guanosine nucleotides degradation III	0.045	Upstream

3.5. Resistance Genes Are Slightly More Abundant Downstream of Wastewater Treatment Plants

In order to measure changes in resistance gene content, we used ShortBRED to identify specific antibiotic resistance associated genes and elements. Alignment of shotgun sequencing data to the hybrid CARD and Lahey databases for antibiotic resistance associated genes and mobile plasmid elements revealed hits to a total of 600 unique terms from all of the samples across all time points. Of the 600 total detected antibiotic resistance associated sequences, nine were more abundant in downstream waters when compared to upstream (statistical tables available from FigShare links in Supplementary Materials). These included carbenicillin and oxacillin beta-lactamases CARB-3 and OXA-1 from the WHO priority pathogen *Pseudomonas aeruginosa*, including two extended spectrum beta-lactamases (Figure 7B and Figure S6). Plasmid-derived sulfonamide resistance for *Vibrio cholerae* species, encoding a dyhydropteroate synthase, was also more abundant in downstream waters. Multiple genes conferring multi-drug resistance (MDR) within the *E. coli* K-12 strain were detected as well, all encoding multiple efflux pump subunits or modulating efflux control. Overall, only two antibiotic resistance elements were significantly different in relative abundance between upstream and downstream sites. A *Streptomyces lividans* methyltransferase was found in significantly higher abundance upstream ($p = 0.044$), while a beta-lactamase from *Pseudomonas aeruginosa* was higher downstream ($p = 0.044$).

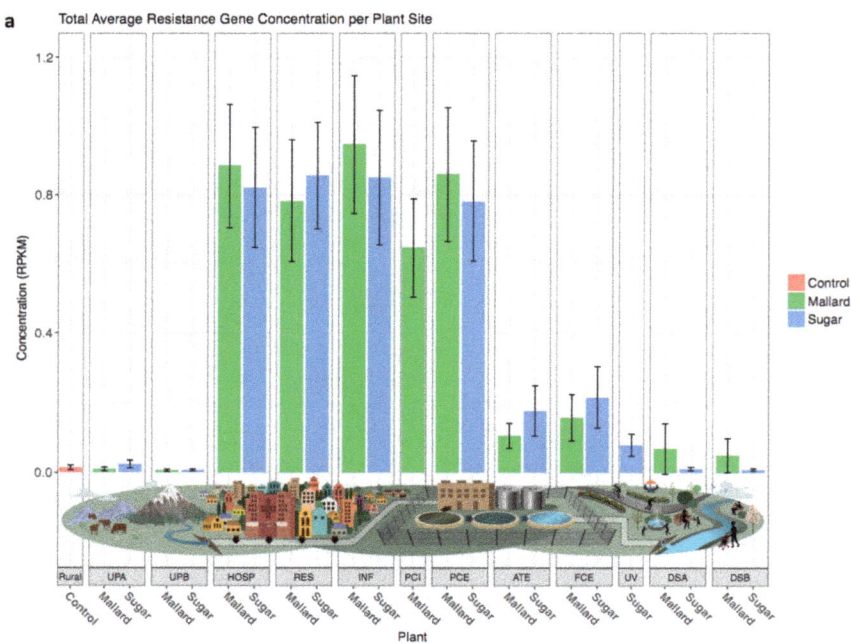

Figure 7. *Cont.*

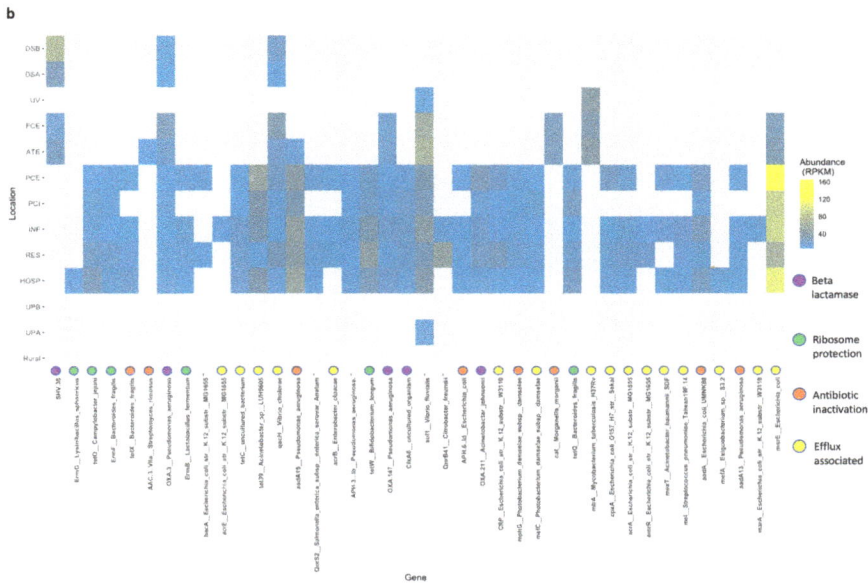

Figure 7. Abundance and differential abundance of antibiotic resistance genes. (**a**) The average number of antibiotic resistance-associated genetic elements across all four time points are reported in RPKM for each different site and each treatment plant. Standard deviations are shown within each bar. Rural sites are displayed in red, with Mallard and Little Sugar Creek sites in green and blue respectively. PCI is exclusive to Mallard, while UV is only in the Sugar Creek facility as denoted in Methods. (**b**) Relative abundance of antibiotic resistance elements. The top 40 average antibiotic resistance genes across all sampling timepoints with an average normalized RPKM of 10 or above are shown for each sampling site. Colored dots indicate common resistance mechanisms conferred by groups of similar resistance genes.

When compared to the relative abundances of ARG sequences within the FCE stage of treatment, 15 of the 600 terms were significantly lower downstream (statistical tables provided in FigShare of Supplementary Materials). An *Exiguobacterium* macrolide efflux pump ($p = 0.043$) was detected in higher relative abundance within partial hospital sewage, while the residential sewage contained significantly more *Escherichia coli* macrolide efflux pump and associated MDR efflux pump membrane proteins ($p = 0.043$). Following treatment in the primary clarifier, no ARGs were altered significantly in abundance from the raw influent. Subsequent digestion in the aeration tank reduced the relative abundance of 84 antibiotic resistance elements, with many falling to undetectable levels. The transition from ATE to FCE did not result in further significant changes to relative abundance of any ARG markers, although a methyltransferase from *Streptomyces lividans* and an aminoglycoside acetyltransferase in *Streptomyces rimosus* were marginally reduced ($p = 0.052$ and 0.051, respectively). The final UV step (implicit in the comparison of FCE to downstream sites DSA and DSB) resulted in the reduction or removal of twelve additional markers, including a *Vibrio cholerae* and *Vibrio fluvialis* sulfonamide dihydropteroate synthase ($p = 0.0004$), an aminoglycoside acetyltransferase in *Serratia marcescens* ($p = 0.014$), a *Pseudomonas aeruginosa* beta-lactamase ($p = 0.054$), a tetracycline efflux pump in *Acinetobacter* ($p = 0.054$), *Photobacterium damselae* macrolide efflux pump ($p = 0.054$), *Escherichia coli* plasmid-encoded efflux pumps for streptogramin, streptomycin, and erythromycin ($p = 0.054$), and suggest elevated levels of a chloramphenicol acetyltransferase in *Morganella morganii*, and an MDR efflux pump for *Mycobacterium tuberculosis* ($p = 0.1$). No significant differences in ARG markers were detected between the two downstream locations. Remote mountain and Uwharrie stream sites and the upstream urban sites were essentially equivalent, with no significant difference in antibiotic resistance markers. Corresponding illustrations for individual antibiotic concentrations and shotgun analysis, ARGs, and all taxa, as well as between site comparisons, are provided in Supplementary Materials (Figures S7–S9).

4. Discussion

In this survey of two Charlotte NC urban waterways and their associated treatment plants, we made four key observations. Each of these suggests targets for future research or innovation in water treatment methodology.

First, we assayed the concentration of 10 antibiotic compounds representing families of commonly used antibiotics, and found that multiple antibiotic concentrations are elevated downstream of wastewater treatment plants. Antibiotic compounds tend to be concentrated inside the plants relative to upstream and influent concentrations, whole compounds persist throughout the treatment cycle, and elevated concentrations of antibiotics are observed in moving waters downstream of the treatment plants. Both the interior sites in the plant and the downstream sites are environments that facilitate the simultaneous exposure of bacteria to multiple antibiotics. Trunk lines handling hospital waste did not have significantly higher antibiotic levels than purely residential trunk lines.

Second, we assayed microbial communities at the same sites within treatment plants and stream watersheds, and found that treated water microbial communities become more like fresh water communities as waste progresses through the treatment process. Some antibiotic resistance terms originating from priority pathogens were observed in low relative abundance in environmental sites and at much higher relative abundance within the influent and in the treatment plants. Surprisingly, influent from hospital-associated and all-residential neighborhoods was not significantly different from the microbial community perspective. We observed an increased relative abundance of expected microbial signatures, such as human gut associated Bacteroidetes and Firmicutes and activated sludge associated taxa, in downstream waters. Levels of these attenuated between the proximal and distal downstream sites and did not significantly change the core microbial community of the streamwater. More downstream sampling locations would be required to accurately model this trend, and it will be of interest to assay antibiotic levels at the endpoint of the Catawba and Yadkin Pee-Dee rivers in southeast coastal waters as well.

Third, we analyzed relative abundance of functional pathways in the stream and treatment plant microbial communities, and found that shifts in stream microbial community function are observed downstream of treated water release points. Here, we focused primarily on the impact of water release on the function of the stream community. When comparing upstream and downstream sites, we observed that some core functions of stream microbes, including many pathway terms for DNA and peptide recycling and biosynthesis, were suppressed relative to upstream sites. This is an interesting finding that we cannot yet explain mechanistically, but it suggests that further investigation of sediments in urban streams under long-term antibiotic stress is necessary to understand the mechanism of impact of this stressor on community function.

Finally, we analyzed presence of antibiotic resistance gene (ARG) signatures in the plant and stream microbial communities, and found that resistance genes are slightly more abundant downstream of wastewater treatment plants. This was not a foregone conclusion of this study, because in other contexts ARG signatures have been observed to increase downstream of treated water release. However, Charlotte Water's treatment process, which includes UV treatment of waters prior to final release, appears to be very effective in reducing ARG relative abundance. While certain ARG signatures, including concerning carbapenem resistance and broad spectrum beta lactamases, originating from organisms, such as *Pseudomonas aeruginosa*, *Vibrio cholerae*, and *Escherichia coli* were detectable in low relative abundance downstream of the plants, the reduction in relative abundance of these signatures relative to influent and internal WWTP locations was dramatic, especially when considering that the testing that Charlotte Water conducts for pathogens in treated water is limited to the standard fecal coliform test. To our knowledge, this is the first study of this scope and resolution to investigate all of these key factors together in one integrated analysis.

Testing for antibiotics and associated resistance genes is not conducted as part of the wastewater treatment and monitoring process at Charlotte Water, nor is this implemented as standard practice in any known treatment facilities in the United States. However, the spread of resistance to common antibiotics is a well-known public health threat, and the presence of antibiotics in sub-lethal concentrations is known to drive microbial evolution [15,16,58]. Based on available knowledge about the dissemination of pharmaceutical compounds in surface waters in North America [59], we hypothesized that either the processing or the release of treated water could create a condition where mixtures of bacteria are simultaneously exposed to multiple antibiotics, and facilitate the spread of antibiotic resistance [60]. We observed that conditions for multiple antibiotic exposure exist inside the water treatment plants as well as at downstream urban locations.

Wastewater treatment plant influent is known to contain significant populations of human gut associated pathogens, and regional sewage microbial profiles are tightly correlated with the microbial profiles of human residents in the region [61]. Although we observe in this study that gut-associated taxa are somewhat elevated in downstream waters, the microbial profile of moving waters in upstream sites in the city is not far removed from the profiles of more remote rural sites, largely devoid of most resistance elements. The addition of treated water to the stream appears as a temporary perturbation in the microbial community that begins to attenuate further away from the release point. This likely allows the stream microbial profile to return to its pre-effluent baseline within a relatively short distance, particularly with the limited downstream population effects that are reported here. Recent works describe similar changes in stream biofilms [34], although the extent of observable species changes was less significant in our samples.

Charlotte's WWTPs accumulate significant quantities and varieties of genetic elements associated with antibiotic resistance, but unlike some other treatment facilities [62], are quite effective in removing them prior to release. Simultaneous exposures of microbes to multiple antibiotics within the plants themselves, where antibiotic concentrations are highest, are relatively short, although a typical residence time of approximately 20 days still encompasses many microbial generations. However, residence times for microbes are much longer in downstream sites, especially in stream sediments and biofilms. We showed that the release of treated water maintains significantly elevated concentrations of multiple antibiotics in the stream for a significant distance downstream of release. The very effective removal of ARG gene signatures by Charlotte Water's treatment process suggests that even though the influent, and the interior of the plant itself, is a diverse hothouse of pathogens and antibiotic resistance, those elements escaping the plant is not likely to be the main mechanism of treated water impact on downstream microbial ecology [63]. Only a small number of recognizable ARGs pass through the water treatment process to end up at detectable levels in the streams. This is not uniformly true in modern water treatment systems, as demonstrated in another recent study [33], and concentrations of specific resistance elements can vary between global processes, as shown from a 2011 study [64]. Rather, we now hypothesize that the persistently elevated concentration of antibiotics downstream of treated water release may be creating its own microenvironment, and that this will impact the function of stream sediment communities and their capacity to provide ecosystem services. Chlorination has been previously shown to be ineffective at ARG removal [65]; however, it is possible that increasing UV treatment intensity and/or exposure time could further reduce dissemination of many ARGs with minimal changes to existing infrastructure [66]. Antibiotic dissemination can potentially be reduced by microfiltration, but these methods are not yet widely implemented and can be potentially costly to use [67]. Surface waters in areas devoid of anthropogenic involvement are known to lack most antibiotic resistance markers, including many mobile elements [68,69]; however, some resistance factors do emerge naturally through the process of recombination and mutation [68], and the conditions with the potential to exacerbate this process do exist in Charlotte's urban streams [70]. Given the tendency of environmental resistance factors to be passed to clinical pathogens [71,72], further study of the impact of maintaining permanent low concentrations of antibiotics in the stream, in the presence of resistance elements from wastewater or environmental sources, is critical for understanding how we can improve water treatment to safeguard both the ecosystem and human health.

Supplementary Materials: The following are available at: http://www.mdpi.com/2073-4441/10/11/1539/s1. NGS datasets created in this study are deposited in the SRA, accession number SRP121672, and scripts are available from GitHub at https://github.com/NCUrbanMicrobiomeProject/InitialStudy. Additional statistical comparisons between sampling sites are available from FigShare as noted in Supplementary Materials. (Table S1) Genera higher in relative abundance in downstream locations also present in at least two wastewater treatment plant locations at a relative abundance of greater than 0.1%. (Figure S1) Concentrations of each antibiotic compound for each sampling location, showing the mean and standard deviation across all four timepoints. (Figure S2) Shared taxa between both plants. (Figure S3) Shared taxa between sample types. (Figure S4) Shannon diversity and Bray-Curtis Beta diversity of all shotgun sample replicates. (Figure S5) Species-level taxonomic classification of differentially abundant clades between sampling locations. (Figure S6) Relative abundance values for antibiotic resistance genes. (Figure S7) Significant differences in antibiotic concentrations between sample site, waterway, and timepoint. (Figure S8) Significant differences in taxa from shotgun sequence analysis between sample site, waterway, and timepoint. (Figure S9) Significant differences between ARGs in collection sites with regards to timepoint, waterway, sample site, and combinations of sample site/timepoint and waterways are shown. Each colored box indicates a significant relative abundance difference between the corresponding sites shown on the x-axis, and the significant resistance term on the y-axis. (File S1) Adjustment values for antibiotic detection and quantification limits. (File S2) Complete linear model statistical results from mass spectrometry data. (File S3) Complete linear model statistical results from shotgun taxonomic classification. (File S4) Complete statistical results from mass spectrometry linear models. (File S5) Standard curve calculations for commercial antibiotic standards. (File S6) Relative abundance values for shotgun taxonomic classification at the genus level.

Author Contributions: K.L., S.S., O.K., S.C., M.R., and C.G. designed the experiment. A.F. and M.T. designed the statistical models, and A.F., M.T., J.J., K.L., and A.L. performed statistical analyses. O.W., A.A.-S., and A.L. conducted antibiotic resistance gene analyses. K.L., A.L., and M.T. performed taxonomic classification. K.L. and C.G. wrote the manuscript with input from all authors.

Funding: This research received no external funding.

Acknowledgments: The authors would like to thank Steven Blanchard (University of North Carolina at Charlotte, UNCC) for assistance troubleshooting scripts and cluster jobs, Douglas A. Shoemaker (UNCC Center for Applied GIS) for assistance in producing geographical maps, Raad Gharaibeh (University of Florida College of Medicine) for advice in establishing our metagenomic pipeline, Jennifer Gibas for the graph axis illustration, Michael Sioda for initializing the metagenomics analysis, and Alicia Sorgen, Taylor Jones, and Kristen Smith for assistance with sample preparation and collection (UNCC). We are grateful to Meredith Bostrom and Randy Crowell of the David H. Murdock Research Institute for sequencing services, Kevin Knagge and Huiyuan Chen for mass spectrometry services (David H. Murdock Research Institute), as well as the technical support and access from Charlotte Water that made this work possible. Research was supported by the UNCC Department of Bioinformatics and Genomics.

Conflicts of Interest: The authors declare no conflict of interest.

References

1. Walsh, C.J.; Roy, A.H.; Feminella, J.W.; Cottingham, P.D.; Groffman, P.M.; Morgan, R.P., II. The urban stream syndrome: Current knowledge and the search for a cure. *J. N. Am. Benthol. Soc.* **2005**, *24*, 706–723. [CrossRef]

2. Yu, S.; Wu, Q.; Li, Q.; Gao, J.; Lin, Q.; Ma, J.; Xu, Q.; Wu, S. Anthropogenic land uses elevate metal levels in stream water in an urbanizing watershed. *Sci. Total Environ.* **2014**, *488–489*, 61–69. [CrossRef] [PubMed]

3. Fisher, J.C.; Newton, R.J.; Dila, D.K.; McLellan, S.L. Urban microbial ecology of a freshwater estuary of Lake Michigan. *Elementa* **2015**, *3*, 000064. [CrossRef] [PubMed]

4. Qu, X.; Ren, Z.; Zhang, H.; Zhang, M.; Zhang, Y.; Liu, X.; Peng, W. Influences of anthropogenic land use on microbial community structure and functional potentials of stream benthic biofilms. *Sci. Rep.* **2017**, *7*, 15117. [CrossRef] [PubMed]

5. Newton, R.J.; McLellan, S.L. A unique assemblage of cosmopolitan freshwater bacteria and higher community diversity differentiate an urbanized estuary from oligotrophic Lake Michigan. *Front. Microbiol.* **2015**, *6*, 1028. [CrossRef] [PubMed]

6. McLellan, S.L.; Fisher, J.C.; Newton, R.J. The microbiome of urban waters. *Int. Microbiol.* **2015**, *18*, 141–149. [PubMed]

7. Kasprzyk-Hordern, B.; Dinsdale, R.M.; Guwy, A.J. The removal of pharmaceuticals, personal care products, endocrine disruptors and illicit drugs during wastewater treatment and its impact on the quality of receiving waters. *Water Res.* **2009**, *43*, 363–380. [CrossRef] [PubMed]

8. Niemuth, N.J.; Jordan, R.; Crago, J.; Blanksma, C.; Johnson, R.; Klaper, R.D. Metformin exposure at environmentally relevant concentrations causes potential endocrine disruption in adult male fish. *Environ. Toxicol. Chem.* **2015**, *34*, 291–296. [CrossRef] [PubMed]
9. Conley, J.M.; Evans, N.; Mash, H.; Rosenblum, L.; Schenck, K.; Glassmeyer, S.; Furlong, E.T.; Kolpin, D.W.; Wilson, V.S. Comparison of in vitro estrogenic activity and estrogen concentrations in source and treated waters from 25 U.S. Drinking water treatment plants. *Sci. Total Environ.* **2017**, *579*, 1610–1617. [CrossRef] [PubMed]
10. Beresford, N.; Baynes, A.; Kanda, R.; Mills, M.R.; Arias-Salazar, K.; Collins, T.J.; Jobling, S. Use of a battery of chemical and ecotoxicological methods for the assessment of the efficacy of wastewater treatment processes to remove estrogenic potency. *J. Vis. Exp.* **2016**, 54243. [CrossRef] [PubMed]
11. Zhang, Y.; Zhang, T.; Guo, C.; Lv, J.; Hua, Z.; Hou, S.; Zhang, Y.; Meng, W.; Xu, J. Drugs of abuse and their metabolites in the urban rivers of Beijing, China: Occurrence, distribution, and potential environmental risk. *Sci. Total Environ.* **2017**, *579*, 305–313. [CrossRef] [PubMed]
12. Li, N.; Sheng, G.P.; Lu, Y.Z.; Zeng, R.J.; Yu, H.Q. Removal of antibiotic resistance genes from wastewater treatment plant effluent by coagulation. *Water Res.* **2017**, *111*, 204–212. [CrossRef] [PubMed]
13. Prasse, C.; Schlüsener, M.P.; Schulz, R.; Ternes, T.A. Antiviral drugs in wastewater and surface waters: A new pharmaceutical class of environmental relevance? *Environ. Sci. Technol.* **2010**, *44*, 1728–1735. [CrossRef] [PubMed]
14. He, K.; Soares, A.D.; Adejumo, H.; McDiarmid, M.; Squibb, K.; Blaney, L. Detection of a wide variety of human and veterinary fluoroquinolone antibiotics in municipal wastewater and wastewater-impacted surface water. *J. Pharm. Biomed. Anal.* **2015**, *106*, 136–143. [CrossRef] [PubMed]
15. Andersson, D.I.; Hughes, D. Evolution of antibiotic resistance at non-lethal drug concentrations. *Drug Resist. Updates* **2012**, *15*, 162–172. [CrossRef] [PubMed]
16. Hughes, D.; Andersson, D.I. Selection of resistance at lethal and non-lethal antibiotic concentrations. *Curr. Opin. Microbiol.* **2012**, *15*, 555–560. [CrossRef] [PubMed]
17. Ma, L.; Xia, Y.; Li, B.; Yang, Y.; Li, L.-G.; Tiedje, J.M.; Zhang, T. Metagenomic assembly reveals hosts of antibiotic resistance genes and the shared resistome in pig, chicken, and human feces. *Environ. Sci. Technol.* **2016**, *50*, 420–427. [CrossRef] [PubMed]
18. Call, D.R.; Matthews, L.; Subbiah, M.; Liu, J. Do antibiotic residues in soils play a role in amplification and transmission of antibiotic resistant bacteria in cattle populations? *Front. Microbiol.* **2013**, *4*, 193. [CrossRef] [PubMed]
19. Wang, F.-H.; Qiao, M.; Su, J.-Q.; Chen, Z.; Zhou, X.; Zhu, Y.-G. High throughput profiling of antibiotic resistance genes in urban park soils with reclaimed water irrigation. *Environ. Sci. Technol.* **2014**, *48*, 9079–9085. [CrossRef] [PubMed]
20. Fahrenfeld, N.; Ma, Y.; O'Brien, M.; Pruden, A. Reclaimed water as a reservoir of antibiotic resistance genes: Distribution system and irrigation implications. *Front. Microbiol.* **2013**, *4*, 130. [CrossRef] [PubMed]
21. Pruden, A. Balancing water sustainability and public health goals in the face of growing concerns about antibiotic resistance. *Environ. Sci. Technol.* **2014**, *48*, 5–14. [CrossRef] [PubMed]
22. Pehrsson, E.C.; Tsukayama, P.; Patel, S.; Mejia-Bautista, M.; Sosa-Soto, G.; Navarrete, K.M.; Calderon, M.; Cabrera, L.; Hoyos-Arango, W.; Bertoli, M.T.; et al. Interconnected microbiomes and resistomes in low-income human habitats. *Nature* **2016**, *533*, 212–216. [CrossRef] [PubMed]
23. LaPara, T.M.; Burch, T.R.; McNamara, P.J.; Tan, D.T.; Yan, M.; Eichmiller, J.J. Tertiary-treated municipal wastewater is a significant point source of antibiotic resistance genes into duluth-superior harbor. *Environ. Sci. Technol.* **2011**, *45*, 9543–9549. [CrossRef] [PubMed]
24. Munck, C.; Albertsen, M.; Telke, A.; Ellabaan, M.; Nielsen, P.H.; Sommer, M.O. Limited dissemination of the wastewater treatment plant core resistome. *Nat. Commun.* **2015**, *6*, 8452. [CrossRef] [PubMed]
25. Christgen, B.; Yang, Y.; Ahammad, S.Z.; Li, B.; Rodriquez, D.C.; Zhang, T.; Graham, D.W. Metagenomics shows that low-energy anaerobic–aerobic treatment reactors reduce antibiotic resistance gene levels from domestic wastewater. *Environ. Sci. Technol.* **2015**, *49*, 2577–2584. [CrossRef] [PubMed]
26. Segata, N.; Waldron, L.; Ballarini, A.; Narasimhan, V.; Jousson, O.; Huttenhower, C. Metagenomic microbial community profiling using unique clade-specific marker genes. *Nat. Methods* **2012**, *9*, 811–814. [CrossRef] [PubMed]

27. R Department Core Team. *R: A Language and Environment for Statistical Computing*; The R Foundation for Statistical Computing: Vienna, Austria, 2017.
28. Wickham, H. *Ggplot2: Elegant Graphics for Data Analysis*; Springer: New York, NY, USA, 2009.
29. Kaminski, J.; Gibson, M.K.; Franzosa, E.A.; Segata, N.; Dantas, G.; Huttenhower, C. High-specificity targeted functional profiling in microbial communities with shortbred. *PLoS Comput. Biol.* **2015**, *11*, e1004557. [CrossRef] [PubMed]
30. McArthur, A.G.; Waglechner, N.; Nizam, F.; Yan, A.; Azad, M.A.; Baylay, A.J.; Bhullar, K.; Canova, M.J.; De Pascale, G.; Ejim, L.; et al. The comprehensive antibiotic resistance database. *Antimicrob. Agents Chemother.* **2013**, *57*, 3348–3357. [CrossRef] [PubMed]
31. Suzek, B.E.; Wang, Y.; Huang, H.; McGarvey, P.B.; Wu, C.H.; the UniProt, C. Uniref clusters: A comprehensive and scalable alternative for improving sequence similarity searches. *Bioinformatics* **2015**, *31*, 926–932. [CrossRef] [PubMed]
32. Benjamini, Y.; Hochberg, Y. Controlling the false discovery rate—A practical and powerful approach to multiple testing. *J. R. Stat. Soc. Ser. B Methodol.* **1995**, *57*, 289–300.
33. Wang, M.; Shen, W.; Yan, L.; Wang, X.H.; Xu, H. Stepwise impact of urban wastewater treatment on the bacterial community structure, antibiotic contents, and prevalence of antimicrobial resistance. *Environ. Pollut.* **2017**, *231*, 1578–7585. [CrossRef] [PubMed]
34. Aubertheau, E.; Stalder, T.; Mondamert, L.; Ploy, M.C.; Dagot, C.; Labanowski, J. Impact of wastewater treatment plant discharge on the contamination of river biofilms by pharmaceuticals and antibiotic resistance. *Sci. Total Environ.* **2017**, *579*, 1387–1398. [CrossRef] [PubMed]
35. Zhang, J.; Yang, Y.; Zhao, L.; Li, Y.; Xie, S.; Liu, Y. Distribution of sediment bacterial and archaeal communities in plateau freshwater lakes. *Appl. Microbiol. Biotechnol.* **2015**, *99*, 3291–3302. [CrossRef] [PubMed]
36. Yang, C.; Zhang, W.; Liu, R.; Li, Q.; Li, B.; Wang, S.; Song, C.; Qiao, C.; Mulchandani, A. Phylogenetic diversity and metabolic potential of activated sludge microbial communities in full-scale wastewater treatment plants. *Environ. Sci. Technol.* **2011**, *45*, 7408–7415. [CrossRef] [PubMed]
37. McLellan, S.L.; Huse, S.M.; Mueller-Spitz, S.R.; Andreishcheva, E.N.; Sogin, M.L. Diversity and population structure of sewage-derived microorganisms in wastewater treatment plant influent. *Environ. Microbiol.* **2010**, *12*, 378–392. [CrossRef] [PubMed]
38. MacRae, J.D.; Smit, J. Characterization of caulobacters isolated from wastewater treatment systems. *Appl. Environ. Microbiol.* **1991**, *57*, 751–758. [PubMed]
39. Lücker, S.; Wagner, M.; Maixner, F.; Pelletier, E.; Koch, H.; Vacherie, B.; Rattei, T.; Damsté, J.S.S.; Spieck, E.; Le Paslier, D.; et al. A metagenome illuminates the physiology and evolution of globally important nitrite-oxidizing bacteria. *Proc. Nat. Acad. Sci. USA* **2010**, *107*, 13479. [CrossRef] [PubMed]
40. Kim, M.K.; Choi, K.M.; Yin, C.R.; Lee, K.Y.; Im, W.T.; Lim, J.H.; Lee, S.T. Odorous swine wastewater treatment by purple non-sulfur bacteria, *Rhodopseudomonas palustris*, isolated from eutrophicated ponds. *Biotechnol. Lett.* **2004**, *26*, 819–822. [CrossRef] [PubMed]
41. Gonzalez-Martinez, A.; Sihvonen, M.; Muñoz-Palazon, B.; Rodriguez-Sanchez, A.; Mikola, A.; Vahala, R. Microbial ecology of full-scale wastewater treatment systems in the polar arctic circle: Archaea, bacteria and fungi. *Sci. Rep.* **2018**, *8*, 2208. [CrossRef] [PubMed]
42. Saunders, A.M.; Albertsen, M.; Vollertsen, J.; Nielsen, P.H. The activated sludge ecosystem contains a core community of abundant organisms. *ISME J* **2016**, *10*, 11–20. [CrossRef] [PubMed]
43. Balcom, I.N.; Driscoll, H.; Vincent, J.; Leduc, M. Metagenomic analysis of an ecological wastewater treatment plant's microbial communities and their potential to metabolize pharmaceuticals. *F1000Research* **2016**, *5*, 1881. [CrossRef] [PubMed]
44. Medhi, K.; Thakur, I.S. Bioremoval of nutrients from wastewater by a denitrifier paracoccus denitrificans istod1. *Bioresour. Technol. Rep.* **2018**, *1*, 56–60. [CrossRef]
45. Pagnier, I.; Croce, O.; Robert, C.; Raoult, D.; La Scola, B. Genome sequence of *Reyranella massiliensis*, a bacterium associated with amoebae. *J. Bacteriol.* **2012**, *194*, 5698. [CrossRef] [PubMed]
46. Pehl, M.J.; Jamieson, W.D.; Kong, K.; Forbester, J.L.; Fredendall, R.J.; Gregory, G.A.; McFarland, J.E.; Healy, J.M.; Orwin, P.M. Genes that influence swarming motility and biofilm formation in *Variovorax paradoxus* eps. *PLoS ONE* **2012**, *7*, e31832. [CrossRef] [PubMed]
47. Kampfer, P.; Witzenberger, R.; Denner, E.B.; Busse, H.J.; Neef, A. *Sphingopyxis witflariensis* sp. Nov., isolated from activated sludge. *Int. J. Syst. Evol. Microbiol.* **2002**, *52*, 2029–2034. [PubMed]

48. Zielinska, M.; Rusanowska, P.; Jarzabek, J.; Nielsen, J.L. Community dynamics of denitrifying bacteria in full-scale wastewater treatment plants. *Environ. Technol.* **2016**, *37*, 2358–2367. [CrossRef] [PubMed]

49. Cai, L.; Ju, F.; Zhang, T. Tracking human sewage microbiome in a municipal wastewater treatment plant. *Appl. Microbiol. Biotechnol.* **2014**, *98*, 3317–3326. [CrossRef] [PubMed]

50. Jiang, K.; Sanseverino, J.; Chauhan, A.; Lucas, S.; Copeland, A.; Lapidus, A.; Del Rio, T.G.; Dalin, E.; Tice, H.; Bruce, D.; et al. Complete genome sequence of *Thauera aminoaromatica* strain mz1t. *Stand. Genom. Sci.* **2012**, *6*, 325–335. [CrossRef] [PubMed]

51. Al Atrouni, A.; Joly-Guillou, M.L.; Hamze, M.; Kempf, M. Reservoirs of non-baumannii acinetobacter species. *Front. Microbiol.* **2016**, *7*, 49. [CrossRef] [PubMed]

52. Chagas, T.P.; Seki, L.M.; Cury, J.C.; Oliveira, J.A.; Davila, A.M.; Silva, D.M.; Asensi, M.D. Multiresistance, beta-lactamase-encoding genes and bacterial diversity in hospital wastewater in rio de janeiro, brazil. *J. Appl. Microbiol.* **2011**, *111*, 572–581. [CrossRef] [PubMed]

53. Carr, E.L.; Eales, K.L.; Seviour, R.J. Substrate uptake by *Gordonia amarae* in activated sludge foams by fish-mar. *Water Sci. Technol.* **2006**, *54*, 39–45. [CrossRef] [PubMed]

54. Hanada, S.; Liu, W.T.; Shintani, T.; Kamagata, Y.; Nakamura, K. *Tetrasphaera elongata* sp. Nov., a polyphosphate-accumulating bacterium isolated from activated sludge. *Int. J. Syst. Evol. Microbiol.* **2002**, *52*, 883–887. [PubMed]

55. Zhang, B.; Xu, X.; Zhu, L. Structure and function of the microbial consortia of activated sludge in typical municipal wastewater treatment plants in winter. *Sci. Rep.* **2017**, *7*, 17930. [CrossRef] [PubMed]

56. Abubucker, S.; Segata, N.; Goll, J.; Schubert, A.M.; Izard, J.; Cantarel, B.L.; Rodriguez-Mueller, B.; Zucker, J.; Thiagarajan, M.; Henrissat, B.; et al. Metabolic reconstruction for metagenomic data and its application to the human microbiome. *PLoS Comput. Biol.* **2012**, *8*, e1002358. [CrossRef] [PubMed]

57. Caspi, R.; Altman, T.; Billington, R.; Dreher, K.; Foerster, H.; Fulcher, C.A.; Holland, T.A.; Keseler, I.M.; Kothari, A.; Kubo, A.; et al. The metacyc database of metabolic pathways and enzymes and the biocyc collection of pathway/genome databases. *Nucleic Acids Res.* **2014**, *42*, D459–D471. [CrossRef] [PubMed]

58. Andersson, D.I.; Hughes, D. Microbiological effects of sublethal levels of antibiotics. *Nat. Rev. Microbiol.* **2014**, *12*, 465–478. [CrossRef] [PubMed]

59. Kolpin, D.W.; Furlong, E.T.; Meyer, M.T.; Thurman, E.M.; Zaugg, S.D.; Barber, L.B.; Buxton, H.T. Pharmaceuticals, hormones, and other organic wastewater contaminants in U.S. Streams, 1999–2000: A national reconnaissance. *Environ. Sci. Technol.* **2002**, *36*, 1202–1211. [CrossRef] [PubMed]

60. Rodriguez-Rojas, A.; Rodriguez-Beltran, J.; Couce, A.; Blazquez, J. Antibiotics and antibiotic resistance: A bitter fight against evolution. *Int. J. Med. Microbiol.* **2013**, *303*, 293–297. [CrossRef] [PubMed]

61. Newton, R.J.; McLellan, S.L.; Dila, D.K.; Vineis, J.H.; Morrison, H.G.; Eren, A.M.; Sogin, M.L. Sewage reflects the microbiomes of human populations. *MBio* **2015**, *6*, e02574. [CrossRef] [PubMed]

62. Yoon, Y.; Chung, H.J.; Wen Di, D.Y.; Dodd, M.C.; Hur, H.G.; Lee, Y. Inactivation efficiency of plasmid-encoded antibiotic resistance genes during water treatment with chlorine, uv, and uv/h2o2. *Water Res.* **2017**, *123*, 783–793. [CrossRef] [PubMed]

63. Laquaz, M.; Dagot, C.; Bazin, C.; Bastide, T.; Gaschet, M.; Ploy, M.C.; Perrodin, Y. Ecotoxicity and antibiotic resistance of a mixture of hospital and urban sewage in a wastewater treatment plant. *Environ. Sci. Pollut. Res. Int.* **2017**, *25*, 9243–9253. [CrossRef] [PubMed]

64. Zhang, X.X.; Zhang, T. Occurrence, abundance, and diversity of tetracycline resistance genes in 15 sewage treatment plants across china and other global locations. *Environ. Sci. Technol.* **2011**, *45*, 2598–2604. [CrossRef] [PubMed]

65. Yuan, Q.B.; Guo, M.T.; Yang, J. Fate of antibiotic resistant bacteria and genes during wastewater chlorination: Implication for antibiotic resistance control. *PLoS ONE* **2015**, *10*, e0119403. [CrossRef] [PubMed]

66. Calero-Caceres, W.; Muniesa, M. Persistence of naturally occurring antibiotic resistance genes in the bacteria and bacteriophage fractions of wastewater. *Water Res.* **2016**, *95*, 11–18. [CrossRef] [PubMed]

67. Watkinson, A.J.; Murby, E.J.; Costanzo, S.D. Removal of antibiotics in conventional and advanced wastewater treatment: Implications for environmental discharge and wastewater recycling. *Water Res.* **2007**, *41*, 4164–4176. [CrossRef] [PubMed]

68. Chen, B.; Yuan, K.; Chen, X.; Yang, Y.; Zhang, T.; Wang, Y.; Luan, T.; Zou, S.; Li, X. Metagenomic analysis revealing antibiotic resistance genes (args) and their genetic compartments in the Tibetan environment. *Environ. Sci. Technol.* **2016**, *50*, 6670–6679. [CrossRef] [PubMed]

69. Pawlowski, A.C.; Wang, W.; Koteva, K.; Barton, H.A.; McArthur, A.G.; Wright, G.D. A diverse intrinsic antibiotic resistome from a cave bacterium. *Nat. Commun.* **2016**, *7*, 13803. [CrossRef] [PubMed]

70. Li, B.; Ju, F.; Cai, L.; Zhang, T. Profile and fate of bacterial pathogens in sewage treatment plants revealed by high-throughput metagenomic approach. *Environ. Sci. Technol.* **2015**, *49*, 10492–10502. [CrossRef] [PubMed]

71. Forsberg, K.J.; Reyes, A.; Wang, B.; Selleck, E.M.; Sommer, M.O.A.; Dantas, G. The shared antibiotic resistome of soil bacteria and human pathogens. *Science* **2012**, *337*, 1107–1111. [CrossRef] [PubMed]

72. Alam, M.Z.; Aqil, F.; Ahmad, I.; Ahmad, S. Incidence and transferability of antibiotic resistance in the enteric bacteria isolated from hospital wastewater. *Braz. J. Microbiol.* **2013**, *44*, 799–806. [CrossRef] [PubMed]

water

MDPI

Article

Distribution and Genotyping of Aquatic *Acinetobacter baumannii* Strains Isolated from the Puzi River and Its Tributaries Near Areas of Livestock Farming

Hsin-Chi Tsai [1,2], Ming-Yuan Chou [3,†], Yi-Jia Shih [4,5,6,†], Tung-Yi Huang [4,†], Pei-Yu Yang [7,†], Yi-Chou Chiu [8], Jung-Sheng Chen [4] and Bing-Mu Hsu [4,9,*]

[1] Department of Psychiatry, School of Medicine, Tzu Chi University, Hualien 970, Taiwan; css30bmw@yahoo.com.tw
[2] Department of Psychiatry, Buddhist Tzu-Chi General Hospital, Hualien 970, Taiwan
[3] Department of Internal Medicine, Cheng Hsin Hospital, Taipei 111, Taiwan; colin73915@hotmail.com
[4] Department of Earth and Environmental Sciences, National Chung Cheng University, Chiayi 621, Taiwan; eja0313@gmail.com (Y.-J.S.); tyhuang27@gmail.com (T.-Y.H.); nicky071214@gmail.com (J.-S.C.)
[5] Department of Biotechnology, Xiamen Ocean Vocational College, Xiamen 361000, China
[6] Fisheries College, Jimei University, Xiamen 361000, China
[7] Department of Laboratory, Show Chwan Memorial Hospital, Changhua 500, Taiwan; peyyuh2900@gmail.com
[8] General Surgery, Surgical Department, Cheng Hsin General Hospital, Taipei 111, Taiwan; ejchiu3@yahoo.com.tw
[9] Center for Innovative on Aging Society (CIRAS), National Chung Cheng University, Chiayi 621, Taiwan
* Correspondence: bmhsu@ccu.edu.tw; Tel.: +886-952840868; Fax: +886-5-2720807
† Ming-Yuan Chou, Yi-Jia Shih, Tung-Yi Huang and Pei-Yu Yang have equal contributions to the first author.

Received: 1 July 2018; Accepted: 28 September 2018; Published: 1 October 2018

check for updates

Abstract: *Acinetobacter baumannii* is an important health care-associated bacterium and a common multidrug-resistant pathogen. The use of antibiotics in the husbandry industry has raised concerns about drug-resistant *A. baumannii* strains, which may affect humans. This study aimed to investigate the seasonal distribution of *A. baumannii* in aquatic environments near areas of livestock farming. The geographic distribution, antibiotic resistance characteristic, and DNA fingerprinting genotype of *A. baumannii* were also studied. The results showed that environmental *A. baumannii* was prevalent during the summer and autumn. The hotspots for *A. baumannii* were found at the sampling sites of livestock wastewater channels (21.4%; 3/14) and the tributaries adjacent to livestock farms (15.4%; 2/13). The prevalence of *A. baumannii* at these locations was significantly higher than those adjacent to the Puzi River. Multidrug-resistant strain of *A. baumannii* was not found in this study, with only one strain (5%; 1/20) being resistant to tetracycline. Of the isolates that were obtained, 10% (2/20) and 20% (4/20) were found to be intermediately resistant to tetracycline and sulphamethoxazole/trimethoprim, respectively. The genotyping patterns and clustering analysis indicated that enterobacterial repetitive intergenic consensus sequence polymerase chain reaction (ERIC-PCR) differentiated *A. baumannii* strains effectively. There were two major clusters that could then be subtyped into 20 *A. baumannii* strains with 15 profiles. The *A. baumannii* strains that were isolated from upstream of the Puzi River and livestock wastewater channels were composed of Cluster I. Cluster II only contained isolates from downstream of the Puzi River area. Furthermore, isolates from adjacent sites were shown to have identical profiles (100%). These results suggest that *A. baumannii* may have spread through free-flowing water in this study. Therefore, we propose that livestock wastewater is one of the sources that contribute to *A. baumannii* pollution in water bodies. In summary, continuous monitoring of antibiotic pollution in livestock wastewater is required.

Keywords: *Acinetobacter baumannii*; antibiotic-resistant strains; aquatic environment; ERIC-PCR

1. Introduction

Acinetobacter baumannii is a Gram-negative, non-fermenting, aerobic bacterium [1–3] which is considered to be ubiquitous as it can be recovered from various environments, including soil or surface water [4–7]. It is also an important health care-associated pathogen, which mostly causes opportunistic infections in immunocompromised individuals [6]. The most prevalent symptoms that are caused by *A. baumannii* in hospitals include urinary tract infection, meningitis, bacteraemia, peritonitis, surgical wound infection, and pneumonia [4,8,9]. Many studies have surveyed the mortality rate of the pathogen, which ranged from 5% in general wards to more than 60% in patients suffering from multidrug-resistant *A. baumannii* infections [8,10]. In the treatment of various *A. baumannii* infections, a reliable method for identification and characterization of the strains is necessary. Numerous methods have been developed for analysis and molecular typing of *A. baumannii*, such as plasmid analysis, ribotyping, multilocus sequence genotyping (MLST), pulsed-field gel electrophoresis (PFGE), and several other sequence-based polymerase chain reaction (PCR) typing techniques [2,4,6,11–13]. Despite MLST and PFGE being highly discriminative genotyping methods, sequence-based DNA fingerprinting PCR techniques have advantages of performance ease and economic viability [2,12,14]. Since the enterobacterial repetitive intergenic consensus polymerase chain reaction (ERIC-PCR) technique has a sound discrimination index (over 94%) [12] and a relatively low cost for large-scale genotyping of *A. baumannii* when compared with other sequence-based PCR techniques, it has been widely adopted in the studies of environmental *A. baumannii* studies.

Multidrug-resistant *A. baumannii* (MDR-AB) is referred to as resistance to three out of four antibiotic classes, namely ceftazidime, ciprofloxacin, gentamicin, and imipenem [15]. As the scale of antibiotic adoption in clinical treatment increases, the presence of MDR-AB also increases, which poses a great challenge for public health; therefore, surveillance of environmental *A. baumannii* is urgent [8,16]. Recently, it has been reported that the incidence of nosocomial MDR-AB infection has elevated significantly. As such, the presence of MDR-AB in the environment could increase the risk of infection [8,17–19].

A. baumannii exists not only in hospitals, but also in its natural habitats, such as soil, surface water, human skin, and inorganic surfaces [6,20–23]. Antimicrobial agents are widely used in livestock and aquaculture to suppress the growth of bacteria and to boost survival rates [24–26]. However, residue antibiotic agents from various sources, such as waste animal feed and animal excrement, can leak into the environment as active ingredients. Indeed, environmental studies have proven that there were substantial amounts of various antibiotic compounds in soils [27]. Long-term antibiotic contamination can change the gene profile of bacteria in livestock and natural habitats. For example, sulfonamide-resistance genes (sul1, sul2, and sul3) have commonly been found in aquatic environments in northern Vietnam, and they have been isolated from the excrement of pigs suffering from diarrhea in Ontario [28,29]. These reports provide evidence that the use of antibiotics in the husbandry industry may promote the emergence of antibiotic-resistant bacteria. The most commonly used livestock-associated antibiotic classes include aminoglycosides, cephalosporins, chloramphenicol, lincosamides, macrolides, penicillins, polyether ionophores, polypeptides, and tetracyclines, with tetracyclines and macrolides combined products as the most frequently used antibiotic agents [24,30–32]. The increased use of antibiotics in livestock and aquaculture could lead to an increase in the population of antibiotic-resistant *A. baumannii* strains, causing the environment to act as a reservoir of these populations [26].

In recent years, many studies have focused on the epidemiology and analysis of antibiotic resistance, resistance mechanisms, and novel treatments of *A. baumannii* [4,6,19,33]. However, evaluations on the impact of antibiotic usage on the environment were less clear. Many studies have revealed that antibiotic residues can be observed in liquid manure from pig husbandries, suggesting that microorganisms surrounding the aquatic environment near these husbandries are affected by these antibiotic residues and result in the risk of natural selection of these antibiotic-resistant pathogens from the environment. The aim of this study was to understand the seasonal variability and antibiotic susceptibility profiles of *A. baumannii* in the Puzi River and its tributaries near areas of livestock farming and to characterise *A. baumannii* in aquatic environments using the ERIC-PCR method.

2. Materials and Methods

2.1. Collection and Concentration of Water Samples

Monthly water samples were collected from the Puzi River from May 2014 to April 2015. Water samples were collected from 32 locations. Subsequently, these 32 locations were separated into three areas based on our previous studies [34] (Figure 1); sites PR01–PR12 (Area A) are located upstream of the Puzi River, sites PR14–PR25 (Area B) are located midstream, and sites PR26–PR34 (Area C) are located downstream. Additionally, water samples from the livestock wastewater channel, household wastewater channel, and tributary were collected at 30 locations around the Puzi River in October 2015. The sampling sites are summarised in Figure 2. At each sampling site, a water sample of approximately 3000 mL was collected for pathogen detection. Each water sample was stored in three sterile one-litre bottles and was transported to the laboratory at 4 °C within 24 h for analysis.

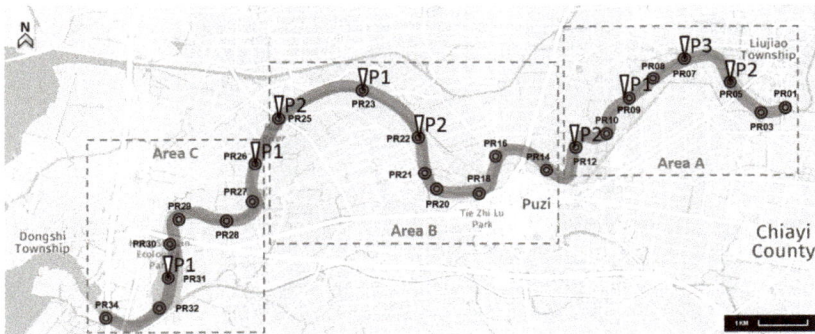

Figure 1. Sampling locations on the Puzi River in this study (P indicates the distribution of *Acinetobacter baumannii*; numbers indicate the number of strains detected).

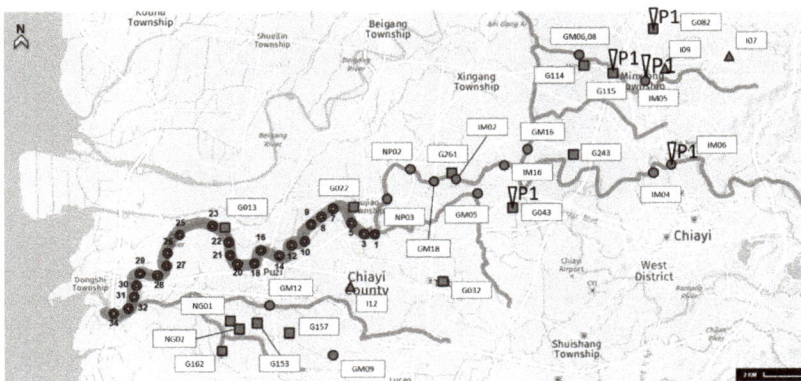

Figure 2. Sampling locations of wastewater surrounding the Puzi River basin in this study (P indicates the distribution of *Acinetobacter baumannii*; numbers indicate the number of isolates detected).

For the detection of specific microbial pathogens, 300 mL of each water sample was filtered through 47 mm GN-6 membranes (Pall, Mexico City, Mexico) with a pore size of 0.45 μm, in a stainless steel filter holder. Subsequently, the membranes were used for sample enrichment of each specific pathogen.

2.2. Enrichment and Identification of A. baumannii

For *A. baumannii* enrichment, the samples that remained on the membranes after filtration were cultured in MacConkey Broth (HIMEDIA, M007, Taipei, Taiwan). Each sample was selectively cultured on CHROMagar™ *Acinetobacter* (CHROMagar, Paris, France) and 5% sheep blood agar (TPM, TPM150M, Taiwan). Subsequently, the agar plates were incubated at 30 °C for 24 h.

DNA extraction was performed using 1 mL of the concentrated pellet from MacConkey Broth (including each suspected isolate from CHROMagar™ *Acinetobacter* agar that underwent re-amplification), using a MagPurix 12s Automated Nucleic Acid Purification System (Zinexts Life Science Corp., New Taipei City, Taiwan) for automated DNA extraction and a MagPurix Viral DNA Extraction Kit ZP02006, according to the manufacturer's manual. Total DNA eluate of *A. baumannii* (2 μL) was mixed with primers (1 μL, 0.4 μM), 5 μL of Fast-Run Taq Master Mix with Dye, and 16 μL of deionised water to yield a final reaction volume of 25 μL. The primers that were used in this study were P-Ab-ITSF: 5'-CAT TAT CAC GGT AAT TAG TG-3' and P-AbI-TSB: 5'-AGA GCA CTG TGC ACT TAA G-3' [35]. The amplification reaction was performed as follows: denaturation for 5 min at 94 °C, followed by 30 cycles of 30 s at 95 °C, 30 s at 52 °C, and 30 s at 72 °C, and a final extension step at 72 °C for 7 min. For *A. baumannii* detection, positive control DNA (*A. baumannii*, ATCC 19606) was also included in each run. The PCR products were electrophoresed on a 2% agarose gel (Biobasic Inc., Markham, ON, Canada), were stained with an ethidium bromide solution, and were visualised under UV light.

2.3. ERIC-PCR for A. baumannii

ERIC-PCR was performed as described by Soni et al. with modifications [36]. The primers ERIC-1 (5'-ATG TAA GCT CCT GGG GAT TCA C-3') and ERIC-2 (5'-AAG TAA GTG ACT GGG GTG AGC G-3') were synthesised to amplify the ERIC-PCR fingerprints of *A. baumannii*. The PCR mixture (25 μL) was comprising of 200 μM of each deoxy-ribonucleotide triphosphate (dNTP), 1.8 U of Taq polymerase (Biolabs, Ipswich, MA, USA), 3 mM of MgCl₂, 10 mM of Tris-HCl (pH = 9.0), 1.0 μM of each primer, and 50 ng of the DNA templates. An additional amount of sterile distilled water was added to attain a volume of 50 μL. The amplification reaction was performed as follows: denaturation for 7 min at 95 °C, followed by 30 cycles of 1 min at 95 °C, 1 min at 52 °C, and 8 min at 65 °C, and a final extension step at 65 °C for 10 min. The ERIC-PCR products were electrophoresed on a 1.5% agarose gel (Biobasic Inc., Markham, Canada) containing Tris-acetate-EDTA (TAE) and 1 μg/mL ethidium bromide at 100 V for 30 min. Subsequently, they were visualised with a UV transilluminator to obtain photographs.

Additionally, the ERIC-PCR patterns were analysed with the Bionumerics software package (Applied Maths, Austin, TX, USA). The relationship between two given isolates was scored using the Jaccard similarity coefficient, and isolates were clustered based on their inter-isolation similarities using the unweighted pair group method with arithmetic averages.

2.4. Antibiotic Susceptibility of A. baumannii

All *A. baumannii* isolates were tested for antibiotic susceptibility by performing the Kirby–Bauer disk diffusion test on Mueller–Hinton agar plates (Becton, Dickinson and Company, Franklin Lakes, NJ, USA) according to the Clinical & Laboratory Standards Institute (CLSI) [37]. The antibiotics and their dosages that were used for testing in this study included: ciprofloxacin (5 μg), cefepime (30 μg), gentamicin (10 μg), imipenem (10 μg), ampicillin/sulbactam (20/10 μg), sulphamethoxazole/trimethoprim (SXT) (23.75/1.75 μg), and tetracycline (30 μg).

3. Results and Discussion

3.1. Presence of A. baumannii in the Aquatic Environment

The observed presence of *A. baumannii* is summarised in Table 1. The results showed that its detection rate was 3.8% annually, and that it was most prevalent in May, July, August, and September 2015. The detection rate was highest in May (20.8%), followed by July (12.5%). The results indicated that this pathogen occurred more frequently during the summer in aquatic environments. This observation is in accordance with the results of previous studies, which have suggested that *A. baumannii* is strictly aerobic and thermotropic (37–42 °C) [6,21,38]. Furthermore, another study suggested that, during the past decade, the prevalence of this pathogen in hospitals was higher between July and October than between January and June [39]. Moreover, Chu et al. found that in Hong Kong, 53% of medical students and new nurses were colonised with *Acinetobacter* during the summer, compared to 32% in winter [40]. The present study is the first to report the annual distribution of *A. baumannii* and its seasonal fluctuations in natural environments.

The Puzi River was separated into three areas (Figure 1). The highest detection rate of *A. baumannii* was found in Area A (5.2%), followed by Area B (4.2%), while the detection rate in Area C was significantly lower (Table 2). The upstream region of the Puzi River (Area A) displayed the highest annual prevalence. Therefore, wastewaters from different sources, including livestock wastewater channels and tributaries of the Puzi River near livestock farming areas, were sampled to investigate the prevalence of *A. baumannii* in October 2014. The results are shown in Table 2. The detection rate of the pathogen was 21.4% (3/14) in livestock wastewater channels and 15.4% (2/13) in the tributaries of the Puzi River. However, the pathogen was not detected in household wastewater. Figure 2 shows the location of livestock farmlands near the Puzi River. The main location of livestock farming was distributed upstream of the Puzi River (57%; 8/14), and the wastewater outlet also flowed into the Puzi River tributaries. These results suggested that *A. baumannii* was transmitted from livestock wastewater into the natural aquatic environment, since the detection rate of this pathogen was higher in wastewater than in the main river. The highest prevalence of *A. baumannii* was in the area upstream of the Puzi River basin. Furthermore, 36% (5/14) of livestock farmland was distributed in the area downstream of the Puzi River basin, which is close to the estuarine environment, where conditions might not be optimal for *A. baumannii* growth.

Table 1. Detection rate of *Acinetobacter baumannii* from the Puzi River per annum.

Sampling Date	Positive Samples	Total Samples	Detection Rate
14 May	5	24	20.8%
14 June	0	24	0
14 July	3	24	12.5%
14 August	1	24	4.2%
14 September	2	24	8.3%
14 October	0	24	0
14 November	0	24	0
14 December	0	24	0
15 January	0	24	0
15 February	0	24	0
15 March	0	24	0
15 April	0	24	0
Total	11	288	3.8%

Table 2. Detection rate of *Acinetobacter baumannii* for different sampling areas.

Location	Sites	Positive Samples	Total Samples	Detection Rate
Puzi River	Area A (PR01-PR12)	5	96	5.2%
	Area B (PR14-PR25)	4	96	4.2%
	Area C (PR26-PR34)	2	96	2.1%
Channels and tributaries	Livestock wastewater channels	3	14	21.4%
	Household wastewater channels	0	3	0%
	Puzi River tributaries	2	13	15.4%

Fernando et al. investigated and isolated *A. baumannii* from a river and nearby dairy farms [38]. Additionally, there are reports that *A. baumannii* isolates, which carried the BLAOXA-23 carbapenemase gene, were isolated from the Seine River in downtown Paris [41] and from the Tietê and Pinheiros rivers in Brazil [42]. Nevertheless, the seasonal distribution and sources of this pathogen remain largely unknown. In previous studies, this organism has primarily been investigated in hospitals and the clinical environment. The results of these studies have consistently indicated that *A. baumannii* is widespread in nosocomial environments [4,17,43]. This organism also prefers to grow in humid conditions. Therefore, the clinical environment acts as a reservoir of *A. baumannii* and leads to opportunistic infections in humans [9,44,45]. However, different places in the natural environment, such as rivers, soil, storage tanks in dairy farms, and manure, can also serve as reservoirs for this pathogen [38]. This study first summarises the observed seasonal prevalence of *A. baumannii* in the aquatic environment. The results show that the hotspot basin environment for this pathogen is livestock wastewater. We suggest that the wastewater from livestock farming is a reservoir for *A. baumannii*. Further, this pathogen spreads into the Puzi River tributary and agricultural irrigation canals through the wastewater that is discharged from livestock wastewater channels. However, the prevalence of *A. baumannii* that was observed in this environmental survey remained lower than that observed in hospital studies. One possible explanation for this result is that the aquatic environment is constantly changing and, consequently, flowing water does not act as a good reservoir for this microorganism.

3.2. Antimicrobial Susceptibility

A total of 20 *A. baumannii*-positive samples were isolated from 318 water samples, which were dispersed in 16 locations out of 32 research sites (Figures 1 and 2). The results of the antibiotic susceptibility tests in *A. baumannii*-positive samples are summarised in Table 3. The results show that the positive control strain (ATCC-16906) was highly resistant to SXT, intermediately resistant to tetracycline, and sensitive to other antibiotic agents (Table 4). In addition, the *A. baumannii* -GI-G11511 strain that was isolated from livestock wastewater was resistant to tetracycline. The *A. baumannii*-GI-IM0611 strain, which was isolated from the Puzi River tributary and agricultural irrigation canals, was also found to be intermediately resistant to tetracycline. Four strains (*A. baumannii* PR07-0531, 09-0531, 22-0511, and 23-0521) were isolated from the Puzi River, and they were found to be intermediately resistant to SXT. Additionally, one strain, *A. baumannii*-PR22-0511, was also found to be intermediately resistant to tetracycline. The other 14 environmental strains were found to be susceptible to all antibiotic agents. These results indicate that most of the isolated strains are quite susceptible to tetracycline. Only 5% (1/20) of the analysed strains were resistant to, and 10% (2/20) intermediately resistant to, tetracycline. Only four strains (20%; 4/20) showed intermediate resistance to SXT. Additionally, no multidrug-resistant *A. baumannii* (MDR-AB) was observed in the aquatic environment. The results for antibiotic susceptibility imply that the *A. baumannii* that is present in the aquatic environment differs from the *A. baumannii* in hospitals, based on the multidrug-resistant outcome. Therefore, we will analyse the homology between the MDR-AB strains from local hospitals and aquatic environments in our future studies.

Table 3. Antibiotic susceptibility of environmental *Acinetobacter baumannii* isolates as determined by Kirby-Bauer disk diffusion tests.

Antibiotics Resistance Phenotype	Number	Resistant	Intermediate	Susceptible
Ciprofloxacin	20	0	0	20 (100%)
Cefepime	20	0	0	20 (100%)
Gentamicin	20	0	0	20 (100%)
Imipenem	20	0	0	20 (100%)
Ampicillin-sulbactam	20	0	0	20 (100%)
Sulphamethoxazole/Trimethoprim	20	0	4 (20%)	16 (80%)
Tetracycline	20	1 (5%)	2 (10%)	17 (85%)

Table 4. Antibiotic susceptibility of different *Acinetobacter baumannii* isolates from the aquatic environment.

No.	Code of Strains	Antibiotics Resistance Phenotype						
		CIP	FEP	G	I	SAM	SXT	T
1	ATCC-16906	S	S	S	S	S	R	I
2	*A. baumannii*-PR07-0531	S	S	S	S	S	I	S
3	*A. baumannii*-PR09-0531	S	S	S	S	S	I	S
4	*A. baumannii*-PR21-0511	S	S	S	S	S	I	I
5	*A. baumannii*-PR23-0521	S	S	S	S	S	I	S
6	*A. baumannii*-PR31-0511	S	S	S	S	S	S	S
7	*A. baumannii*-PR07-0711	S	S	S	S	S	S	S
8	*A. baumannii*-PR07-0721	S	S	S	S	S	S	S
9	*A. baumannii*-PR22-0711	S	S	S	S	S	S	S
10	*A. baumannii*-PR26-0721	S	S	S	S	S	S	S
11	*A. baumannii*-PR25-0811	S	S	S	S	S	S	S
12	*A. baumannii*-PR25-0821	S	S	S	S	S	S	S
13	*A. baumannii*-PR05-0911	S	S	S	S	S	S	S
14	*A. baumannii*-PR05-0912	S	S	S	S	S	S	S
15	*A. baumannii*-PR12-0911	S	S	S	S	S	S	S
16	*A. baumannii*-PR12-0912	S	S	S	S	S	S	S
17	*A. baumannii*-GI-IM0511	S	S	S	S	S	S	S
18	*A. baumannii*-GI-IM0611	S	S	S	S	S	S	I
19	*A. baumannii*-GI-G04311	S	S	S	S	S	S	S
20	*A. baumannii*-GI-G08211	S	S	S	S	S	S	S
21	*A. baumannii*-GI-G11511	S	S	S	S	S	S	R

Notes: CIP: Ciprofloxacin, FEP: Cefepime, GEN: Gentamicin, IPM: Imipenem, SAM: Ampicillin-sulbactam, SXT: Sulphamethoxazole/Trimethoprim, T: Tetracycline, R: Resistant, I: Intermediately resistant, S: Susceptible.

To conclude the results, it can be said that the environmental *A. baumannii* strains in this study showed much lower antibiotic resistance than hospital *A. baumannii*. The results also indicated that the strains that were present in water samples were quite susceptible to, and/or intermediately resistant to, certain antibiotics, including SXT and tetracycline. Previous studies have demonstrated that tetracyclines and SXT are the common classes of antimicrobials that are used in livestock [24,26]. Our observations indicate potential mechanisms by which the frequent use of agricultural antibiotics may lead to the formation of antibiotic-resistant bacteria strains. Such strains also pose potential risks to humans, through the direct transmission of the resistant bacteria by water sources or the transfer of resistance genes from antibiotic-resistant bacteria in the agricultural environment into human pathogens [42,46]. Therefore, it is necessary to understand the mechanism of transmission of MDR-AB strains into aquatic environments. In future studies, we will not only focus on understanding the association between agricultural antibiotic usage conditions and the environmental prevalence of MDR-AB, but also on the analysis of homology between MDR-AB strains from local hospitals and aquatic environments.

3.3. ERIC-PCR Fingerprint Analysis

A total of 20 *A. baumannii* strains were further characterised for strain genotyping and were compared with a reference strain (ATCC 16906) using ERIC-PCR analysis. Five isolates were collected from livestock wastewater and the Puzi River tributaries near livestock farming areas, and 15 isolates were collected from the Puzi River basin. The standard that was used to determine the degree of similarity between different ERIC-PCR fingerprints was based on similarities between the reference strains and the samples from the different *A. baumannii* colonies at the same sampling sites (Site No. PR05, 07, 12, and 25). According to the ERIC-PCR analysis, the Jaccard similarity coefficient between the two reference strains was 85%. The isolates from different colonisation areas at the four sampling sites (*A. baumannii*-PR12-0911 vs. *A. baumannii*-PR12-0912, *A. baumannii*-PR05-0911 vs. *A. baumannii*- PR05-0912, *A. baumannii*-PR25-0811 vs. *A. baumannii*-PR25-0821, and *A. baumannii* -PR07-0711 vs. *A. baumannii*-PR07-0721) showed a Jaccard similarity coefficient of 100%, which was used to confirm the genotyping (Figure 3). Two major clusters were observed at a similarity level of less than 20%, and 20 *A. baumannii* strains were subtyped into 15 profiles. The results also showed that the strains from Area A of the Puzi River were similar to the livestock wastewater and belonged to cluster I (bootstrap value (*p*-value) = 90%). Further, cluster II contained isolates from the other Puzi River area (*p*-value = 90%) (Figure 3). Maleki et al. (2016) found that the diversity of genetic patterns of *A. baumannii* that were observed by ERIC-PCR analysis was due to the wide distribution in hospitals [2]. Here, we postulate that this high diversity is due to the distributions of sampling sites with different sources. Strains *A. baumannii*-PR22-0511 and *A. baumannii*-PR23-0521 from two adjacent sampling sites showed an identical profile (100% similarity). This result suggests that one mechanism of *A. baumannii* transmission is through free-flowing water, which leads to spreading to other aquatic environments.

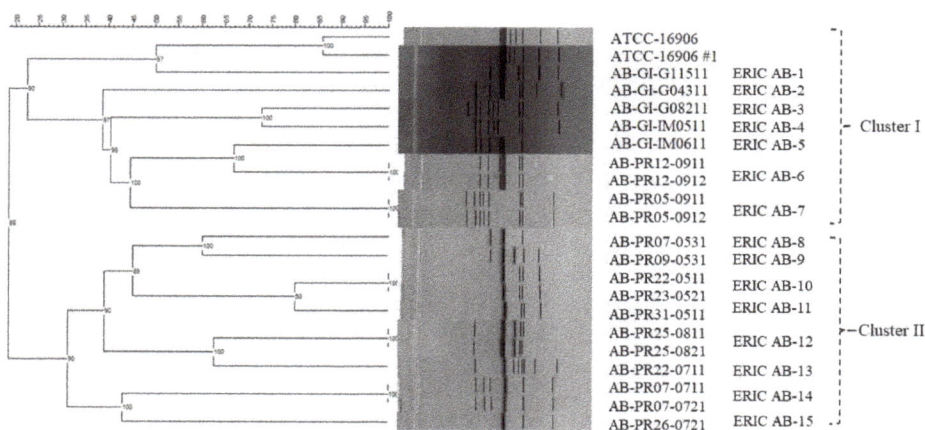

Figure 3. Amplification clustering patterns of *Acinetobacter baumannii* (AB) by enterobacterial repetitive intergenic consensus PCR (ERIC-PCR).

We regard the genotyping of *A. baumannii* isolates as a necessary means to control the epidemic that has been caused by this organism. Therefore, different DNA fingerprinting techniques have been developed for the quick and accurate classification of *A. baumannii* isolates. Molecular genotyping methods, including plasmid profiling, ribotyping, PFGE, MLST, and PCR-based typing methods, have been evaluated as potential methods to characterise *A. baumannii* isolates [6]. Despite MLST and PFGE being highly discriminative genotyping methods, PCR-based DNA fingerprinting techniques have advantages of performance ease and economic viability [2,6,12]. To date, there have been many studies describing different PCR-based methods to type MDR-AB [47–49]. In general, the ERIC-PCR method is a common, easy, and quick fingerprinting technique for characterising *A. baumannii* isolates [2,48,49]. The results of this study indicate that the ERIC-PCR method is useful for the analysis of genetic variation among environmental *A. baumannii* isolates.

4. Conclusions

In this study, we first investigated the seasonal temporal distribution and antibiotic resistance of *A. baumannii* in natural aquatic environments. In conclusion, the seasonal prevalence of *A. baumannii* and the percentage of antibiotic-resistant *A. baumannii* isolates in water bodies were found to be lower than those in known nosocomial environments. However, we observed that the highest detection rate of *A. baumannii* occurred in livestock wastewater. We also observed one tetracycline-resistant strain. In addition, four strains were found to be intermediately resistant to SXT, and one was intermediately resistant to tetracycline. These results indicate the necessity of monitoring on the use of antimicrobials in livestock. Further, livestock wastewater is a potential source of *A. baumannii* contamination. This is an important issue for the transmission pathway of *A. baumannii* between the environment and hospitals, or even long-term care facilities, and is worth further exploration.

Author Contributions: Conceptualization, B.-M.H and H.-C.T; Methodology, Y.-J.S; Software, M.-Y.C; Validation, P.-Y.Y, J.-S.C and Y.-C.C; Formal Analysis, J.-S.C and Y.-J.S; Investigation, J.-S.C.; Resources, M.-Y.C; Data Curation, J.-S.C and Y.-J.S; Writing-Original Draft Preparation, H.-C.T, M.-Y.C and Y.-J.S; Writing-Review & Editing, P.-Y.Y, T.-Y.H and B.-M.H; Visualization, B.-M.H; Supervision, B.-M.H, P.-Y.Y, H.-C.T and J.-S.C; Project Administration, B.-M.H and H.-C.T; Funding Acquisition, B.-M.H, M.-Y.C, P.-Y.Y and H.-C.T.

Funding: This research was funded by the Ministry of Science and Technology of Taiwan (MOST 106-2116-M-194-013), the Centers for Disease Control, Taiwan, R.O.C. (MOHW105-CDC-C-114-112601 and 106-CDC-C-114-122601), the Buddhist Tzu Chi Hospital, R.O.C., Show Chwan Health Care System and Cheng Hsin General Hospital (TCRD106-37, RD107023, CHGH106-01). This research was also supported by the Center for Innovative Research on Aging Society (CIRAS) from The Featured Areas Research Center Program within the framework of the Higher Education Sprout Project by the Ministry of Education (MOE) in Taiwan.

Conflicts of Interest: The authors declare that they have no conflict of interest.

References

1. Alaei, N.; Aziemzadeh, M.; Bahador, A. Antimicrobial resistance profiles and genetic elements involved in carbapenem resistance in acinetobacter baumannii isolates from a referral hospital in Southern Iran. *J. Glob. Antimicrob. Resist.* **2016**, *5*, 75–79. [CrossRef] [PubMed]

2. Maleki, A.; Vandyousefi, J.; Mirzaie, Z.; Ghafourian, S.; Kazemian, H.; Sadeghifard, N. Molecular analysis of the isolates of acinetobacter baumannii isolated from tehran hospitals using ERIC-PCR method. *Mod. Med. Lab. J.* **2016**, *1*, 12–16. [CrossRef]

3. Rosa, R.; Depascale, D.; Cleary, T.; Fajardo-Aquino, Y.; Kett, D.H.; Munoz-Price, L.S. Differential environmental contamination with acinetobacter baumannii based on the anatomic source of colonization. *Am. J. Infect. Control.* **2014**, *42*, 755–757. [CrossRef] [PubMed]

4. Zhang, H.-Z.; Zhang, J.-S.; Qiao, L. The acinetobacter baumannii group: A systemic review. *World J. Emerg. Med.* **2013**, *4*, 169. [CrossRef] [PubMed]

5. Taitt, C.R.; Leski, T.; Stockelman, M.G.; Craft, D.W.; Zurawski, D.V.; Kirkup, B.C.; Vora, G.J. Antimicrobial resistance determinants in acinetobacter baumannii isolates taken from military treatment facilities. *Antimicrob. Agents Chemother.* **2014**, *58*, 767–781. [CrossRef] [PubMed]

6. Peleg, A.Y.; Seifert, H.; Paterson, D.L. Acinetobacter baumannii: Emergence of a successful pathogen. *Clin. Microbiol. Rev.* **2008**, *21*, 538–582. [CrossRef] [PubMed]

7. Jawad, A.; Snelling, A.; Heritage, J.; Hawkey, P. Exceptional desiccation tolerance of acinetobacter radioresistens. *J. Hosp. Infect.* **1998**, *39*, 235–240. [CrossRef]

8. Fournier, P.E.; Richet, H.; Weinstein, R.A. The epidemiology and control of acinetobacter baumannii in health care facilities. *Clin. Infect. Dis.* **2006**, *42*, 692–699. [CrossRef] [PubMed]

9. Denton, M.; Wilcox, M.; Parnell, P.; Green, D.; Keer, V.; Hawkey, P.; Evans, I.; Murphy, P. Role of environmental cleaning in controlling an outbreak of acinetobacter baumannii on a neurosurgical intensive care unit. *J. Hosp. Infect.* **2004**, *56*, 106–110. [CrossRef] [PubMed]

10. Garnacho-Montero, J.; Ortiz-Leyba, C.; Jimenez-Jimenez, F.; Barrero-Almodovar, A.; Garcia-Garmendia, J.; Bernabeu-Wittell, M.; Gallego-Lara, S.; Madrazo-Osuna, J. Treatment of multidrug-resistant acinetobacter baumannii ventilator-associated pneumonia (VAP) with intravenous colistin: A comparison with imipenem-susceptible vap. *Clin. Infect. Dis.* **2003**, *36*, 1111–1118. [CrossRef] [PubMed]

11. Bartual, S.G.; Seifert, H.; Hippler, C.; Luzon, M.A.D.; Wisplinghoff, H.; Rodríguez-Valera, F. Development of a multilocus sequence typing scheme for characterization of clinical isolates of acinetobacter baumannii. *J. Clin. Microbiol.* **2005**, *43*, 4382–4390. [CrossRef] [PubMed]

12. Bou, G.; Cervero, G.; Dominguez, M.; Quereda, C.; Martinez-Beltran, J. PCR-based DNA fingerprinting (REP-PCR, AP-PCR) and pulsed-field GEL electrophoresis characterization of a nosocomial outbreak caused by imipenem-and meropenem-resistant acinetobacter baumannii. *Clin. Microbiol. Infect.* **2000**, *6*, 635–643. [CrossRef] [PubMed]

13. Seifert, H.; Gerner-Smidt, P. Comparison of ribotyping and pulsed-field gel electrophoresis for molecular typing of acinetobacter isolates. *J. Clin. Microbiol.* **1995**, *33*, 1402–1407. [PubMed]

14. Vila, J.; Marcos, M.; De Anta, M.J. A comparative study of different pcr-based DNA fingerprinting techniques for typing of the acinetobacter calcoaceticus-A. Baumannii complex. *J. Med. Microbiol.* **1996**, *44*, 482–489. [CrossRef] [PubMed]

15. Karlowsky, J.A.; Draghi, D.C.; Jones, M.E.; Thornsberry, C.; Friedland, I.R.; Sahm, D.F. Surveillance for antimicrobial susceptibility among clinical isolates of pseudomonas aeruginosa and acinetobacter baumannii from hospitalized patients in the United States, 1998 to 2001. *Antimicrob. Agents Chemother.* **2003**, *47*, 1681–1688. [CrossRef] [PubMed]

16. Fournier, P.-E.; Vallenet, D.; Barbe, V.; Audic, S.; Ogata, H.; Poirel, L.; Richet, H.; Robert, C.; Mangenot, S.; Abergel, C. Comparative genomics of multidrug resistance in acinetobacter baumannii. *PLoS Genet.* **2006**, *2*, e7. [CrossRef] [PubMed]

17. Falagas, M.; Kopterides, P. Risk factors for the isolation of multi-drug-resistant acinetobacter baumannii and pseudomonas aeruginosa: A systematic review of the literature. *J. Hosp. Infect.* **2006**, *64*, 7–15. [CrossRef] [PubMed]

18. Hsueh, P.-R.; Teng, L.-J.; Chen, C.-Y.; Chen, W.-H.; Ho, S.-W.; Luh, K.-T. Pandrug-resistant acinetobacter baumannii causing nosocomial infections in a university hospital, Taiwan. *Emerg. Infect. Dis.* **2002**, *8*, 827. [CrossRef] [PubMed]

19. Dijkshoorn, L.; Nemec, A.; Seifert, H. An increasing threat in hospitals: Multidrug-resistant acinetobacter baumannii. *Nat. Rev. Microbiol.* **2007**, *5*, 939–951. [CrossRef] [PubMed]

20. Visca, P.; Seifert, H.; Towner, K.J. Acinetobacter infection—An emerging threat to human health. *IUBMB Life* **2011**, *63*, 1048–1054. [CrossRef] [PubMed]

21. Karumathil, D.P.; Yin, H.-B.; Kollanoor-Johny, A.; Venkitanarayanan, K. Effect of chlorine exposure on the survival and antibiotic gene expression of multidrug resistant acinetobacter baumannii in water. *Int. J. Environ. Res. Public Health* **2014**, *11*, 1844–1854. [CrossRef] [PubMed]

22. Francey, T.; Gaschen, F.; Nicolet, J.; Burnens, A.P. The role of acinetobacter baumannii as a nosocomial pathogen for dogs and cats in an intensive care unit. *J. Vet. Intern. Med.* **2000**, *14*, 177–183. [CrossRef] [PubMed]

23. Vaneechoutte, M.; Devriese, L.A.; Dijkshoorn, L.; Lamote, B.; Deprez, P.; Verschraegen, G.; Haesebrouck, F. Acinetobacter baumannii-infected vascular catheters collected from horses in an equine clinic. *J. Clin. Microbiol.* **2000**, *38*, 4280–4281. [PubMed]

24. Landers, T.F.; Cohen, B.; Wittum, T.E.; Larson, E.L. A review of antibiotic use in food animals: Perspective, policy, and potential. *Public Health Rep.* **2012**, *127*, 4–22. [CrossRef] [PubMed]

25. McEwen, S.A.; Fedorka-Cray, P.J. Antimicrobial use and resistance in animals. *Clin. Infect. Dis.* **2002**, *34*, S93–S106. [CrossRef] [PubMed]

26. Widyasari-Mehta, A.; Hartung, S.; Kreuzig, R. From the application of antibiotics to antibiotic residues in liquid manures and digestates: A screening study in one european center of conventional pig husbandry. *J. Environ. Manag.* **2016**, *177*, 129–137. [CrossRef] [PubMed]

27. Kemper, N. Veterinary antibiotics in the aquatic and terrestrial environment. *Ecol. Indic.* **2008**, *8*, 1–13. [CrossRef]

28. Boerlin, P.; Travis, R.; Gyles, C.L.; Reid-Smith, R.; Janecko, N.; Lim, H.; Nicholson, V.; McEwen, S.A.; Friendship, R.; Archambault, M. Antimicrobial resistance and virulence genes of escherichia coli isolates from swine in ontario. *Appl. Environ. Microbiol.* **2005**, *71*, 6753–6761. [CrossRef] [PubMed]

29. Phuong Hoa, P.T.; Nonaka, L.; Hung Viet, P.; Suzuki, S. Detection of the SUL1, SUL2, and SUL3 genes in sulfonamide-resistant bacteria from wastewater and shrimp ponds of north vietnam. *Sci. Total Environ.* **2008**, *405*, 377–384. [CrossRef] [PubMed]

30. Sarmah, A.K.; Meyer, M.T.; Boxall, A.B. A global perspective on the use, sales, exposure pathways, occurrence, fate and effects of veterinary antibiotics (VAS) in the environment. *Chemosphere* **2006**, *65*, 725–759. [CrossRef] [PubMed]

31. Cai, Y.; Cai, Y.e.; Cheng, J.; Mou, S.; Yiqiang, L. Comparative study on the analytical performance of three waveforms for the determination of several aminoglycoside antibiotics with high performance liquid chromatography using amperometric detection. *J. Chromatogr. A* **2005**, *1085*, 124–130. [CrossRef] [PubMed]

32. Butaye, P.; Devriese, L.A.; Haesebrouck, F. Antimicrobial growth promoters used in animal feed: Effects of less well known antibiotics on gram-positive bacteria. *Clin. Microbiol. Rev.* **2003**, *16*, 175–188. [CrossRef] [PubMed]

33. Moradi, J.; Hashemi, F.B.; Bahador, A. Antibiotic resistance of acinetobacter baumannii in Iran: A systemic review of the published literature. *Osong Public Health Res. Perspect.* **2015**, *6*, 79–86. [CrossRef] [PubMed]

34. Tsai, H.C.; Chou, M.Y.; Wu, C.C.; Wan, M.T.; Kuo, Y.J.; Chen, J.S.; Huang, T.Y.; Hsu, B.M. Seasonal distribution and genotyping of antibiotic resistant strains of listeria innocua isolated from a river basin categorized by ERIC-PCR. *Int. J. Environ. Res. Public Health* **2018**, *15*, 1559. [CrossRef] [PubMed]

35. Chen, T.L.; Siu, L.K.; Wu, R.C.; Shaio, M.F.; Huang, L.Y.; Fung, C.P.; Lee, C.M.; Cho, W.L. Comparison of one-tube multiplex PCR, automated ribotyping and intergenic spacer (ITS) sequencing for rapid identification of acinetobacter baumannii. *Clin. Microbiol. Infect.* **2007**, *13*, 801–806. [CrossRef] [PubMed]

36. Soni, D.K.; Singh, M.; Singh, D.V.; Dubey, S.K. Virulence and genotypic characterization of listeria monocytogenes isolated from vegetable and soil samples. *BMC Microbiol.* **2014**, *14*, 241. [CrossRef] [PubMed]

37. Clinical and Laboratory Standards Institute. *Methods for Dilution Antimicrobial Susceptibility Tests for Bacteria That Grow Aerobically, Approved Standard M7-A7*, 7th ed.; Clinical and Laboratory Standards Institute: Wayne, PA, USA, 2006.

38. Fernando, D.M.; Khan, I.U.; Patidar, R.; Lapen, D.R.; Talbot, G.; Topp, E.; Kumar, A. Isolation and characterization of acinetobacter baumannii recovered from campylobacter selective medium. *Front. Microbiol.* **2016**, *7*, 1871. [CrossRef] [PubMed]

39. McDonald, L.C.; Banerjee, S.N.; Jarvis, W.R.; System, N.N.I.S. Seasonal variation of acinetobacter infections: 1987–1996. *Clin. Infect. Dis.* **1999**, *29*, 1133–1137. [CrossRef] [PubMed]

40. Chu, Y.W.; Leung, C.M.; Houang, E.T.S.; Ng, K.C.; Leung, C.B.; Leung, H.Y.; Cheng, A.F.B. Skin carriage of acinetobacters in Hong Kong. *J. Clin. Microbiol.* **1999**, *37*, 2962–2967. [PubMed]

41. Girlich, D.; Poirel, L.; Nordmann, P. First isolation of the BLAOXA-23 carbapenemase gene from an environmental acinetobacter baumannii isolate. *Antimicrob. Agents Chemother.* **2010**, *54*, 578–579. [CrossRef] [PubMed]

42. Turano, H.; Gomes, F.; Medeiros, M.; Oliveira, S.; Fontes, L.C.; Sato, M.I.; Lincopan, N. Presence of high-risk clones of OXA-23-producing acinetobacter baumannii (ST79) and SPM-1-producing pseudomonas aeruginosa (ST277) in environmental water samples in Brazil. *Diagn. Microbiol. Infect. Dis.* **2016**, *86*, 80–82. [CrossRef] [PubMed]

43. Gestal, M.C.; Zurita, J.; Gualpa, G.; Gonzalez, C.; Mino, A.P. Early detection and control of an acinetobacter baumannii multi-resistant outbreak in a hospital in quito, Ecuador. *J. Infect. Dev. Ctries.* **2016**, *10*, 1294–1298. [CrossRef] [PubMed]

44. Aygün, G.; Demirkiran, O.; Utku, T.; Mete, B.; Ürkmez, S.; Yılmaz, M.; Yaşar, H.; Dikmen, Y.; Öztürk, R. Environmental contamination during a carbapenem-resistant acinetobacter baumannii outbreak in an intensive care unit. *J. Hosp. Infect.* **2002**, *52*, 259–262. [CrossRef] [PubMed]

45. Umezawa, K.; Asai, S.; Ohshima, T.; Iwashita, H.; Ohashi, M.; Sasaki, M.; Kaneko, A.; Inokuchi, S.; Miyachi, H. Outbreak of drug-resistant acinetobacter baumannii ST219 caused by oral care using tap water from contaminated hand hygiene sinks as a reservoir. *Am. J. Infect. Control.* **2015**, *43*, 1249–1251. [CrossRef] [PubMed]

46. Ferreira, A.E.; Marchetti, D.P.; De Oliveira, L.M.; Gusatti, C.D.S.; Fuentefria, D.B.; Corcao, G. Presence of OXA-23-producing isolates of acinetobacter baumannii in wastewater from hospitals in Southern Brazil. *Microbiol. Drug Resist.* **2011**, *17*, 221–227. [CrossRef] [PubMed]

47. Ece, G.; Erac, B.; Cetin, H.Y.; Ece, C.; Baysak, A. Antimicrobial susceptibility and clonal relation between acinetobacter baumannii strains at a tertiary care center in Turkey. *Jundishapur J. Microbiol.* **2015**, *8*, e15612. [CrossRef] [PubMed]

48. Aljindan, R.; Alsamman, K.; Elhadi, N. ERIC-PCR genotyping of acinetobacter baumannii isolated from different clinical specimens. *Saudi J. Med. Med. Sci.* **2018**, *6*, 13.

49. Heidari, H.; Halaji, M.; Taji, A.; Kazemian, H.; Shamsabadi, M.S.; Sisakht, M.T.; Ebrahim-Saraie, H.S. Molecular analysis of drug-resistant acinetobacter baumannii isolates by ERIC-PCR. *Meta Gene* **2018**, *17*, 132–135. [CrossRef]

water

MDPI

Article

Presence of Antibiotic-Resistant *Escherichia coli* in Wastewater Treatment Plant Effluents Utilized as Water Reuse for Irrigation

Asli Aslan *, Zachariah Cole, Anunay Bhattacharya and Oghenekpaobor Oyibo

Department of Epidemiology and Environmental Health Sciences, Georgia Southern University, Statesboro, GA 30460, USA; zc00856@georgiasouthern.edu (Z.C.); ab10694@georgiasouthern.edu (A.B.); oo00806@georgiasouthern.edu (O.O.)
* Correspondence: aaslan@georgiasouthern.edu; Tel.: +1-912-478-2565

Received: 20 May 2018; Accepted: 15 June 2018; Published: 18 June 2018

check for
updates

Abstract: Providing safe water through water reuse is becoming a global necessity. One concern with water reuse is the introduction of unregulated contaminants to the environment that cannot be easily removed by conventional wastewater treatment plants (WWTP). The occurrence of ampicillin, sulfamethoxazole, ciprofloxacin, and tetracycline-resistant *Escherichia coli* through the treatment stages of a WWTP (raw sewage, post-secondary, post-UV and post-chlorination) was investigated from January to May 2016. The highest concentrations of antibiotic resistant *E. coli* in the effluent were detected in April after rainfall. Ampicillin-resistant *E. coli* was the most common at the post UV and chlorination stages comprising 63% of the total *E. coli* population. The minimum inhibitory concentration (MIC) analysis showed that one in five isolates was resistant to three or more antibiotics, and the majority of these *E. coli* were resistant to ampicillin, followed by sulfamethoxazole and ciprofloxacin. The highest MIC was detected at the finished water after application of multiple disinfection methods. Tetracycline resistance was the least observed among others, indicating that certain drug families may respond to wastewater treatment differently. Currently, there are no policies to enforce the monitoring of antibiotic-resistant pathogen removal in WWTP. Better guidelines are needed to better regulate reuse water and prevent health risk upon exposure to antibiotic-resistant bacteria.

Keywords: antibiotic resistance; chlorination; *Escherichia coli*; fecal indicator bacteria; reuse water; UV-disinfection

1. Introduction

The discovery of antibiotics has been one of the significant successes in human history. Most of these pharmaceuticals, however, become irrelevant to the disease they were intended to treat over time, as microorganisms have rapidly developed resistance mechanisms to fight back this once lifesaving intervention. Today, over 20,000 potential resistance genes in genome sequencing databases have been discovered since the first antibiotic resistance reported in the late 1930s, right after its medicinal use [1].

Microorganisms harboring resistance genes end up in water [2] and soil [3]. Wastewater [4], agricultural runoff [5], and hospital waste [6] have been reported as sources of antibiotic resistance in the aquatic environment. Water contaminated with antibiotic-resistant bacteria (ARB) has the potential to affect aquatic biodiversity [7,8] and human health adversely. These organisms are introduced to our drinking water resources [9,10] and food systems through irrigation [11,12]. This issue has become a global concern, and the World Health Organization has recently declared ARB as an emerging pollutant in water [13].

It is necessary to address problems arising from water reuse due to water scarcity issues worldwide. One potential concern with reuse is that chemical and biological contaminants in the WWTP effluents can be introduced to the environment. In an earlier study, [14] detected several pharmaceutical and personal care products in surface water and water 30 cm beneath the soil where turf-grass fields were irrigated with reuse water. Today, with increasing water scarcity, WWTP in some states such as California, Texas, and Arizona have been using reclaimed water for irrigation purposes. According to the USEPA, in 2012, 30 states and one U.S. territory have adopted water reuse regulations [15]. Recycled water has been monitored by targeting fecal coliform bacteria [16], but antibiotic-resistant indicator bacteria have not been a part of the monitoring efforts.

The purpose of this study was to determine the antibiotic removal efficacy of a conventional WWTP whose effluents were utilized to irrigate recreational landscapes. *Escherichia coli*, as the fecal indicator bacteria, were targeted for antibiotic resistance in the WWTP. Ampicillin, sulfamethoxazole, ciprofloxacin and tetracycline-resistant *E. coli* were cultured to calculate the removal rates and variability in the resistant population from inflow to the effluent (reuse water). Furthermore, the impact of multiple disinfection steps on the removal of these *E. coli* populations was compared to provide a more detailed assessment of antibiotic resistance in reuse water.

2. Materials and Methods

2.1. Sample Collection

A WWTP serving a small urban community with a population of 25,000 people in Georgia was used for the study. The plant did not receive any industrial discharge. Samples were collected during morning hours (before 9 am), and sampling was repeated five times in 2016 (January–May). Triplicates of grab samples (1 L) were collected from the inflow, post-secondary, post-UV, and post-chlorination stages. All samples ($n = 60$) were transported to the laboratory on ice and processed within 6 h of collection. Seven days' cumulative precipitation data (total precipitation of the day of sampling and previous six days) were obtained from the University of Georgia Weather Network (www.georgiaweather.net).

2.2. Antibiotic Resistant Escherichia coli Culture Collection

2.2.1. Escherichia coli Isolation

Serial dilutions from 10^1 to 10^6 were prepared for the influent and secondary effluent by using sterile phosphate saline water. Triplicates of diluted influent, secondary effluent, undiluted UV-treated effluent, and chlorinated reclaimed water were filtered through a sterile membrane filtration system using 0.45 μm sterile filters. The chlorinated samples were neutralized with 10% sodium thiosulfate prior to analysis. Presumptive *Escherichia coli* were grown on mI agar at 35 ± 0.5 °C. for 18 h [15]. The mI medium contained cefsulodin (final concentration 5 μg/mL) and has been reported to inhibit the growth of gram-positive organisms and non-coliform gram-negative bacteria in the literature [15,16].

The antibiotics tested for resistance in *E. coli* were selected from among the most commonly used and clinically relevant pharmaceuticals (Table 1). Antibiotic-resistant presumptive *E. coli* were enumerated by culturing a separate set of filters on mI Agar plates with an antibiotic of concern—tetracycline (final concentration 16 μg/L), ampicillin (final concentration 32 μg/L), sulfamethoxazole (final concentration 350 μg/L) and ciprofloxacin (final concentration 4 μg/L)—at 37 ± 0.5 °C. All antibiotic concentrations were based on the Clinical & Laboratory Standards Institute (CLSI, Wayne, PA, USA) breakpoints [17]. A control set of filters were also incubated on media without antibiotics at the same conditions, and antibiotic resistance was calculated as percentages by using the formula [18]:

$$\% \ intermediate \ or \ resistant \ = \ \frac{(Presumptive \ E. \ coli) \ on \ antibiotic \ plate}{(Presumptive \ E. \ coli) on \ control \ plate} \times 100 \qquad (1)$$

Table 1. Antibiotics used in the study.

Antibiotic	Abbreviation	Drug Family
Ampicillin	AM	β-lactam penicillin
Ciprofloxacin	CI	Fluoroquinolone
Sulfamethoxazole/Trimethoprim	ST	Folic acid synthesis inhibitor
Tetracycline	TC	Tetracycline

2.2.2. Isolate Identification

Colonies were randomly picked from the mI plates in aseptic conditions, transferred into typtic soy broth (TSB) and grown in a shaker incubator at 37 ± 0.5 °C for 18 h. These cultures were then washed with phosphate saline solution three times and finally stored in cryovials containing TSB with 50% glycerol at -80 °C for further analysis.

The species of these isolates were further confirmed by real time polymerase chain reaction. Briefly, cultures grown overnight were lysed in a bead mill for 60 s at 5000 rpm and the debris was removed by centrifugation [19]. The DNA concentration at the end of the crude extraction was measured using a UV-spectrophotometer (NanoDrop™ 2000, Thermo Scientific, Wilmington, DE, USA). Each DNA extract was analyzed in duplicate by the EC23S857 assay for *E. coli* [20]. The reaction mixture contained 12.5 μl of Environmental Master Mix 2.0, 2.5 μL of 2 mg/mL bovine serum albumin, 1 μM of each primer, 2 μL of DNA-free water and 5 μL of the DNA extracts for a total reaction volume of 25 μL; and the thermal cycling protocols were 10 min at 95 °C, followed by 40 cycles of 15 s at 95 °C and 60 s at 56 °C. A positive control (*E. coli*, ATCC® 25922™) and a no template control were also run during the analysis for quality control.

2.3. Antibiotic Susceptibility Testing of the Isolates

The *E. coli* isolates ($n = 96$) were further tested for antibiotic susceptibility by Epsilometer test (ETEST®) (bioMérieux, Marcy l'Etoile, France) as described by the manufacturer. Briefly, overnight cultures of isolates were streaked on Mueller Hinton Agar plates and antibiotic strips were placed on these plates after they were completely dry. These plates were then inverted and placed in a 35 ± 0.5 °C incubator for 18 h. Each isolate was tested for ampicillin, ciprofloxacin, sulfamethoxazole and tetracycline susceptibility. At the end of the incubation, the minimum inhibitory concentration (MIC) values for each isolate and each antibiotic were recorded as described by the manufacturer. The CLSI breakpoints were used to interpret the data [13] and were reported as susceptible (S), intermediate (I) or resistant (R).

2.4. Data Analysis

The data were imported and the analysis was performed using the SAS (Statistical Analysis Software) 9.4. Univariate analysis for each variable was performed to assess the normality and distribution. The difference between the mean of the different antibiotic resistant bacteria at different stages of wastewater treatment was assessed using the Wilcoxon rank sum test (NPAR1WAY).

3. Results

3.1. Presumptive E. coli Growth on Plates Supplemented with Antibiotics

3.1.1. Growth on the Control Plates

The presumptive *E. coli* entering the WWTP ($2.5 \times 10^7 \pm 1.36 \times 10^7$ CFU/100 mL) were removed significantly during UV disinfection with an average 5.2 log (Table 2), and an additional 1.1 log reduction was achieved with chlorination before the treated effluent was released for irrigation purposes. Two out of five sampling events (February and April) had the highest number of the presumptive *E. coli* (1.1×10^2 and 3.2×10^2 CFU/100 mL respectively) (Figure 1) at the disinfected effluent stage.

Table 2. Log removal of *Escherichia coli* in different treatment stages.

Media	Influent to Secondary	Secondary to UV	Influent to UV	Influent to UV + Chlorination
Control	−2.68	−2.56	−5.24	−6.33
ST	−2.55	−2.61	−5.17	−6.15
CI	−3.11	−2.17	−5.29	−5.95
TC	−2.36	−2.90	−5.26	−5.79
AM	−2.50	−2.83	−5.33	−6.26

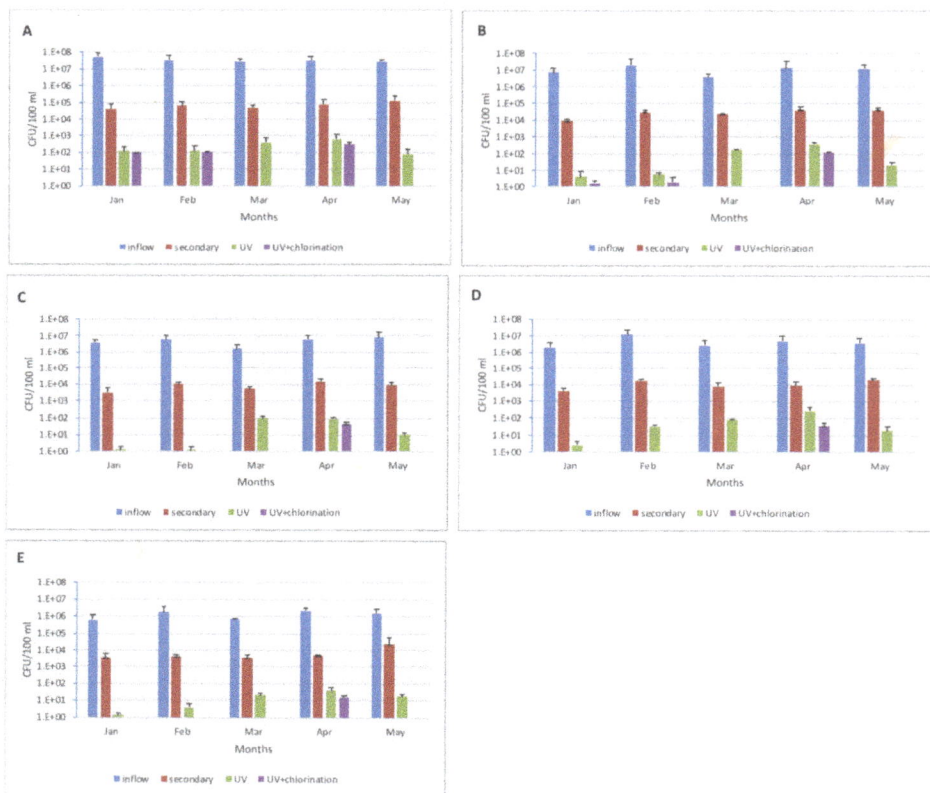

Figure 1. *Escherichia coli* growth on control (**A**), ampicillin (**B**), ciprofloxacin (**C**), sulfamethoxazole (**D**) and tetracycline (**E**) supplemented media.

3.1.2. *E. coli* Growth on Antibiotic Supplemented Media

The mean concentrations of ampicillin resistant *E. coli* that grew on AM/MI agar were higher than the *E. coli* detected on any other antibiotic supplemented media. The mean concentrations of *E. coli* were $1.2 \times 10^7 \pm 1.04 \times 10^7$ CFU/100 mL at the inflow and decreased down to $2.4 \times 10^1 \pm 5.1 \times 10^1$ CFU/100 mL at the reclaimed water stage. Almost half (47%) of *E. coli* entering the WWTP were able to grow on the AM/MI agar (Table 3). An additional 0.9 logs of *E. coli* were removed with the chlorination after the conventional treatment and UV disinfection (Table 2). At the effluent, the percentage of the presumptive *E. coli* that were able to grow on the AM/MI agar increased significantly (Table 3). *E. coli* was detected post-chlorination in January, February and April (2×10^0, 4×10^0 and 1.2×10^2 CFU/100 mL respectively) on this medium. April was also the month with the highest amount of precipitation (total of 1.5 inches in seven days prior to the sampling) followed by January and February (Figure 2).

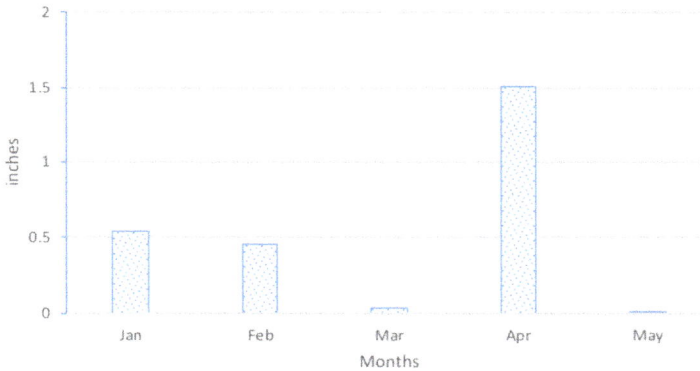

Figure 2. Cumulative precipitation (7 days total) before each sampling event.

Table 3. The percentage of *E. coli* growth on media supplemented with antibiotics through the treatment stages.

Stage	AM	ST	CI	TC
Inflow	47	24	19	6
Secondary	42	17	12	12
UV	47	29	16	5
UV + chlorination	63	21	25	9

At the inflow, *E. coli* growth on sulfamethoxazole supplemented MI media (ST/MI) reached $6.1 \times 10^6 \pm 6.95 \times 10^6$ CFU/100 mL. These concentrations decreased to $7.2 \times 10^1 \pm 1.07 \times 10^2$ CFU/100 mL and $8 \times 10^0 \pm 1.57 \times 10^1$ CFU/100 mL post UV and chlorination stages, respectively. Similar to the control and AM/MI, *E. coli* growth on ST/MI media was the highest in April (2.6×10^2 CFU/100 mL).

E. coli concentrations on ciprofloxacin supplemented media (CI/MI agar) ranked third among other antibiotic conditions ($4.7 \times 10^6 \pm 3.59 \times 10^6$ CFU/100 mL). Almost one fifth (19%) of the presumptive *E. coli* population at the inflow could grow on CI/MI agar. The number of bacteria decreased to $9.6 \times 10^0 \pm 1.92 \times 10^1$ CFU/100 mL post-chlorination, and 25% of the *E. coli* were able to grow on the CI/MI agar at this last stage (Table 3).

E. coli growth on tetracycline supplemented media had the lowest occurrence at the inflow with a mean of $1.4 \times 10^6 \pm 8.99 \times 10^5$ CFU/100 mL (Figure 1). Only 6% of the *E. coli* entering the WWTP grew on the TI/MI agar. However, the percentage of *E. coli* doubled post-secondary stage (12%). Chlorination followed by UV disinfection removed some of these bacteria, and 9% of the *E. coli* were able to grow on the TC/MI agar at the end of the treatment process.

The nonparametric analysis using the Wilcoxon rank sum test yielded significant results for all the antibiotics. The difference between inflow and UV, inflow and UV + chlorination, and secondary and UV for ampicillin, ciprofloxacin, sulfamethoxazole, and tetracycline were strongly significant ($p < 0.001$). The difference was moderately significant for ampicillin, ciprofloxacin, and sulfamethoxazole between UV and chlorination ($p < 0.05$). Tetracycline concentrations, on the other hand, were not significant between these two disinfection stages ($p > 0.05$).

3.1.3. Antibiotic Susceptibility in *E. coli* Isolates

All isolates were confirmed as *E. coli* based on PCR validation. The cycle thresholds for isolates ranged from 18 to 28. Antibiotic resistance was observed in *E. coli* isolates obtained from different stages of treatment (Table 4). The highest resistance was to ampicillin (85% R and 1% I). Among these isolates, sulfamethoxazole resistance ranked second (Table 4). Ciprofloxacin and tetracycline resistant *E. coli* were the least observed isolates (10% R and 20% I for CI and 30% R for TC, respectively).

Table 4. Percentage of antibiotic resistant *E. coli* isolates.

Antibiotic	MIC Interpretive Criteria (μg/mL)			Percent Resistant [†] (*n* = 96)
	S	I	R	
AM	≤8	16	≥32	95
CI	≤1	2	≥4	30
ST	≤2	-	≥4	70
TC	≤4	8	≥16	30

[†] For all antibiotics, any 'intermediate' resistance was included with resistant.

Resistance to three or more antibiotics (multidrug-resistant) was observed in 21% of the *E. coli* isolates. Based on the minimum inhibitory concentrations, resistance to ampicillin was widespread among the multidrug resistant *E. coli* and four of these isolates had ampicillin MIC > 256 μg/mL (Table 5). Three of these high MICs (EC5, EC9 and EC14) were isolated from the effluent where both UV and chlorination was applied to the finished water. Two of these isolates were also resistant to tetracycline with MIC > 256 μg/mL (EC5 and EC9). Ampicillin, ciprofloxacin, and trimethoprim–sulfamethoxazole multidrug resistance was observed in 15 of the *E. coli* isolates (75%). One isolate (EC12) was resistant to all four antibiotics.

Table 5. The minimum inhibitory concentration (MIC) values of multidrug-resistant *E. coli*.

Isolate	Location	AM	CI	ST	TC
EC1	inflow	64	4	6	2
EC2	secondary	64	2	4	3
EC3	secondary	48	3	>32	0.5
EC4	inflow	48	8	6	0.38
EC5	UV + chlorination	>256	0.5	>32	>256
EC6	UV + chlorination	48	2	>32	2
EC7	UV + chlorination	64	24	>32	4
EC8	UV	48	3	4	1
EC9	UV + chlorination	>256	3	0.047	>256
EC10	UV + chlorination	128	6	4	4
EC11	secondary	>256	3	4	4
EC12	UV	64	2	12	15
EC13	inflow	64	4	0.38	24
EC14	UV + chlorination	>256	6	>32	3
EC15	inflow	48	>32	>32	4
EC16	inflow	16	0.016	4	32
EC17	inflow	24	4	0.047	>256
EC18	UV	24	3	24	0.5
EC19	UV	48	4	6	2
EC20	secondary	48	3	>32	2

4. Discussion

In the study, the occurrence of antibiotic-resistant *E. coli* in a conventional WWTP in which effluents have been utilized to irrigate recreational landscapes was investigated. The results show that reclaimed water harbors *E. coli* resistant to a suite of commonly used antibiotics in medicine (ampicillin, sulfamethoxazole, ciprofloxacin, and tetracycline). The resistance to these antibiotics was also observed among *E. coli* isolates from irrigation waters in other studies [21–23]. Culturable fecal indicator bacteria such as *E. coli* have been instrumental in monitoring the impact of effluents on the environment [24]. According to the National Pollutant Discharge Elimination System (NPDES), a WWTP in the USA can only discharge effluents below fecal indicator bacteria guideline values to ensure minimum adverse environmental impact [25]. However, there are currently no guidelines to monitor antibiotic resistant *E. coli* in reuse water. The results of our study provide baseline information on the occurrence of these antibiotic-resistant indicator bacteria.

Multiple disinfection methods (UV followed by chlorination) applied during the reclaimed water production significantly decreased the number of *E. coli*. In an earlier study, [26] estimated 3.9 log removal of fecal indicator bacteria in a WWTP utilized for reuse purposes; approximately 2 logs lower than our findings. This may be a result of different treatment methods among WWTPs. Earlier studies showed that chlorine-related disinfection by-products might potentially induce antibiotic resistance [27], and UV may not entirely remove antibiotics or antibiotic resistant genes from the effluent [28]. In addition, [29] showed that rapid sand filtration used for wastewater treatment failed to remove *E. coli* cells. The USEPA states in the guidelines of water reuse that the reclaimed water programs vary with the intensity of treatment based on the anticipated human exposure to the effluent [30]. The variability in resistance is not only limited to the treatment type, and environmental conditions may have a significant impact on the occurrence of antibiotic-resistant *E. coli* even within the same plant. Our study showed that the antibiotic-resistant *E. coli* concentrations were high at the effluent when there was rain prior to the sampling event. Precipitation events are often a burden on the treatment efficacy. Earlier studies showed an association between heavy rainfall and pathogen removal [31]. A metagenomics study in wastewater treatment plants also showed that the diversity in the microbial community significantly increased after rainfall events [32], suggesting poor disinfection due to increased flow and short retention time. Further research is needed with a study designed specifically to assess the impact of rain on the removal efficiency of antibiotic-resistant bacteria.

Another important finding was that one in every five *E. coli* isolated from the WWTP had multidrug resistance. Similarly, [33] found that *E. coli* O157:H7 plasmids were resistant to seven different antibiotics including ampicillin, ciprofloxacin, and tetracycline. It is well known that *E. coli* can survive long-term and proliferate in the environment [34], and that the long-term persistence of antibiotic-resistant *E. coli* can cause potential public health outcomes upon exposure. Therefore, the ecology of *E. coli*—in particular, persistence and seasonality—plays a significant role in the health risks [35,36]. Investigating the culturable fraction of antibiotic-resistant bacteria through WWTP stages in our study provided information that can further be used to assess health risk upon exposure. A majority of the studies on the antibiotic resistance from WWTP have investigated antibiotic resistance genes, which provide vast knowledge on the horizontal gene transport and fate of these genes during treatment. However, these genes may exist in the wastewater as naked DNA. In addition, antibiotic-resistant genes can be naturally found in the environment as these genes have been detected in pristine environments dating back thousands of years, before the antibiotic era [14,37,38]. These genes may also persist in the environment longer than the cells, which may cause an overestimation of the health risk. Earlier studies showed that *E. coli* genes decayed slower than cultivated *E. coli* in water (T_{99} = 5.65 days and 2.02 days, respectively) [39–41]. Therefore, investigating the culturable fraction of antibiotic resistance can help to fill some of these knowledge gaps to estimate health risk.

The *E. coli* that are resistant to certain antibiotics which have been used for a long time in medicine could still be detected at the effluent even after multiple disinfection steps. For example, approximately 50% of the *E. coli* entering the WWTP were resistant to ampicillin, and the percentage increased to 63% at the effluent before leaving the plant after disinfecting with both UV and chlorine. Moreover, three out of four isolates of *E. coli* with multidrug resistance had minimum inhibition concentrations that were above the detection limits. These results suggest that the total *E. coli* community may have serotypes that are resistant to antibiotics and disinfection at the same time. Similar to our findings, [42] suggested that the disinfection byproducts promoted the evolution of resistant *E. coli* strains. Similarly, [43] showed that the dose applied for UV disinfection can create a selective environment for antibiotic resistant *E. coli* to survive better than other serotypes within the population. These findings suggest that better treatment processes are needed to take over old technologies to mitigate antibiotic resistance in the environment. Studies have shown that membrane bioreactor systems are capable of achieving better removal of microorganisms than conventional activated sludge systems [44]. Alternatively, tertiary treatment with filtration followed by disinfection was also reported to be effective in antibiotic resistance removal [45]. Approaches such as these relatively new technologies may reduce the antibiotic-resistant bacteria load entering aquatic environment.

5. Conclusions

The findings of this study show that using conventional methods of wastewater treatment to produce reclaimed water may pose challenges to removing antibiotic-resistant bacteria. Factors impacting the treatment efficacy such as the microbial community composition entering and leaving the plant, physicochemical factors impacting treatment, and extreme weather events that can adversely affect the flow and overall plant capacity need to be addressed while tackling the contribution of WWTP to antibiotic resistance in the environment. Our results show that multiple disinfection methods such as UV and chlorination may remove fecal indicator bacteria to acceptable levels for reuse, but the remaining cells in the effluent exhibit multidrug resistance phenotypes. The presence of these strains in the effluent needs to be considered while developing new regulations for water reuse. Further research is required in order to evaluate the health risks of using reclaimed water harboring antibiotic-resistant bacteria for drinking, agricultural, and recreational purposes.

Author Contributions: A.A. designed the study and wrote the manuscript; Z.C. contributed to sample collection; Z.C. and O.O. performed the laboratory analysis; Z.C. and A.B. analyzed the data.

Funding: This research was funded by Georgia Southern University Office of Research and Development.

Conflicts of Interest: The authors declare no conflict of interest.

References

1. Davies, J.; Davies, D. Origins and evolution of antibiotic resistance. *Microbiol. Mol. Biol. Rev.* **2010**, *74*, 417–433. [CrossRef] [PubMed]
2. Stoll, C.; Sidhu, J.P.S.; Tiehm, A.; Toze, S. Prevalence of clinically relevant antibiotic resistance genes in surface water samples collected from Germany and Australia. *Environ. Sci. Technol.* **2012**, *46*, 9716–9726. [CrossRef] [PubMed]
3. Burch, T.R.; Sadowsky, M.J.; LaPara, T.M. Fate of antibiotic resistance genes and class 1 integrons in soil microcosms following the application of treated residual municipal wastewater solids. *Environ. Sci. Technol.* **2014**, *48*, 5620–5627. [CrossRef] [PubMed]
4. Ben, W.; Wang, J.; Cao, R.; Yang, M.; Zhang, Y.; Qiang, Z. Distribution of antibiotic resistance in the effluents of ten municipal wastewater treatment plants in China and the effect of treatment processes. *Chemosphere* **2017**, *172*, 392–398. [CrossRef] [PubMed]
5. Li, X.; Watanabe, N.; Xiao, C.; Harter, T.; McCowan, B.; Liu, Y.; Atwill, E.R. Antibiotic-resistant *E. coli* in surface water and groundwater in dairy operations in Northern California. *Environ. Monit. Assess.* **2014**, *186*, 1253–1260. [CrossRef] [PubMed]
6. Carnelli, A.; Mauri, F.; Demarta, A. Characterization of genetic determinants involved in antibiotic resistance in *Aeromonas* spp. and fecal coliforms isolated from different aquatic environments. *Res. Microbiol.* **2017**, *168*, 461–471. [CrossRef] [PubMed]

7. Port, J.A.; Cullen, A.C.; Wallace, J.C.; Smith, M.N.; Faustman, E.M. Metagenomic frameworks for monitoring antibiotic resistance in aquatic environments. *Environ. Health Perspect.* **2014**, *122*, 222–228. [CrossRef] [PubMed]

8. Manaia, C.M.; Macedo, G.; Fatta-Kassinos, D.; Nunes, O.C. Antibiotic resistance in urban aquatic environments: Can it be controlled? *Appl. Microbiol. Biotechnol.* **2016**, *100*, 1543–1557. [CrossRef] [PubMed]

9. Zhang, S.; Han, B.; Gu, J.; Wang, C.; Wang, P.; Ma, Y.; Cao, J.; He, Z. Fate of antibiotic resistant cultivable heterotrophic bacteria and antibiotic resistance genes in wastewater treatment processes. *Chemosphere* **2015**, *135*, 138–145. [CrossRef] [PubMed]

10. Frey, S.K.; Topp, E.; Khan, I.U.H.; Ball, B.R.; Edwards, M.; Gottschall, N.; Sunohara, M.; Lapen, D.R. Quantitative *Campylobacter* spp., antibiotic resistance genes, and veterinary antibiotics in surface and ground water following manure application: Influence of tile drainage control. *Sci. Total Environ.* **2015**, *532*, 138–153. [CrossRef] [PubMed]

11. Wang, F.-H.; Qiao, M.; Lv, Z.-E.; Guo, G.-X.; Jia, Y.; Su, Y.-H.; Zhu, Y.-G. Impact of reclaimed water irrigation on antibiotic resistance in public parks, Beijing, China. *Environ. Pollut.* **2014**, *184*, 247–253. [CrossRef] [PubMed]

12. Blaustein, R.A.; Shelton, D.R.; Van Kessel, J.A.; Karns, J.S.; Stocker, M.D.; Pachepsky, Y.A. Irrigation waters and pipe-based biofilms as sources for antibiotic-resistant bacteria. *Environ. Monit. Assess.* **2016**, *188*, 56. [CrossRef] [PubMed]

13. World Health Organization. *Antimicrobial Resistance: An emerging Water, Sanitation and Hygiene Issue Briefing Note*; WHO/FWC/WSH/14.07; World Health Organization: Geneva, Switzerland, 2014. [CrossRef]

14. Xu, J.; Chen, W.; Wu, L.; Green, R.; Chang, A.C. Leachability of some emerging contaminants in reclaimed municipal wastewater-irrigated turf grass fields. *Environ. Toxicol. Chem.* **2009**, *28*, 1842–1850. [CrossRef] [PubMed]

15. United States Environmental Protection Agency. *Method 1604: Total Coliforms and Escherichia coli in Water by Membrane Filtration Using a Simultaneous Detection Technique (MI Medium)*; Standard Methods, EPA-821-R-02-024; United States Environmental Protection Agency: Washington, DC, USA, 2002.

16. Brenner, K.P.; Rankin, C.C.; Roybal, Y.R.; Stelma, G.N.; Scarpino, P.V; Dufour, A.P. New medium for the simultaneous detection of total coliforms and *Escherichia coli* in water. *Appl. Environ. Microbiol.* **1993**, *59*, 3534–3544. [PubMed]

17. Clinical and Laboratory Standards Institute. *Performance Standards for Antimicrobial Susceptibility Testing*, 27th ed.; CLSI Supplement M100; Clinical and Laboratory Standards Institute: Wayne, PA, USA, 2017.

18. Watkinson, A.J.; Micalizzi, G.R.; Bates, J.R.; Costanzo, S.D. Novel method for rapid assessment of antibiotic resistance in *Escherichia coli* isolates from environmental waters by use of a modified chromogenic agar. *Appl. Environ. Microbiol.* **2007**, *73*, 2224–2229. [CrossRef] [PubMed]

19. United States Environmental Protection Agency. *Method 1609: Enterococci in Water by TaqMan®Quantitative Polymerase Chain Reaction (qPCR) with Internal Amplification Control (IAC) Assay*; Standard Methods, EPA-820-R-15-099; U.S. Environmental Protection Agency: Washington, DC, USA, 2015.

20. Chern, E.C.; Siefring, S.; Paar, J.; Doolittle, M.; Haugland, R.A. Comparison of quantitative PCR assays for *Escherichia coli* targeting ribosomal RNA and single copy genes. *Lett. Appl. Microbiol.* **2011**, *52*, 298–306. [CrossRef] [PubMed]

21. Roe, M.T.; Vega, E.; Pillai, S.D. Antimicrobial resistance markers of Class 1 and Class 2 Integron-bearing *Escherichia coli* from irrigation water and sediments. *Emerg. Infect. Dis.* **2003**, *9*, 822–826. [CrossRef] [PubMed]

22. Vital, P.G.; Zara, E.S.; Paraoan, C.E.M.; Dimasupil, M.A.Z.; Abello, J.J.M.; Santos, I.T.G.; Rivera, W.L. Antibiotic resistance and extended-spectrum beta-lactamase production of *Escherichia coli* isolated from irrigation waters in selected urban farms in metro Manila, Philippines. *Water* **2018**, *10*, 548. [CrossRef]

23. Aijuka, M.; Charimba, G.; Hugo, C.J.; Buys, E.M. Characterization of bacterial pathogens in rural and urban irrigation water. *J. Water Health* **2015**, *13*, 103–117. [CrossRef] [PubMed]

24. Ishii, S.; Sadowsky, M.J. *Escherichia coli* in the environment: Implications for water quality and human health. *Microbes Environ.* **2008**, *23*, 101–108. [CrossRef] [PubMed]

25. Smith, J.E.; Perdek, J.M. Assessment and management of watershed microbial contaminants. *Crit. Rev. Environ. Sci. Technol.* **2004**, *34*, 109–139. [CrossRef]

26. Al-Jassim, N.; Ansari, M.I.; Harb, M.; Hong, P.Y. Removal of bacterial contaminants and antibiotic resistance genes by conventional wastewater treatment processes in Saudi Arabia: Is the treated wastewater safe to reuse for agricultural irrigation? *Water Res.* **2015**, *73*, 277–290. [CrossRef] [PubMed]

27. Kim, I.; Yamashita, N.; Tanaka, H. Performance of UV and UV/H$_2$O$_2$ processes for the removal of pharmaceuticals detected in secondary effluent of a sewage treatment plant in Japan. *J. Hazard. Mater.* **2009**, *166*, 1134–1140. [CrossRef] [PubMed]

28. Xi, C.; Zhang, Y.; Marrs, C.F.; Ye, W.; Simon, C.; Foxman, B.; Nriagu, J. Prevalence of antibiotic resistance in drinking water treatment and distribution systems. *Appl. Environ. Microbiol.* **2009**, *75*, 5714–5718. [CrossRef] [PubMed]

29. Srinivasan, S.; Aslan, A.; Xagoraraki, I.; Alocilja, E.; Rose, J.B. *Escherichia coli, Enterococci,* and *Bacteroides thetaiotaomicron* qPCR signals through wastewater and septage treatment. *Water Res.* **2011**, *45*, 2561–2572. [CrossRef] [PubMed]

30. United States Environmental Protection Agency. *Guidelines for Water Reuse*; EPA/600/R-12/61; U.S. Environmental Protection Agency: Washington, DC, USA, 2012.

31. Carducci, A.; Verani, M. Effects of bacterial, chemical, physical and meteorological variables on virus removal by a wastewater treatment plant. *Food Environ. Virol.* **2013**, *5*, 69–76. [CrossRef] [PubMed]

32. McLellan, S.L.; Huse, S.M.; Mueller-Spitz, S.R.; Andreishcheva, E.N.; Sogin, M.L. Diversity and population structure of sewage-derived microorganisms in wastewater treatment plant influent. *Environ. Microbiol.* **2010**, *12*, 378–392. [CrossRef] [PubMed]

33. Chigor, V.N.; Umoh, V.J.; Smith, S.I.; Igbinosa, E.O.; Okoh, A.I. Multidrug resistance and plasmid patterns of *Escherichia coli* O157 and other *E. coli* isolated from diarrhoeal stools and surface waters from some selected sources in Zaria, Nigeria. *Int. J. Environ. Res. Public Health* **2010**, *7*, 3831–3841. [CrossRef] [PubMed]

34. Winfiel, M.D.; Groisman, E.A. Role of nonhost enviroments in the lifestyles of *Salmonella* and *E. coli*. *Appl. Environ. Microbiol.* **2003**, *69*, 3687–3694. [CrossRef]

35. McLain, J.E.T.; Williams, C.F. Assessing environmental impacts of treated wastewater through monitoring of fecal indicator bacteria and salinity in irrigated soils. *Environ. Monit. Assess.* **2012**, *184*, 1559–1572. [CrossRef] [PubMed]

36. Jiang, Y.; Xu, C.; Wu, X.; Chen, Y.; Han, W.; Gin, K.Y.H.; He, Y. Occurrence, seasonal variation and risk assessment of antibiotics in Qingcaosha reservoir. *Water* **2018**, *10*, 115. [CrossRef]

37. Bhullar, K.; Waglechner, N.; Pawlowski, A.; Koteva, K.; Banks, E.D.; Johnston, M.D.; Barton, H.A.; Wright, G.D. Antibiotic resistance is prevalent in an isolated cave microbiome. *PLoS ONE* **2012**, *7*. [CrossRef] [PubMed]

38. Dcosta, V.M.; King, C.E.; Kalan, L.; Morar, M.; Sung, W.W.L.; Schwarz, C.; Froese, D.; Zazula, G.; Calmels, F.; Debruyne, R.; et al. Antibiotic resistance is ancient. *Nature* **2011**, *477*, 457–461. [CrossRef] [PubMed]

39. Jin, G.; Englande, A.J.; Bradford, H.; Englande, A.J. Comparison of *E. coli, Enterococci,* and fecal coliform as indicators for brackish water quality assessment. *Water Environ. Res.* **2004**, *76*, 245–255. [CrossRef] [PubMed]

40. Dick, L.K.; Stelzer, E.A; Bertke, E.E.; Fong, D.L.; Stoeckel, D.M. Relative decay of *Bacteroidales* microbial source tracking markers and cultivated *Escherichia coli* in freshwater microcosms. *Appl. Environ. Microbiol.* **2010**, *76*, 3255–3262. [CrossRef] [PubMed]

41. Martinez, J.L. Environmental pollution by antibiotics and by antibiotic resistance determinants. *Environ. Pollut.* **2009**, *157*, 2893–2902. [CrossRef] [PubMed]

42. Li, D.; Zeng, S.; He, M.; Gu, A.Z. Water disinfection byproducts induce antibiotic resistance-role of environmental pollutants in resistance phenomena. *Environ. Sci. Technol.* **2016**, *50*, 3193–3201. [CrossRef] [PubMed]

43. Zhang, C.-M.; Xu, L.-M.; Wang, X.-C.; Zhuang, K.; Liu, Q.-Q. Effects of ultraviolet disinfection on antibiotic-resistant *Escherichia coli* from wastewater: Inactivation, antibiotic resistance profiles and antibiotic resistance genes. *J. Appl. Microbiol.* **2017**, *123*, 295–306. [CrossRef] [PubMed]

44. Zhang, K.; Farahbakhsh, K. Removal of native coliphages and coliform bacteria from municipal wastewater by various wastewater treatment processes: Implications to water reuse. *Water Res.* **2007**, *41*, 2816–2824. [CrossRef] [PubMed]

45. Quach-Cu, J.; Herrera-Lynch, B.; Marciniak, C.; Adams, S.; Simmerman, A.; Reinke, R.A. The effect of primary, secondary, and tertiary wastewater treatment processes on antibiotic resistance gene (ARG) concentrations in solid and dissolved wastewater fractions. *Water* **2018**, *10*, 37. [CrossRef]

water

MDPI

Article

Antibiotic Resistance and Extended-Spectrum Beta-Lactamase Production of *Escherichia coli* Isolated from Irrigation Waters in Selected Urban Farms in Metro Manila, Philippines

Pierangeli G. Vital [1,2,*], Enrico S. Zara [1], Cielo Emar M. Paraoan [1], Ma. Angela Z. Dimasupil [1], Joseth Jermaine M. Abello [1], Iñigo Teodoro G. Santos [1] and Windell L. Rivera [1,2]

[1]   Institute of Biology, College of Science, University of the Philippines Diliman, Quezon City 1101, Philippines;
      enrico.s.zara@gmail.com (E.S.Z.); paraoan_cielo@yahoo.com (C.E.M.P.);
      angela.dimasupil@gmail.com (M.A.Z.D.); josethabello08@gmail.com (J.J.M.A.);
      itgsantos@yahoo.com.ph (I.T.G.S.); wlrivera@gmail.com (W.L.R.)
[2]   Natural Sciences Research Institute, University of the Philippines Diliman, Quezon City 1101, Philippines
*   Correspondence: pierangeli.vital@upd.edu.ph or piervital@hotmail.com

Received: 31 January 2018; Accepted: 20 April 2018; Published: 25 April 2018

check for updates

Abstract: Highly-polluted surface waters are increasingly used for irrigation in different agricultural settings because they have high nutrient content and are readily available. However, studies showed that they are reservoirs for the emergence and dissemination of antibiotic-resistant bacteria in the environment. In this study, the resistance of 212 *Escherichia coli* isolates from irrigation water, soil, and vegetables in selected urban farms in Metro Manila, Philippines was evaluated. Results showed that antibiotic resistance was more prevalent in water (67.3%) compared to soil (56.4%) and vegetable (61.5%) isolates. Resistance to tetracycline was the highest among water (45.6%) and vegetable (42.3%) isolates while ampicillin resistance was the highest among soil isolates (33.3%). Multidrug-resistant (MDR) isolates were also observed and they were more prevalent in water (25.3%) compared to soil (2.8%) and vegetable (8.4%) isolates. Interestingly, there are patterns of antibiotic resistance that were common to isolates from different samples. Extended-spectrum beta-lactamase production (ESBL) was also investigated and genes were observed to be present in 13 isolates. This provides circumstantial evidence that highly-polluted surface waters harbor antibiotic-resistant and MDR *E. coli* that may be potentially transferred to primary production environments during their application for irrigation purposes.

Keywords: antibiotic resistance; ESBL; *Escherichia coli*; irrigation water; gastrointestinal infections

1. Introduction

Agricultural productivity heavily relies on the use of irrigation water in irrigating agricultural plants, applying fertilizers and pesticides, and processing of farm products. In urban agricultural areas where there is limited water supply and high demand for clean water, irrigation water is usually sourced from surface waters contaminated with agricultural runoff, livestock and wildlife fecal material, wastewater discharge, and septic leakage [1,2]. Although surface waters are practical to use because of their availability and high nutrient content, they can be a potential source of fecal contaminants and pathogenic microorganisms that may be transferred to farm products, such as fresh produce, during irrigation [1–3].

The problem of microbial contamination of fresh produce is often compounded by the emergence of antibiotic resistance among pathogenic microorganisms. Antibiotic resistance is caused by the widespread and increasing use of antibiotics [4–6] and it is increasingly becoming a global health concern because it limits the available therapeutic options, resulting in higher treatment and hospitalization costs and increased rates of mortality and morbidity [7–9]. A principal method by which bacteria resist antibiotics is through the production of β-lactamases, enzymes that hydrolyze β-lactam antibiotics. β-lactam antibiotics are the most commonly-used antibiotics, including penicillins, cephalosporins, and carbapenems. One group of β-lactamase enzymes, the extended spectrum β-lactamases (ESBLs), produced by *Escherichia coli*, is of particular significance. These are able to target a wider range of antibiotics and plasmids that contain genes for ESBLs which often carry genes for resistance to various other antibiotics [10–12].

Surface waters have been considered as an important source of microbial antibiotic resistance [13–15]. A considerable fraction of antibiotics used in clinical, agricultural, and household settings usually end up in aquatic environments and serve as regulatory and signaling molecules among bacteria [16–18]. The presence of low concentrations of antibiotics in aquatic environments imposes a selection pressure that promotes antibiotic tolerance and emergence of antibiotic resistance in aquatic bacteria [15]. Therefore, surface waters used for irrigation purposes may serve as pools of antibiotic-resistant bacteria that have profound influence on the microbial quality of primary production environments.

In highly-urbanized and densely-populated areas, such as Metro Manila, Philippines, the use of highly-polluted surface waters for irrigation is widely practiced, especially as urban agriculture is gaining popularity as a tool to increase household income and meet subsistence food needs [2,19]. This necessitates the assessment and monitoring of the quality of surface waters used for irrigation.

The present study aims to evaluate the resistance of 212 *E. coli* isolates from irrigation water, soil, and vegetables in selected urban farms in Metro Manila, Philippines against nine commonly-used antibiotics and to ascertain the patterns of antibiotic resistance among the isolates. Further, the isolates were screened for ESBL production and detection of genes encoding ESBLs to assess one of the mechanisms for antibiotic resistance. This information may be useful to raise awareness on the importance of prevention measures to be taken, and it also alerts for the widespread use of antimicrobials, especially in agricultural sectors.

2. Materials and Method

2.1. Bacterial Isolates

A total of 212 culture-positive *E. coli* isolates that were previously collected from irrigation water, soil, and vegetable samples from six urban agricultural farms in Metro Manila, Philippines were used in this study [2]. The farms were small scale urban farms in Quezon City, Marikina City and Pasig City that are situated near residential areas and cultivate different vegetables that are sold in nearby wet markets. The farms were chosen because the irrigation waters used in these farming sites are derived from highly-polluted surface waters.

The isolates were sub-cultured into Tryptic Soy Broth (TSB) (Merck, Darmstadt, Germany) and incubated at 35 °C for 24 h, after which 20% glycerol was added prior to storage at −20 °C for further analysis.

2.2. Antibiotic Susceptibility Test

Antibiotic susceptibility testing was patterned after the Antimicrobial Susceptibility Manual of the American Society for Microbiology [20]. In its procedure, pure isolates were sub-cultured into Trypticase Soy Agar (TSA) plates (BD BBL™, Franklin Lakes, NJ, USA) and incubated at 35 °C for 16 to 18 h. The isolates were inoculated into 3 mL 0.85% saline solution and the turbidity of the suspension was standardized to that of 0.5% McFarland standard. Using a sterile cotton swab, the standardized suspension was swabbed and inoculated evenly on the entire surface of Mueller Hinton agar (MHA) plates (Hi-Media Laboratories, India). Then, nine antibiotics which include: tetracycline (30 μg), ciprofloxacin (5 μg), cefotaxime (30 μg), chloramphenicol (30 μg), nalidixic acid (30 μg), streptomycin

(10 µg), ampicillin (10 µg), cephalotin (30 µg), and trimethoprim (30 µg) were placed on the surface of inoculated MHA plates and incubated at 35 °C for 16 to 18 h. The antibiotics used in this study were selected as they represent each major class of antibiotics that is used in the treatment of human and animal *E. coli* infections. Subsequently the clearing zones on the plates were measured using a caliper. The diameters of the clearing zones were interpreted in accordance to the Clinical and Laboratory Standards Institute manual (CLSI) [21]. The test was performed in triplicate and *E. coli* ATCC®25922 was used as the negative control.

2.3. Screening for ESBL Production

Initial screening of *E. coli* isolates for ESBL production was performed using the disk diffusion method of the CLSI manual [21]. The antibiotics used were ceftazidime (30 µg) and cefotaxime (30 µg) (BBL Sensi-Disc, BD Diagnostics, MD, USA). Briefly, overnight cultures of the isolates were inoculated into 3 mL 0.85% saline solution and the turbidity of the suspension was adjusted to that of 0.5% McFarland standard. The suspension was then spread evenly on the entire surface of MHA plates, after which, ceftazidime and cefotaxime disks were placed. After 16 to 18 h of incubation, inhibition zones were measured and interpreted based CLSI manual [21].

Isolates screened as ESBL producers were subjected to a confirmatory test through the double disk synergy test [12,22]. Suspected ESBL-producing isolates were inoculated into 3 mL 0.85% saline solution to match the turbidity of 0.5% McFarland standard. The suspension was spread evenly into MHA plates and ceftazidime and cefotaxime disks were placed 20 to 30 mm from a disk containing amoxicillin/clavulanic acid (20 µg/10 µg). An increase in zone diameter of 5 mm or greater for either ceftazidime or cefotaxime, when combined with clavulanic acid, compared with their zone diameters alone, indicated ESBL production by the isolate [23].

2.4. Detection of ESBL Genes

ESBL genes were identified from the confirmed ESBL producers through PCR amplification using primers targeting different ESBL genes. Two multiplex sets including Multiplex I TEM, SHV and OXA-1 like, and Multiplex II CTX-M group 1, 2, and 9 were used in this study. Additionally, a simplex set CTX-M group 8/25 was used in this study, following the method of Dallenne et al. [23].

2.5. Statistical Analysis

All data were analyzed statistically using Statistical Package for Social Sciences ver. 20.0 (IBM Corp., Armonk, NY, USA). One-way analysis of variance (ANOVA) was used to evaluate significant differences in antibiotic resistance among the isolates recovered from the three different samples. Data that were found to be significant were further analyzed using Tukey's HSD comparison of means and a *p* value less than 0.05 was accepted as significant. The chi-square test was also employed to compare the frequency of resistance of the isolates from different samples to each antibiotic. Data were considered as statistically significant based on a $p < 0.05$.

3. Results

3.1. Antimicrobial Resistance

A total of 212 culture-positive *E. coli* consisting of 147 irrigation water, 39 soil, and 26 vegetable isolates were obtained from 190 water, 91 soil, and 92 vegetable samples. The identity of the isolates was confirmed in a previous study [2] using a PCR assay that amplifies the ß-glucoronidase (*uidA*) gene which encodes an acid hydrolase that catalyzes the cleavage of a wide variety of 3-glucuronidases used by of *E. coli* [24].

The resistance of 212 isolates was tested against nine antibiotics. The percentages of isolates that were susceptible, intermediate and resistant to each antibiotic are presented in Table 1. As shown, there are more antibiotic-resistant isolates from irrigation water compared to soil and vegetables.

Resistance to tetracycline was the highest among water (45.6%) and vegetable (42.3%) isolates, whereas ampicillin resistance was the highest among soil isolates (33.3%). Meanwhile, resistance to nalidixic acid was the lowest among the water isolates (2.7%) and resistance to ciprofloxacin (2.6%) and nalidixic acid (2.6%) was the lowest among the soil isolates. Finally, there was no resistance to nalidixic acid and streptomycin observed among the vegetable isolates.

Table 1. Percentages of *E. coli* isolates from agricultural irrigation water in Metro Manila, Philippines that were susceptible (S), intermediate (I) and resistant (R) to antibiotics. Resistance breakpoints were based on CLSI standards.

Antibiotic (μg)	Resistance Breakpoint (mm)	Water $n = 147$			Soil $n = 39$			Vegetables $n = 26$		
		S	I	R	S	I	R	S	I	R
Tetracycline (30)	≤14	53.7	0.7	45.6	69.2	0.0	30.8	57.7	0.0	42.3
Ciprofloxacin (5)	≤15	87.1	6.1	6.8	97.4	0.0	2.6	92.3	3.8	3.8
Cefotaxime (30)	≤22	84.4	2.0	13.6	94.9	0.0	5.1	96.2	0.0	3.8
Chloramphenicol (30)	≤12	70.7	12.9	16.3	94.9	0.0	5.1	92.3	3.8	3.8
Nalidixic acid (30)	≤13	90.5	6.8	2.7	92.3	5.1	2.6	92.3	7.7	0.0
Streptomycin (10)	≤11	77.6	9.5	12.9	87.2	5.1	7.7	88.5	11.5	0.0
Ampicillin (10)	≤13	66.0	0.0	34.0	64.1	2.6	33.3	65.4	0.0	34.6
Cephalothin (30)	≤14	70.1	20.4	9.5	76.9	10.3	12.8	65.4	23.1	11.5
Trimethoprim (30)	≤10	79.6	0.0	20.4	84.6	0.0	15.4	80.8	0.0	19.2

Overall, the resistance to nine antibiotics was highest among the water isolates (67.3%) compared to soil (56.4%) and vegetable (61.5%) isolates. ANOVA revealed that there was significant difference among the antibiotic resistance of isolates from the three different samples ($p = 0.008$). Tukey's HSD, a post hoc analysis showed that the antibiotic resistance of water and vegetable isolates against each antibiotic was significantly different from each other ($p = 0.000$). Comparison of the frequency of resistance of the isolates from different samples to a single antibiotic showed significant differences ($p < 0.05$) for tetracycline, ciprofloxacin, chloramphenicol, nalidixic acid, streptomycin, ampicillin, cephalothin, and trimethoprim. Meanwhile, no significant difference was observed among the frequency of resistance of the isolates from the different samples to cefotaxime ($p = 0.075$).

Among the 212 *E. coli* isolates, 36.5% were found to be resistant to at least three antibiotics and were considered as multidrug resistant (MDR). Table 2 summarizes the prevalence of MDR isolates from irrigation water, soil, and vegetables. MDR isolates was more prevalent in water (25.3%) compared to soil (2.8%) and vegetables (8.4%). Intriguingly, resistance to eight different antibiotics was observed among the water isolates (0.7%).

Table 2. Percentages of multidrug resistant *E. coli* isolates from agricultural irrigation water in Metro Manila, Philippines.

Number of Antibiotics to Which Isolates are Resistant	Water $n = 147$	Soil $n = 39$	Vegetables $n = 26$
3	10.89	0.7	7.7
4	7.5	0.7	0.7
5	4.8	1.4	0.0
6	1.4	0.0	0.0
8	0.7	0.0	0.0
Total	25.3	2.8	8.4

The patterns of antibiotic resistance among the *E. coli* isolates were also evaluated in this study. As shown in Table 3, the resistance pattern most prevalent among the isolates is resistance to a combination of tetracycline and ampicillin (2.8%), followed by resistance to a combination of ampicillin and cephalothin (2.4%) and a combination of tetracycline and chloramphenicol (2.4%). These antibiotic resistance patterns all occurred among the water isolates. Of the MDR isolates, however, the most prevalent pattern is resistance to a combination of ciprofloxacin, chloramphenicol, nalidixic acid, and ampicillin (2.4%), and this occurred among water isolates.

Although the patterns of antibiotic resistance greatly differ among water, soil, and vegetable isolates, there were antibiotic resistance patterns that are common to the isolates from different sources. For instance, resistance patterns Amp-Tmp and T-Amp were common to isolates from the three different samples, while resistance patterns Amp-Kf and T-Tmp were common to both water and soil isolates. Lastly, resistance pattern T-Amp-Tmp was common to water and vegetable isolates.

3.2. Initial Screening for ESBL Production

Out of 200 *E. coli* isolates evaluated for possible ESBL production through antimicrobial susceptibility testing with ceftazidime and cefotaxime, 27 isolates tested positive. Thirteen isolates (16.3%) were potential or suspected ESBL producers, having zones of 22 mm or less and 27 mm or less for ceftazidime and cefotaxime, respectively. This included two of the 48 isolates (4.2%) from Diliman and 25 of the 32 isolates (34.4%) from Marikina. Of the potential ESBL producers, one showed resistance and two showed intermediate resistance to ceftazidime, while two showed resistance and nine showed intermediate resistance to cefotaxime.

3.3. Confirmatory Testing for ESBL Production

ESBL production was phenotypically confirmed through the double-disk synergy test in 27 (8.75%) of all *E. coli* isolates evaluated, as shown in Figure 1. Of the 27 suspected ESBL producers, ESBL production was confirmed in one (1.2%) of the isolates from Diliman, and in six (18.8%) of the isolates from Marikina. Molecular testing through PCR amplification of target genes only detected *blaTEM*, *CTX-M* groups 1, 2, and 9 and *CTX-M* group 8/25. Results showed that from the 27 isolates, only 13 were positive for *blaTEM*, and five and eight isolates for *blaCTX-M* 1 and 2 respectively. At the same time, no amplicons corresponding to *blaSHV-1* and *blaOXA* were observed.

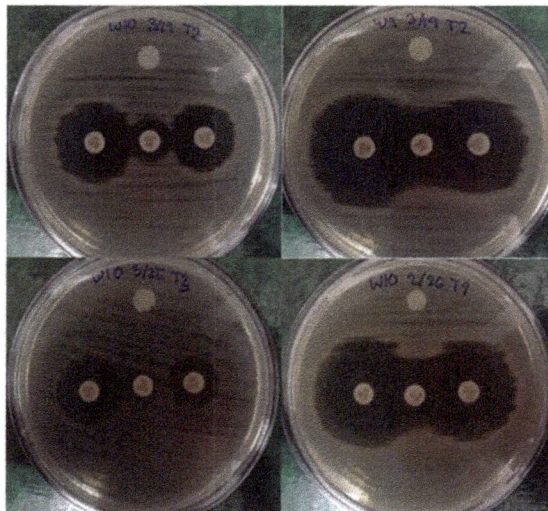

Figure 1. Negative (**left**) and positive (**right**) results for confirmatory screening of *E. coli* isolates from agricultural irrigation waters in Quezon City and Marikina City, Philippines showing zone extension for ESBL-producing isolates.

Table 3. Antibiotic resistance pattern among *E. coli* isolates from agricultural irrigation water in Metro Manila, Philippines.

Source	Number of Antibiotic to Which Isolates Are Resistant	Antibiotic Resistance Pattern	Frequency
Water n = 147	0	-	48
	1	Amp; Na; S; T; Tmp	8; 3; 2; 19; 3
	2	Amp-Kf; Amp-Tmp; S-Tmp; T-Amp; T-C; T-Na; T-Tmp	5; 3; 2; 6; 5; 2; 3
	3	Cip-Na-Amp; Cip-Na-S; S-Amp-Tmp; T-Amp-Kf, T-Amp-Tmp; T-C-Amp;	1; 1; 1; 2; 2; 2;
		T-Cip-Amp; T-C-Na; T-C-Tmp; T-Na-Amp; T-S-Amp	1; 3; 1; 1
	4	Cip-C-Na-Amp; T-Cip-C-Tmp; T-C-S-Amp; T-Na-Amp-Kf;	1; 1; 1; 1;
		T-Na-S-Amp;	1; 5
	5	T-Na-S-Tmp; T-S-Amp-Tmp	1; 1; 1; 1;
		Na-S-Amp-Kf-Tmp; T-Cip-C-Na-Ctx; T-Cip-C-Na-Tmp; T-Cip-Na-Amp-Kf;	1; 1; 1
	6	T-C-Na-Amp-Tmp; T-Ctx-Amp-Kf-Tmp; T-S-Amp-Kf-Tmp	1; 1
	8	Cip-Na-Ctx-Amp-Kf-Tmp; T-C-Na-S-Amp-Tmp T-C-Na-Ctx-S-Amp-Kf-Tmp	1
Soil n = 39	0	-	17
	1	Amp; Ctx; S; T; Tmp	2; 1; 1; 3; 1
	2	Amp-Kf; Amp-Tmp; T-Amp; T-Tmp	3; 1; 4; 2
	3	T-Cip-Na	1
	4	S-Amp-Kf-Tmp	1
	5	T-C-Amp-Kf-Tmp; T-C-Na-S-Amp	1; 1
Vegetables n = 26	0	-	10
	1	Kf; T	1; 5
	2	Amp-Tmp; Cip-Na; T-Amp	2; 1; 3
	3	Amp-Kf-Tmp; T-Amp-Tmp	1; 2
	4	T-C-Amp-Kf	1

Key: T = tetracycline; Cip = ciprofloxacin; Ctx = cefotaxime; C = chloramphenicol; Na = nalidixic acid; S = streptomycin; Amp = ampicillin; Kf = cephalothin; Tmp = trimethoprim.

4. Discussion

Several studies have shown that surface waters used for irrigation purposes harbor antibiotic-resistant bacteria. In this study, results showed that irrigation water contain the highest antibiotic resistant-bacteria, followed by vegetable and soil samples. The higher prevalence of antibiotic-resistant bacteria in irrigation water implies that it harbors a sufficient number of bacteria that may be potentially transferred to the primary production environment such as agricultural soils and vegetables during irrigation. Similar results were observed in numerous studies where irrigation waters from different sources are reported to contain elevated numbers of antibiotic resistant *E. coli* [25–30]. Among the *E. coli* isolates from irrigation water used in this study, the highest resistance was observed in tetracycline, followed by resistance to ampicillin. The resistance to these antibiotics was also observed among *E. coli* isolates from irrigation waters in several studies [25,26,28]. The presence of tetracycline-resistant *E. coli* in irrigation waters is particularly disturbing, as tetracycline is widely available and extensively used in developing countries as a first-line drug in the treatment of gastrointestinal infections [15,25,31].

Importantly, MDR *E. coli* isolates were also observed. Multidrug resistance is defined as resistance of an isolate to antibiotics belonging to at least three different classes of antibiotics [32,33]. In this study, MDR isolates were more prevalent in irrigation water isolates compared to soil and vegetable isolates. Similar results were obtained by Paraoan et al. [30] where 46 *E. coli* isolates (58.22%) from the agricultural irrigation waters in Bulacan, Philippines were found to be MDR. Further, the results of this study were in agreement with the results of other studies which documented the presence of MDR *E. coli* from irrigation waters. For instance, Roe et al. [25] showed that the Rio Grande River, a major source of irrigation water for both the USA and Mexico, harbors MDR *E. coli* with a prevalence rate of 32%. Another important source of irrigation water in Mexico is the San Pedro River which was found to be contaminated by MDR *E. coli* with a prevalence rate of 30.6% [27]. In a study of Chigor et al. [26], surface waters used for irrigation in Zaria, Nigeria were found to be contaminated with pathogenic *E. coli* O157:H7, which are also MDR. While pathogenic *E. coli* strains were not identified in our study, our results nevertheless contribute to the growing body of evidence showing that irrigation waters serve as reservoirs of MDR bacteria. Curiously, there are certain patterns of antibiotic resistance that are common to *E. coli* isolates obtained from three different samples and from two different samples, such as water and soil isolates and water and vegetable isolates. One plausible explanation for this observation is the horizontal transmission of antibiotic resistant and MDR *E. coli* across the samples. Taken together, these results indicate that highly-polluted surface waters used for irrigation in selected urban farms in Metro Manila, Philippines harbor antibiotic-resistant and MDR *E. coli*.

The emergence of antibiotic resistance in aquatic environments, such as surface waters, and its subsequent dissemination has been widely documented [13–15,34]. Antibiotics of various origins enter aquatic milieus through different routes. As most antibiotics are poorly metabolized and absorbed by the body [35–39], antibiotic residues used in clinical and domestic settings are released from patients' urine and feces and discharged as wastewater effluents [36]. Similarly, outdated antibiotic remainders used in domestic settings are disposed of deliberately in household drains [16]. Meanwhile, antibiotics used in poultry and livestock domestication are released from animals' urine and feces and combine with agricultural runoffs that are usually introduced in nearby aquatic systems, such as surface waters and wastewater treatment plants [16,17,37]. Likewise, antibiotics used in aquaculture are deliberately introduced as feed additives into aquatic farms.

Although these antibiotics are usually diluted and degraded in aquatic environments, resulting to relatively low concentrations, they may act as regulatory substances and signaling molecules in bacteria [18]. In addition, trace and sub-inhibitory concentrations of antibiotics create a mutant selection window that significantly increases the tolerance of bacteria to antibiotics and potentially promotes and preserves antibiotic resistant bacteria through adaptive mutations. Consequently, antibiotic resistance can be acquired by other bacteria in aquatic environments through horizontal gene transfers, such as conjugation, transduction, and transformation.

In urban agricultural farms of Metro Manila, Philippines, irrigation waters are typically derived from small bodies of water that are linked to Pasig River, Laguna Lake, and Manila Bay, which form a complex water system within Metro Manila [39]. Interestingly, these aquatic environments are contaminated by antibiotic residues of sulfamethoxazole, sulfamethazine, trimethoprim, and lincomycin that are presumably derived from human and animal use. Additionally, sulfamethoxazole-resistant genes, such as *sul1*, *sul2*, and *sul3*, were also observed [18]. The antibiotic residues found in water systems of Metro Manila are also detected in major rivers of China and Vietnam [18].

The application of these waters for irrigation purposes raises public health concern as they may potentially contaminate and disseminate antibiotic-resistant and multidrug resistant *E. coli* in the primary production environment, such as agricultural soils and fresh produce [26,28]. The presence of antibiotic resistant and MDR *E. coli* in fresh produce exposes humans to serious health hazards as there are vegetables that do not undergo microbial inactivation or preservation treatment prior to consumption [6,29]. Further, washing of vegetables does not completely eliminate the presence of these bacteria as they can be internalized in natural apertures of the vegetables, such as stomata, or localized in artificial crevices, cracks, and cuts [40]. Eventually, consumption of contaminated fresh produce may potentially cause foodborne gastrointestinal infections in humans. With regard to different ESBL gene types, a number of recent investigations highlighted the emergence and prevalence of CTX-M ESBLs as the most common type worldwide [41]. Before considering this, however, it is important to take into account the shifts in prevalence of ESBL genes across recent years and decades. In the 1990s, TEM- and SHV-type ESBLs were the dominant ESBL types, most often encountered in *K. pneumoniae* in hospitals. In the following decade, *E. coli* began to be recognized as the main source of ESBLs, with an increase in CTX-M ESBLs being described and blaCTX-M genes being recognized [41]. Therefore, different ESBL types, while not currently prevalent, may occur in many different locations. It is not unexpected, however, that the obtained result of the blaTEM gene's prevalence contrasts with the reported prevalence of CTX-M ESBLs, since the dominant ESBL type may vary between countries [42].

In the Philippines, there have also been studies to characterize molecularly ESBLs found in different bacterial isolates. In a study by Cabrera and Rodriguez (2009) [43], SHV-12 was found to be the dominant ESBL in Enterobacteriaceae tested, with some isolates carrying blaTEM-1. A year after, another study found CTX-M ESBLs as the predominant type of ESBLs in 95% of isolates of Enterobacteriaceae obtained from the same location, indicating a shift in ESBL genes similar to those in other countries [44]. Two further studies had a similar result, finding CTX-M ESBLs to be most prevalent in ESBL-producing bacteria and supporting the trend observed in many surveys [12,45]. Like other research, however, the above studies were focused on clinical isolates, and it appears that there is no data on ESBL-producing bacteria in environmental samples from surface waters in the Philippines. Hence, our study is the first to report the presence of ESBL-producing bacteria in surface waters used for irrigation in the Philippines.

The results of this study may circumstantially indicate horizontal transmission of antibiotic-resistant and MDR *E. coli* from irrigation water to agricultural soils and vegetables. However, the results need to be interpreted with caution, as there was an unequal distribution of the number of *E. coli* isolates from the samples used in this study. The results also cannot conclusively determine the occurrence and direction of lateral transfer of antibiotic resistant and MDR *E. coli*. Hence, the possibility that antibiotic resistant and MDR *E. coli* originated from other sources cannot be discounted. The presence of antibiotic-resistant and MDR *E. coli* in fresh produce may be naturally inherent in vegetables due to the ubiquity of antibiotic resistant bacteria [4,35]. It may also be due to some other contaminating sources, such as poor handling of fresh produce during cultivation, harvesting and marketing and unsanitary production equipment and conditions [2,4].

The results of this study were also limited by the absence of sequencing to identify the ESBL genes and by clonality testing to determine the genetic relatedness of *E. coli* isolates used in this study. Therefore, the molecular identification of ESBL genes and genetic association of *E. coli* isolates used in this study are areas that warrant further investigation. With these findings, it is important that awareness regarding antibiotic resistance and its generation is raised among the public and in healthcare settings. Release of contaminants into the environment should be controlled in order to prevent emergence of antibiotic resistances [11].

5. Conclusions

The findings of this study showed that highly-polluted surface waters used for irrigation in selected urban farms of Metro Manila, Philippines are contaminated with antibiotic-resistant and MDR *E. coli*. Soil and vegetables obtained from the sampling sites likewise contain antibiotic-resistant and MDR isolates, albeit to a lesser extent. Additionally, certain patterns of antibiotic resistance were common to isolates obtained from different samples. This provides circumstantial evidence that surface waters harbor antibiotic-resistant and MDR bacteria that may be transferred to the primary production environment when used for irrigation, and may potentially cause foodborne gastrointestinal infections. Further research is warranted to unequivocally establish the occurrence of the horizontal transmission of antibiotic-resistant and MDR *E. coli* across the samples.

Author Contributions: Enrico S. Zara, Cielo Emar M. Paraoan, Ma. Angela Z. Dimasupil, Joseth Jermaine M. Abello, and Iñigo Teodoro G. Santos performed the experiments, analyzed the data, and prepared the manuscript; and Pierangeli G. Vital and Windell L. Rivera conceptualized the study, headed the project, and were in charge of the preparation of the paper.

Acknowledgments: This work was supported by the Natural Sciences Research Institute, University of the Philippines Diliman and the Department of Agriculture—Bureau of Agricultural Research of the Philippine Government.

Conflicts of Interest: The authors declare no conflict of interest.

References

1. Steele, M.; Odumeru, J. Irrigation water as source of foodborne pathogens on fruit and vegetables. *J. Food Prot.* **2004**, *67*, 2839–2849. [CrossRef] [PubMed]
2. Garcia, B.C.B.; Dimasupil, M.A.Z.; Vital, P.G.; Widmer, K.W.; Rivera, W.L. Fecal contamination in irrigation water and microbial quality of vegetable primary production in urban farms of Metro Manila, Philippines. *J. Environ. Sci. Health Part B* **2015**, *50*, 734–743. [CrossRef] [PubMed]
3. Food and Agriculture Organization of the United Nations World Health Organization. *Microbiological Hazards in Fresh Fruits and Vegetables—Microbiological Risks Assessment Series*; Pre-Publication Version; FAO: Rome, Italy; WHO: Geneva, Switzerland, 2008.
4. Holvoet, K.; Sampers, I.; Callens, B.; Dewulf, J.; Uyttendaele, M. Moderate prevalence of antimicrobial resistance in *Escherichia coli* isolates from lettuce, irrigation water and soil. *Appl. Environ. Microbiol.* **2013**, *79*, 6677–6683. [CrossRef] [PubMed]
5. Blaustein, R.A.; Shelton, D.R.; Van Kessel, J.A.S.; Karns, J.S.; Stocker, M.D.; Pachepsky, Y.A. Irrigation waters and pipe-based biofilms as sources for antibiotic-resistant bacteria. *Environ. Monit. Assess.* **2015**, *188*, 1–12. [CrossRef] [PubMed]
6. Levy, S.B. Factors impacting on the problem of antibiotic resistance. *J. Antimicrob. Chemother.* **2002**, *49*, 25–30. [CrossRef] [PubMed]
7. Levy, S.B.; Marshall, B. Antibacterial resistance worldwide: Causes, challenges and responses. *Nat. Med.* **2004**, *10*, 122–129. [CrossRef] [PubMed]
8. Huddleston, J.R. Horizontal gene transfer in the human gastrointestinal tract: Potential spread of antibiotic resistance genes. *Infect. Drug Resist.* **2014**, *7*, 167–176. [CrossRef] [PubMed]
9. Cosgrove, S.E. The relationship between antimicrobial resistance and patient outcomes: Mortality, length of hospital stay, and health care costs. *Clin. Infect. Dis.* **2006**, *42*, S82–S89. [CrossRef] [PubMed]
10. Tissera, S.; Lee, S.M. Isolation of Extended Spectrum β-lactamase (ESBL) Producing Bacteria from Urban Surface Waters in Malaysia. *Malays. J. Med. Sci.* **2013**, *20*, 14–22. [PubMed]
11. Lucena, M.A.H.; Metillo, E.B.; Oclarit, J.M. Prevalence of CTX-M Extended Spectrum β-lactamase-producing Enterobacteriaceae at a Private Tertiary Hospital in Southern Philippines. *Philipp. J. Sci.* **2012**, *141*, 117–121.

12. Cruz, M.C.; Bacani, C.S.; Mendoza, A.B.; Hedreyda, C.T. Evaluation of extended-spectrum beta-lactamase production in *Escherichia coli* clinical isolates from three hospitals in Luzon, Philippines. *Philipp. Sci. Lett.* **2014**, *7*, 438–444.

13. Schwartz, T.; Kohnen, W.; Jansen, B.; Obst, U. Detection of antibiotic-resistant bacteria and their resistance genes in wastewater, surface water, and drinking water biofilms. *FEMS Microbiol. Ecol.* **2003**, *43*, 325–335. [CrossRef] [PubMed]

14. Watkinson, A.J.; Micalizzi, G.B.; Graham, G.M.; Bates, J.B.; Costanzo, S.D. Antibiotic-resistant *Escherichia coli* in wastewaters, surface waters, and oysters from an urban riverine system. *Appl. Environ. Microbiol.* **2007**, *73*, 5667–5670. [CrossRef] [PubMed]

15. Borjesson, S. *Antibiotic Resistance in Wastewater: Methicillin-Resistant Staphylococcus Aureus (MRSA) and Antibiotic Resistance Genes*; LiU-Tryck: Linköping, Sweden, 2009; ISBN 978-91-7393-629-3.

16. Amaya, E.; Reyes, D.; Paniagua, M.; Calderon, S.; Rashid, M.U.; Colque, P.; Kuhn, I.; Mollby, R.; Weintraub, A.; Nord, C.E. Antibiotic resistance patterns of *Escherichia coli* isolates from different aquatic environmental sources in Leon, Nicaragua. *Clin. Microbiol. Infect.* **2012**, *18*, E347–E354. [CrossRef] [PubMed]

17. Carey, D.E.; McNamara, P.J. The impact of triclosan on the spread of antibiotic resistance in the environment. *Front. Microbiol.* **2014**, *5*, 780. [CrossRef] [PubMed]

18. Suzuki, S.; Ogo, M.; Miller, T.W.; Shimizu, A.; Takada, H.; Siringan, M.A.T. Who possesses drug resistance genes in the aquatic environment?: Sulfamethoxazole (smx) resistance genes among the bacterial community in water environment of Metro Manila, Philippines. *Front. Microbiol.* **2013**, *4*, 102. [CrossRef] [PubMed]

19. Campilan, D.; Boncodin, R.; de Guzman, C. *Multi-Sectoral Initiatives for Urban Agriculture in Metro Manila, Philippines CIP Program Report 1999–2000*; International Potato Center: Lima, Peru, 2001.

20. Coyle, M.B. *Antimicrobial Susceptibility Testing*; Manual; American Society for Microbiology Press: Washington, DC, USA, 2005.

21. Clinical and Laboratory Standards Institute. *Performance Standards for Antimicrobial Susceptibility Testing*; CLSI: Wayne, PA, USA, 2012; p. 188.

22. Jarlier, V.; Nicolas, M.H.; Fournier, G.; Philippon, A. Extended Broad-Spectrum β-Lactamases Conferring Transferable Resistance to Newer β-Lactam Agents in Enterobacteriaceae: Hospital Prevalence and Susceptibility Patterns. *Rev. Infect. Dis.* **1988**, *10*, 867–878. [CrossRef] [PubMed]

23. Dallenne, C.; Da Costa, A.; Decre, D.; Favier, C.; Arlet, G. Development of a set of multiplex PCR assays for the detection of genes encoding important β-lactamases in Enterobacteriaceae. *J. Antimicrob. Chemother.* **2010**, *65*, 490–495. [CrossRef] [PubMed]

24. Takahashi, H.; Kimura, B.; Tanaka, Y.; Shinozaki, J.; Suda, T.; Fujii, T. Real-time PCR and enrichment culture for sensitive detection and enumeration of *Escherichia coli*. *J. Microbiol. Methods* **2009**, *79*, 124–127. [CrossRef] [PubMed]

25. Roe, M.T.; Vega, E.; Pillai, S.D. Antimicrobial resistance markers of class 1 and class 2 integron-bearing *Escherichia coli* from irrigation water and sediments. *Emerg. Infect. Dis.* **2003**, *9*, 822–826. [CrossRef] [PubMed]

26. Chigor, V.N.; Umoh, V.J.; Smith, S.I.; Igbinosa, E.O.; Okoh, A.I. Multidrug resistance and plasmid patterns of *Escherichia coli* O157 and other *E. coli* isolated from diarrhoeal stools and surface waters from some selected sources in Zaria, Nigeria. *Int. J. Environ. Res. Public Health* **2010**, *7*, 3831–3841. [CrossRef] [PubMed]

27. Ramirez Castillo, F.Y.; Avelar Gonzalez, F.J.; Garneau, P.; Márquez Díaz, F.; Guerrero Barrera, A.L.; Harel, J. Presence of multidrug resistant pathogenic *Escherichia coli* in the San Pedro River located in the State of Aguascalientes, Mexico. *Front. Microbiol.* **2013**, *4*, 147. [CrossRef] [PubMed]

28. Lamprecht, C.; Romanis, M.; Huisamen, N.; Carinus, A.; Schoeman, N.; Gunnar, O.S.; Trevor, J.B. *Escherichia coli* with virulence factors and multidrug resistance in the Plankenburg River. *S. Afr. J. Sci.* **2014**, *110*, 1–6. [CrossRef]

29. Njage, P.M.K.; Buys, E.M. Pathogenic and commensal *Escherichia coli* from irrigation water show potential in transmission of extended spectrum and Ampc β-lactamases determinants to isolates from lettuce. *Microb. Biotechnol.* **2015**, *8*, 462–473. [CrossRef] [PubMed]

30. Paraoan, C.E.M.; Rivera, W.L.; Vital, P.G. Detection of Class I and II integrons for the assessment of antibiotic and multidrug resistance among *Escherichia coli* isolates from agricultural irrigation waters in Bulacan, Philippines. *J. Environ. Sci. Health Part B* **2017**, *52*, 306–313. [CrossRef] [PubMed]

31. Chopra, I.; Roberts, M. Tetracycline antibiotics: Mode of action, applications, molecular biology, and epidemiology of bacterial resistance. *Microbiol. Mol. Biol. Rev.* **2001**, *65*, 232–260. [CrossRef] [PubMed]

32. Magiorakos, A.P.; Srinivasan, A.; Carey, R.B.; Carmeli, Y.; Falagas, M.E.; Giske, C.G.; Harbarth, S.; Hindler, J.F.; Kahlmeter, G.; Olsson-Liljequist, B. Multidrug-resistant, extensively drug-resistant and pandrug-resistant bacteria: An international expert proposal for interim standard definitions for acquired resistance. *Clin. Microbiol. Infect.* **2012**, *18*, 268–281. [CrossRef] [PubMed]

33. Shakya, P.; Barrett, P.; Diwan, V.; Marothi, Y.; Shah, H.; Chhari, N.; Tamhankar, A.J.; Pathak, A.; Lundborg, C.S. Antibiotic resistance among *Escherichia coli* isolates from stool samples of children aged 3 to 14 years from Ujjain, India. *BMC Infect. Dis.* **2013**, *13*, 1–6. [CrossRef] [PubMed]

34. Marti, E.; Variatza, E.; Balcazar, J.L. The role of aquatic ecosystems as reservoirs of antibiotic resistance. *Trend Microbiol.* **2013**, *22*, 36–41. [CrossRef] [PubMed]

35. Taylor, D.N. Poorly absorbed antibiotics for the treatment of traveler' diarrhea. *Clin. Infect. Dis.* **2005**, *41*, S564–S570. [CrossRef] [PubMed]

36. Diwan, V.; Tamhankar, A.J.; Khandal, R.K.; Sen, S.; Aggarwal, M.; Marothi, Y.; Iyer, R.V.; Sundblad-Tonderski, K.; Stalsby-Lundborg, C. Antibiotics and antibiotic-resistant bacteria in waters associated with a hospital in Ujjain, India. *BMC Public Health* **2010**, *10*, 414. [CrossRef] [PubMed]

37. Zhu, Y.G.; Johnson, T.A.; Su, J.Q.; Qiao, M.; Guo, G.X.; Stedtfeld, R.D.; Hashsham, S.A.; Tiedje, J.M. Diverse and abundant antibiotic resistance genes in chinese swine farms. *Proc. Natl. Acad. Sci. USA* **2013**, *110*, 3435–3440. [CrossRef] [PubMed]

38. Wang, Q.; Mao, D.; Mu, Q.; Luo, Y. Enhanced horizontal transfer of antibiotic resistance genes in freshwater microcosms induced by an ionic liquid. *PLoS ONE* **2015**, *10*, e0126784. [CrossRef] [PubMed]

39. UN Center for Human Settlements. *Guidelines for Settlement Planning in Areas Prone to Flood Disasters*; United Nations Center for Human Settlements: Nairobi, Kenya, 1995.

40. Sela, S.; Manulis-Sasson, S. What else can we do to mitigate contamination of fresh produce by foodborne pathogens? *Microb. Biotechnol.* **2015**, *8*, 29–31. [CrossRef] [PubMed]

41. Canton, R.; Coque, T.M. The CTX-M β-lactamase pandemic. *Curr. Opin. Microbiol.* **2006**, *9*, 466–475. [CrossRef] [PubMed]

42. Hawkey, P.M. Prevalence and clonality of extended-spectrum β-lactamases in Asia. *Clin. Microbiol. Infect.* **2008**, *14*, 159–165. [CrossRef] [PubMed]

43. Cabrera, E.C.; Rodriguez, R.D. First report on the occurrence of SHV-12 extended-spectrum beta-lactamase-producing Enterobacteriaceae in the Philippines. *J. Microbiol. Immunol. Infect.* **2009**, *42*, 74–85. [PubMed]

44. Tian, G.; Garcia, J.; Adams-Haduch, J.M.; Evangelista, J.P.; Destura, R.V.; Wang, H.; Doi, Y. CTX-M as the predominant extended-spectrum β-lactamases among Enterobacteriaceae in Manila, Philippines. *J. Antimicrob. Chemother.* **2010**, *65*, 584–586. [CrossRef] [PubMed]

45. Kanamori, H.; Navarro, R.B.; Yano, H.; Sombrero, L.T.; Capeding, M.R.Z.; Lupisan, S.P.; Olveda, R.M.; Arai, K.; Kunishima, H.; Hirakata, Y.; et al. Molecular characteristics of extended-spectrum β-lactamases in clinical isolates of Enterobacteriaceae from the Philippines. *Acta Trop.* **2011**, *120*, 140–145. [CrossRef] [PubMed]

water

MDPI

Article

An Increase of Abundance and Transcriptional Activity for *Acinetobacter junii* Post Wastewater Treatment

Muhammad Raihan Jumat [1], Muhammad Fauzi Haroon [2], Nada Al-Jassim [1], Hong Cheng [1] and Pei-Ying Hong [1,*

[1] Water Desalination and Reuse Center (WDRC), Biological and Environmental Science & Engineering Division (BESE), King Abdullah University of Science and Technology (KAUST), Thuwal 23955-6900, Saudi Arabia; raihan.jumat@kaust.edu.sa (M.R.J.); nada.aljassim@kaust.edu.sa (N.A.-J.); hong.cheng@kaust.edu.sa (H.C.)

[2] Department of Organismic and Evolutionary Biology, Harvard University, Cambridge, MA 02138, USA; fauziharoon@gmail.com

* Correspondence: peiying.hong@kaust.edu.sa; Tel.: +966-(0)-808-2218

Received: 29 January 2018; Accepted: 28 March 2018; Published: 6 April 2018

Abstract: A membrane bioreactor (MBR)-based wastewater treatment plant (WWTP) in Saudi Arabia is assessed over a five-month period in 2015 and once in 2017 for bacterial diversity and transcriptional activity using metagenomics, metatranscriptomics and real time quantitative polymerase chain reaction (RT-qPCR). *Acinetobacter* spp. are shown to be enriched in the chlorinated effluent. Members of the *Acinetobacter* genus are the most abundant in the effluent and chlorinated effluent. At the species level, *Acinetobacter junii* have higher relative abundances post MBR and chlorination. RNA-seq analysis show that, in *A. junii*, 288 genes and 378 genes are significantly upregulated in the effluent and chlorinated effluent, respectively, with 98 genes being upregulated in both. RT-qPCR of samples in 2015 and 2017 confirm the upregulation observed in RNA-seq. Analysis of the 98 genes show that majority of the upregulated genes are involved in cellular repair and metabolism followed by resistance, virulence, and signaling. Additionally, two different subpopulations of *A. junii* are observed in the effluent and chlorinated effluent. The upregulation of cellular repair and metabolism genes, and the formation of different subpopulations of *A. junii* in both effluents provide insights into the mechanisms employed by *A. junii* to persist in the conditions of a WWTP.

Keywords: *Acinetobacter junii*; wastewater treatment plant; antibiotic resistance; metal resistance genes; persistence

---

1. Introduction

It is predicted that, by 2050, many countries worldwide will bear the brunt of global water scarcity [1]. To relieve the pressure on finite sources of fresh water, treated wastewater has been purported as an alternative source. If adequately treated, recycled wastewater may serve for irrigation and other non-potable uses. Guidelines are typically put in place in most countries to guide the level of treatment required for the wastewaters prior to reuse. For example, Saudi Arabia requires a fecal coliform count of <1000 CFU/100 mL for reuse in restricted irrigation or <2.2 CFU/100 mL for unrestricted irrigation [2]. In general, most studies show that wastewater treatment plants (WWTPs) are successful in reducing bacterial diversity and abundance [3–5]. However, WWTPs have repeatedly been shown to be hotbeds of antibiotic resistance genes (ARGs), antibiotic-resistant bacteria (ARB) and pathogen enrichment [6–8]. In particular, virulent and resistant pathogens are constantly detected in the effluents of WWTPs [4,9–11]. The subsequent dissemination of pathogens from WWTPs is a significant cause of concern for public health.

To circumvent the problems associated with conventional WWTPs, WWTPs worldwide have recently coupled microfiltration or ultrafiltration membranes to their activated sludge processes. This configuration is typically referred to as the aerobic membrane bioreactor (MBR). The membrane separates biomass from wastewater by filtration, resulting in higher water quality [12]. MBRs are typically operated with a longer sludge retention time compared to the conventional activated sludge processes. This results in lower sludge production rates due to biodegradation of the biomass and lower growth yield, likely arising from the depletion of available substrates within the sludge biomass. The effluent from the MBR is then subjected to disinfection, typically by chlorine, which inactivates any viable microorganisms that may be present [13,14]. Chlorine is an oxidizing agent and achieves its disinfection efficacy by oxidizing nucleic acids, proteins and destabilizing cell walls of microorganisms present in the WWTP effluent [15]. Chlorine further reacts with water to form hypochlorous acid and hypochlorite, which is a type of reactive oxygen species (ROS) that can generate oxidative stress towards microorganisms. Oxidative stress can further trigger the formation of intracellular ROS which are highly reactive molecules that can interfere with normal functions of bacteria [16–18]. Collectively, the MBR and chlorination were shown to be responsible in reducing the microbial load and diversity of wastewater in a WWTPs situated in Saudi Arabia [19,20], possibly due to imposition of harsh conditions that do not favor bacterial proliferation and growth.

However, several antibiotic-resistant strains, which were absent in the influent, were detected in the effluent and chlorinated effluent [4]. Among these were members of the *Acinetobacter* genus which were found to be resistant to ampicillin, kanamycin, erythromycin, tetracycline and chloramphenicol (Tables S1 and S2). Similarly, antibiotic-resistant *Acinetobacter* spp. have been previously isolated from other WWTPs, suggesting that antibiotic resistance confers tolerance to the harsh WWTP conditions [21,22]. An earlier study also showed that *Acinetobacter* spp. undergo a shift in metabolism in activated sludge, which might be another mechanism adopted by this genus to survive the wastewater treatment process [23]. These studies, although enlightening, used conventional polymerase chain reaction (PCR), which is skewed towards known genes and dependent on primer design, to detect for changes in *Acinetobacter* spp. activity.

Acinetobacter spp. are Gram negative, coccobacilli bacteria belonging to the *Moraxellaceae* family. *Acinetobacter* spp. have been shown to thrive in hospital settings, being responsible for up to 9% of nosocomial infections [24]. Among members of this genus, *Acinetobacter junii* has been shown to be an opportunistic pathogen causing disease in individuals with known malignancies, or in those who had received prior antibiotic treatment or invasive procedures. The symptoms of infection with *A. junii* include septicemia, bacteremia, empyema, peritonitis and keratitis [25–27]. Members of this genus have also been shown to facilitate uptake of ARGs from *E. coli* through horizontal gene transfer at higher rates than other studied bacteria [28].

Given the ubiquitous detection of members of the genera *Acinetobacter* in diverse environments including activated sludge [29], different stages of WWTPs [30], hospital effluent [31], clinical samples [32] and environmental samples [21,33,34], we hypothesize that *Acinetobacter* spp. may utilize several strategies to facilitate their survival and persistence. In this study, we further utilized *A. junii*, which was detected to be ubiquitously present in samples collected from different stages of the WWTP, as a model bacterium to address this hypothesis. It was hypothesized that *A. junii* upregulates specific genes to allow for its survival in the harsh environments of a WWTP. To determine this, *A. junii* from different stages of a WWTP serving a university campus is studied by RNA-seq. This overcomes the biasness of PCR and captures global changes in transcriptional activity of *Acinetobacter* spp. before and after wastewater treatment. Results indicate an increase in relative abundance of *Acinetobacter* spp. down the treatment process with a concomitant increase in transcriptional activity. Majority of the upregulated genes correspond to cellular resistance and metabolism. Other upregulated genes encode for metal-resistance, efflux pumps and mobility proteins, all which have been implicated in conferring antibiotic resistance and increased survivability in harsh environmental conditions.

2. Materials and Methods

2.1. Wastewater Sampling

A WWTP located on the campus of King Abdullah University of Science and Technology (KAUST), Thuwal, Saudi Arabia was sampled for the purposes of this study. The operation of the WWTP has been previously described [19]. Briefly, large screens prevent entry of insoluble and bulky solids into the primary clarifier. Wastewater was retained inside the primary clarifier for ca. 3 h to allow settle-able suspended solids to settle at the bottom of clarifier in the presence of coagulants. The clarified sewage then undergoes biological treatment in an anoxic sludge tank. A second oxic sludge tank is equipped with a 0.4 μm pore sized-submerged microfiltration membrane which forms the membrane bioreactor. Both sludge tanks have a total capacity of 1600 m^3. Mixed liquor suspended solids (MLSS) concentration of the activated sludge was maintained at 16 g/L. The membranes were operated at a flux of 15.5 L/(m^2·h) with a trans-membrane pressure below 20 kPa. These membranes have been cleaned monthly with 5% ClO$^-$ solution throughout their 7 years of operation. The MBR is operated at a 4-h hydraulic retention time and a 40-day sludge retention time (SRT). The MBR produces 4000 m^3 of effluent daily. Up to 60 L of the MBR effluent was sampled each time, and referred to as "effluent" throughout this study. In this WWTP, MBR effluent is stored in an 8500 m^3 holding tank and mixed with 400 m^3/d of blow-down from a nearby seawater cooling tower [20]. This mixed stream is disinfected with a free chlorine residual of 0.5 mg/L at a contact time of 2.5 h, thrice daily resulting in a contact time (CT) of 75 mg·min/L. While Saudi Arabia currently does not have any regulations on the minimum required chlorine dose for disinfection, the operators voluntarily chose to adhere to the recommended minimum value of 30 mg·min/L as suggested by the US-EPA for disinfection of waters for reuse purposes [35]. The chlorinated water from this holding tank was sampled and referred to as "chlorinated effluent" throughout this study. Influent, effluent and chlorinated effluent were sampled in July 2015, October 2015, November 2015 and October 2017. For each sampling expedition, 20 L of influent and 60 L of both effluents were collected.

2.2. Tangential Flow Filtration

To concentrate any microbial matter, each wastewater sample was passed through a tangential flow filtration (TFF) system fitted with a T-Series cassette with a 100 kDa cut-off, which should retain any matter >0.005 μm (Pall Corporation, Port Washington, NY, USA). The recovery efficiency of the TFF system was previously described using enterovirus as model contaminant, with an efficiency ranging from 47.8% to 86.9% [19]. Influent samples were centrifuged at 7500× g for 10 min to obtain 12 L of clarified influent prior to TFF concentration. Samples were concentrated to 5 mL of retentate. These concentrated samples were denoted as Influent-Retentate, Effluent-Retentate and Chlorinated Effluent-Retentate. To ensure recovery of microbial matter that was adhered to the cassette, the system was washed with 50 mL of 1 × Phosphate Buffered Saline (PBS) containing 0.01% Tween 60 (Wash Buffer). The wash buffer was recirculated through the system several times before it was concentrated to 5 mL. These samples are denoted as Influent-Wash, Effluent-Wash and Chlorinated Effluent-Wash. Each of these concentrated samples were stored in −80 °C prior to nucleic acid extraction. The filter was washed with 0.1 N NaOH between samples, as per manufacturer's instructions.

2.3. Nucleic Acid Extraction and Metagenomics Sequence Analysis

DNA and RNA were extracted from equal volumes of the retentate and wash concentrates using the DNeasy Blood and Tissue and RNeasy Midi kits (Qiagen, Hilden, Germany), respectively. Concentrations of the extracted DNA and RNA were quantified using the Qubit 2.0 fluorometer (Thermo Fisher Scientific, Carlsbard, CA, USA). To obtain 1 μg of DNA and RNA for metagenomic and transcriptomic analysis, DNA and RNA extracts were pooled from July 2015, October 2015 and November 2015 samples. DNA libraries were constructed using the TruSeq DNA LT kit and sequenced by the HiSeq2000 system (Illumina, San Diego, CA, USA). RNA libraries were generated by using the TruSeq Stranded mRNA plus Ribo-Zero Epidemiology kit and sequenced by the Illumina NextSeq 500 system. For quality control, 1 µL of the resultant libraries was loaded onto an Agilent Technologies 2100 Bioanalyzer (Santa Clara, CA, USA). All the sequenced samples yielded a single band at approximately 260 bp. All sequencing reactions were carried out at the KAUST Bioscience Core Laboratory. Raw metagenomic reads were trimmed using Trimmomatic v.0.3.2 [36]. Trimmed reads from each sample were concatenated together. To profile the metagenomes for the most abundant genus among samples collected throughout the WWTP, MetaPhlAn v2.0 [37] was performed using default parameters. Output tables were merged together using "merge_metaphlan_tables.py" and heat maps were created using "metaphlan_hclust_heatmap.py". Initial results of this analysis show a large proportion of the assembled contigs belonging to the species *Acinetobacter junii*. To further characterize *A. junii* down the treatment process of this WWTP, RNA-seq analysis was carried out as described below.

RNA-seq data were further analyzed by CLC Genomics Workbench version 8.0.1 from CLC Bio (Cambridge, MA, USA). The complete genome of *A. junii* strain 65 was downloaded from NCBI GenBank (accession number NZ_CP019041.1). RAST annotated genomic DNA sequences were used as reference to map the RNA-seq. Reads were only assembled if the fraction of the read which aligned to the reference genome was greater than 0.9 and if the read matched other regions of the reference genome at fewer than 10 nucleotide positions. Mapped RNA-seq files were only considered if statistically significant ($p < 0.05$) according to the Baggerly proportion-based test [38]. Fold changes of gene expression were calculated with reference to RNA expression in the influent wastewater samples. Only genes with a fold change of more than 2 or less than -2 were considered as significantly upregulated or downregulated, respectively. All sequencing data generated for this study were deposited in the European Nucleotide Archive (ENA) under study accession number PRJEB15519.

2.4. cDNA Synthesis

RNA extracted from the wastewater samples were used as template to synthesize complementary DNA (cDNA) for confirmatory qRT-PCR experiments. cDNA was synthesized according to Invitrogen's SuperScript III First-Strand Synthesis for RT-PCR. Briefly, each 6 µL of RNA sample was incubated with 1 µL of random hexamers and 1 µL of Annealing Buffer at 65 °C for 5 min. The tubes were incubated on ice for 1 min followed by the addition of 10 µL of 2 × First-Strand Reaction Mix and 2 µL of SuperScript III/RNaseOUT Enzyme mix. The samples were incubated at 25 °C for 10 min, 50 °C for 50 min and terminated at 85 °C for 5 min. The resulting cDNA was stored at -20 °C, prior to RT-qPCR.

2.5. RT-qPCR

RT-qPCR was employed to confirm the upregulations observed in the RNA-seq. Primers were designed for 16 of the genes showing significant upregulation in either or both effluents. Table S3 lists the gene names, primer sequences and gene categorization (Table S3). qPCR standards were amplified from lab grown *A. junii* strains (DSMZ 14968, 1532). PCR products were cloned into pCR2.1 vectors (Thermo Fisher Scientific, Carlsbad, CA, USA). Plasmids carrying the PCR products were extracted from transformed TOP10 competent cells (Thermo Fisher Scientific, Carlsbad, CA, USA) by using PureYield™ Plasmid Miniprep System (Promega, Madison, WI, USA). Plasmids were sent to the KAUST Genomics Core Lab for Sanger sequencing to determine the presence of the right primer sequences. Plasmid copy numbers were then determined based on the concentration, insert and vector sizes. Plasmids were subsequently diluted from 10^2 to 10^8 copies/µL to attain 8-points qPCR standard curves for each gene. Quantities of gene expression were calculated by the relative standard curve method. RT-qPCR were carried out using 10 µL of Applied Biosystsems' Fast SYBR™ Green

Master Mix (Thermo Fisher Scientific, Carlsbad, CA, USA), 0.4 µL of each primer (10 µM), 8.2 µL of molecular-biology grade water and 1 µL of template. RT-qPCR was performed using Applied Biosystem®7900HT Fast Real-Time OCR system with 96-well block module (Thermo Fisher Scientific, Carlsbad, CA, USA). A melting curve analysis was performed with a dissociation cycle that included an increment of temperature from 60 to 95 °C, at an interval of 0.5 °C for 5 s. After optimization, copy numbers of the gene were determined and normalized against *rpoB* gene expression.

2.6. ICP-MS

The metal content (^{24}Mg, ^{27}Al, ^{47}Ti, ^{53}Cr, ^{55}Mn, ^{56}Fe, ^{57}Fe, ^{59}Co, ^{60}Ni, ^{63}Cu, ^{66}Zn, ^{90}Zr, ^{107}Ag, ^{111}Cd, ^{202}Hg, and ^{208}Pb) of October 2017 wastewater samples were measured on an inductively coupled plasma mass spectrometry (ICP-MS) (Agilent 7500). Wastewater samples were filtered through 0.22 µm Whatman™ Puradisc 23-mm syringe filters (GE Healthcare, Little Chalfort, Buckinhamshire, UK) prior to measurement. Measurements were made against commercially available standards at 0, 1, 5, 10, 50 and 100 parts per billion in 2% HNO_3 (CMS-3, CMS-4, CMS-5, and MSHG) (Inorganic Ventures, Christiansburg, VA, USA). Heavy metal ion concentrations were displayed as averaged values from two reads on the ICP-MS.

3. Results

3.1. Metagenomics Analysis Revealed Enrichment of Acinetobacter spp. Post Wastewater Treatment

To determine the bacterial composition of the wastewater samples, samples collected in July 2015, October 2015 and November 2015 were pooled and their extracted nucleic acids were sequenced by omics-based sequencing. The sequencing reads were compared against a collated bacterial database. Figure 1 shows the detection of microorganisms throughout the WWTP at the genus and species levels. Further classification of microorganisms at the kingdom, phylum, class and family levels were listed in the Supplementary Materials (Figures S2–S5).

While viruses and archaea decreased in relative abundance down the treatment process, bacteria increased in relative abundance in the effluent and chlorinated effluent compared to the influent (Figure S2). A similar observation was seen at the phylum level with bacteria belonging to Proteobacteria compared to the other phyla (Figure S3). Out of the 22 classes of bacteria detected in this study, members of the Gammaproteobacteria and Actinobacteria increased in relative abundance in the effluent and continued to increase in the chlorinated effluent (Figure S4). Members of Gammaproteobacteria accounted for the highest relative abundance within the total microbial community in the chlorinated effluent. A further examination at the lower taxonomical levels revealed that the *Moraxellaceae* accounted for the highest relative abundance of the detected Gammaproteobacteria class (Figure S5). Specifically, the only genus detected in this WWTP belonging to the *Moraxellaceae* was *Acinetobacter*. Members of this genera were detected in increasing relative abundances in the effluent and the chlorinated effluent, similar to the trend observed at the family level.

Out of the 50 most detected species, five belonged to the *Acinetobacter* genus (Figure 1A). Two of these strains were detected at higher relative abundances in the chlorinated effluent than in the influent (*A. venetians* and *A. junii*).

A.

Genera

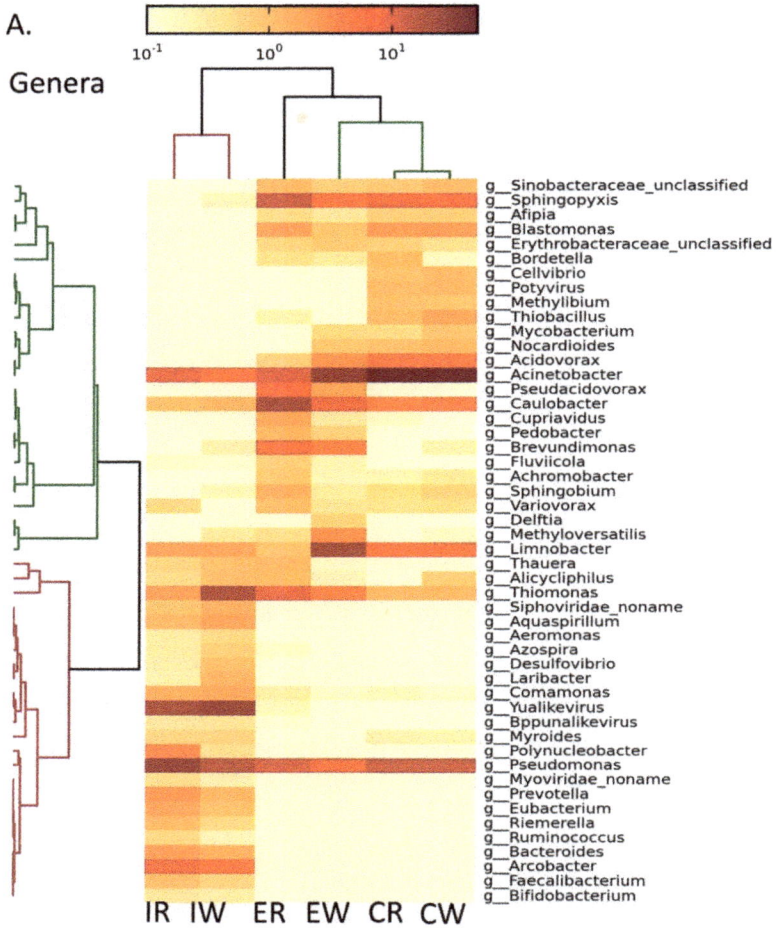

Figure 1. *Cont.*

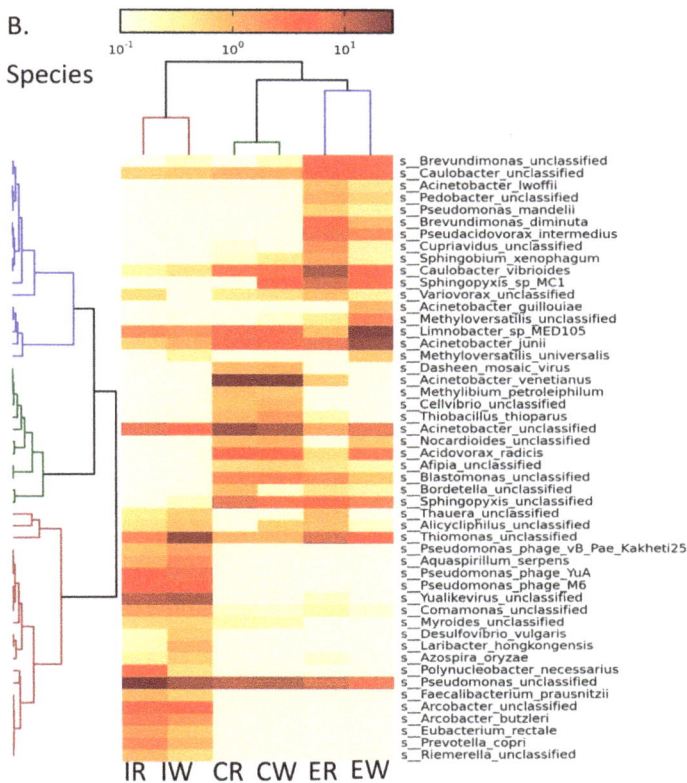

Figure 1. Heat map representing relative abundances of genera (**A**) and species (**B**), detected in the wastewater samples collected in July, October and November 2015. Colored scale bar represents the relative abundances for each heat map. IR: Influent-retentate, IW: Influent-wash, ER: Effluent-retentate, EW: Effluent-wash, CR: Chlorinated Effluent-retentate, CW: Chlorinated Effluent-wash.

3.2. Gene Expression Analysis Revealed an Increase in Gene Expression of Acinetobacter junii Post MBR and Chlorination

Acinetobacter junii has been shown to an opportunistic pathogen [26,39] and since the chlorinated effluent is being introduced into the environment, further analysis of this species from the 2015 wastewater samples were carried out using metatranscriptomics approach.

A total of 288 genes (8.9% of genome) were found to be significantly upregulated in the effluent. This number increased to 378 (11.7%) in the chlorinated effluent. The effluent and chlorinated effluent shared 98 genes (3.1%) which were upregulated (Figure S1A). Thirty-five of these 98 genes were not classified by RAST. The largest category of the upregulated genes belonged to the cellular repair ($n = 25$) and metabolism ($n = 15$) categories. The remaining genes were involved in resistance ($n = 9$), signaling ($n = 7$), virulence ($n = 5$), stress response ($n = 4$), regulation of transcription and translation ($n = 3$) (Table 1). All 288 and 378 genes that were upregulated in the effluent and the chlorinated effluent are listed in Tables S4 and S5.

A. junii displayed 13 (0.4%) and 121 (3.8%) downregulated genes in the effluent and chlorinated effluent. Only 6 (0.19%) of these downregulated genes were shared between the effluent and chlorinated effluent (Figure S1B). Two of these 6 genes were involved in cellular repair with the rest not being categorized (Table 2).

Table 1. Genes of *Acinetobacter junii* that were upregulated in both the effluent and chlorinated effluent. The gene expression values of the effluent-retentate and chlorinated effluent-retentate were compared to the gene expression values of the influent-retentate. The gene expression values of the effluent-wash and chlorinated effluent-wash were compared to the gene expression values of the influent-wash. Each of these upregulations were included if they were more than two-fold change and were statistically significant ($p < 0.05$) on the Baggerly proportion-based test. Genes without any known classifications were omitted. More intense red shading signified a high fold induction.

Locus Tag	Gene Product	Fold Change				Category
		Effluent Retentate	Effluent Wash	Chlorinated Effluent Retentate	Chlorinated Effluent Wash	
BVL33_15990	50S ribosomal protein L10	2.06	31.09	2.51	80.82	Cellular Repair
BVL33_10015	L-threonine dehydrogenase	2.94	12.93	2.32	3.36	Cellular Repair
BVL33_00375	F0F1 ATP synthase subunit epsilon	2.04	12.66	3.30	16.98	Cellular Repair
BVL33_12120	proteheme IX farnesyltransferase	2.11	7.08	3.26	4.70	Cellular Repair
BVL33_06310	outer membrane protein assembly factor BamE	2.45	5.89	2.96	3.42	Cellular Repair
BVL33_12235	cytochrome b	2.71	5.14	2.45	4.75	Cellular Repair
BVL33_02910	bacterioferritin	37.79	5.09	57.84	7.09	Cellular Repair
BVL33_13450	RNA-binding protein	4.75	4.91	7.36	4.48	Cellular Repair
BVL33_10260	adenosine kinase	3.72	4.74	3.67	3.52	Cellular Repair
BVL33_01915	disulfide bond formation protein B	3.58	4.74	9.65	6.64	Cellular Repair
BVL33_02390	recombinase RecB	5.06	4.66	8.33	6.01	Cellular Repair
BVL33_10760	NAD kinase	2.99	4.37	4.90	3.95	Cellular Repair
BVL33_01480	iron-sulfur cluster insertion protein ErpA	2.18	4.10	2.47	3.01	Cellular Repair
BVL33_01875	short-chain dehydrogenase	6.64	3.61	2.28	3.27	Cellular Repair
BVL33_14300	gamma-glutamylcyclotransferase	2.71	2.90	4.87	5.64	Cellular Repair
BVL33_08930	dehydratase	2.25	2.62	9.67	4.05	Cellular Repair
BVL33_15970	DNA transfer protein p32	250.50	2.56	493.00	21.17	Cellular Repair
BVL33_14275	folate-binding protein YgfZ	2.26	2.53	4.89	4.89	Cellular Repair
BVL33_06815	phosphoribosylglycinamide formyltransferase	2.11	2.49	4.90	8.94	Cellular Repair
BVL33_11785	DNA-binding protein HU	16.05	2.23	9.16	4.32	Cellular Repair
BVL33_01730	lipoprotein-34 precursor (NlpB)	5.43	2.03	2.46	2.84	Cellular Repair
BVL33_14490	SPOR domain-containing protein	329.60	2.13	425.30	3.80	Cellular Repair & Metabolism
BVL33_14010	acetyltransferase	2.35	3.04	2.19	4.41	Cellular Repair & Metabolism
BVL33_13315	acyl-CoA thioesterase	2.47	3.04	2.11	7.39	Cellular Repair & Metabolism
BVL33_13885	acyltransferase	117.20	2.01	421.20	8.79	Cellular Repair & Metabolism
BVL33_03425	30S ribosomal protein S17	2.27	46.72	3.25	7.78	Metabolism
BVL33_03420	50S ribosomal protein L29	2.32	39.53	3.50	8.58	Metabolism
BVL33_03375	30S ribosomal protein S10	2.08	30.42	2.28	5.55	Metabolism
BVL33_03435	50S ribosomal protein L24	2.42	24.00	4.37	3.22	Metabolism
BVL33_03500	30S ribosomal protein S4	3.41	13.28	2.47	16.66	Metabolism
BVL33_14915	acyl-CoA dehydrogenase	4.44	7.56	2.46	3.19	Metabolism
BVL33_04785	integration host factor subunit alpha	12.70	5.81	6.56	11.07	Metabolism
BVL33_14350	VOC family protein	2.87	4.87	3.28	4.64	Metabolism
BVL33_14925	2,4-dienoyl-CoA reductase	3.46	4.51	4.17	5.50	Metabolism

Table 1. *Cont.*

Locus Tag	Gene Product	Fold Change				Category
		Effluent Retentate	Effluent Wash	Chlorinated Effluent Retentate	Chlorinated Effluent Wash	
BVL33_04625	3-hydroxyisobutyrate dehydrogenase	3.14	3.10	2.44	3.25	Metabolism
BVL33_03235	GTP cyclohydrolase II	6.02	2.98	6.53	3.89	Metabolism
BVL33_13355	SCP-2 sterol transfer family protein	3.01	6.90	3.02	13.07	Metabolite Transport
BVL33_02100	IS4 family transposase	15.00	4.70	77.00	6.11	Resistance
BVL33_15710	Na+/H+ antiporter subunit C	5.83	8.50	4.87	33.77	Resistance
BVL33_10090	MBL fold metallo-hydrolase	7.89	5.56	14.22	8.98	Resistance
BVL33_13770	efflux transporter periplasmic adaptor subunit	4.53	4.02	5.41	2.33	Resistance
BVL33_13780	adenosine deaminase	6.49	3.83	3.28	2.04	Resistance
BVL33_02685	sodium:proton antiporter	27.50	2.57	144.50	7.71	Resistance
BVL33_01605	efflux transporter periplasmic adaptor subunit	4.76	2.33	2.46	3.82	Resistance
BVL33_11410	glutathione S-transferase	3.46	2.77	3.26	2.52	Resistance & Signalling
BVL33_11715	glutathione S-transferase	4.52	2.51	2.46	3.09	Resistance & Signalling
BVL33_14240	response regulator	2.51	9.69	3.70	60.47	Signalling
BVL33_02040	ion channel protein Tsx	3.37	7.82	9.72	2.61	Signalling
BVL33_13200	co-chaperone GroES	2.70	6.16	2.82	4.36	Signalling
BVL33_14235	response regulator	5.73	3.49	4.93	2.63	Signalling
BVL33_14220	hybrid sensor histidine kinase/response regulator	4.35	2.94	2.00	2.01	Signalling
BVL33_04800	thiol reductase thioredoxin	4.42	9.02	4.23	4.60	Stress Response
BVL33_12425	superoxide dismutase	4.44	6.13	3.76	19.06	Stress Response
BVL33_00220	NADPH-dependent FMN reductase	2.24	5.24	19.19	27.53	Stress Response
BVL33_12940	DNA starvation/stationary phase protection protein	823.90	3.50	179.50	2.17	Stress Response
BVL33_13710	transcriptional regulator	63.30	4.46	226.20	7.05	Transcriptional Regulator
BVL33_07060	serine protease	3.54	2.27	2.11	3.06	Transcriptional Regulator
BVL33_04615	ribosome silencing factor RsfS	2.76	9.01	2.47	17.79	Translation Regulator
BVL33_14205	entericidin, EcnA/B family	5.45	11.83	2.47	12.14	Virulence, Disease & Defense
BVL33_14745	preprotein translocase subunit YajC	14.79	8.03	4.87	10.43	Virulence, Disease & Defense
BVL33_04950	protein FliA	17.07	5.88	4.21	9.87	Virulence, Disease & Defense
BVL33_15450	HopJ type III effector protein	146.90	4.97	792.20	4.59	Virulence, Disease & Defense
BVL33_09420	bifunctional adenosylcobinamide kinase/adenosylcobinamide-phosphate guanylyltransferase	63.40	2.24	170.20	13.80	Virulence, Disease & Defense

Table 2. Genes of *Acinetobacter junii* that were downregulated in both the effluent and chlorinated effluent. The gene expression values of the effluent-retentate and chlorinated effluent-retentate were compared to the gene expression values of the influent-retentate. The gene expression values of the effluent-wash and chlorinated effluent-wash were compared to the gene expression values of the influent-wash. Each of these downregulations were included if they were less than -2-fold change and were statistically significant ($p < 0.05$) on the Baggerly proportion-based test. The magnitude of downregulation is indicated by the intensity of green shading.

Locus Tag	Gene Product	Fold Change				Category
		Effluent Retentate	Effluent Wash	Chlorinated Effluent Retentate	Chlorinated Effluent Wash	
BVL33_10620	thiamine biosynthesis protein ThiS	−2.16	−30.91	−90.70	−139.10	Cellular Repair
BVL33_11020	(4Fe-4S)-binding protein	−2.16	−50.30	−72.30	−150.90	Cellular Repair
BVL33_15925	hypothetical protein	−4.31	−104.94	−192.00	−2.43	None
BVL33_00180	hypothetical protein	−88.10	−90.14	−88.10	−126.20	None
BVL33_03255	hypothetical protein	−2.16	−4.66	−83.20	−3.89	None
BVL33_06630	amino acid transport protein	−88.10	−6.40	−88.10	−6.40	None

3.3. Cellular Repair Genes Upregulated in Both Effluent and Chlorinated Effluent

The largest category of genes upregulated in both effluents were involved in cellular repair (Table 1). Two of the most upregulated genes in this category are BVL33_14490 and BVL33_15970 which encode for a SPOR domain-containing protein and a DNA transfer protein, p32, respectively. These genes are highly upregulated in the retentate samples (329.6 and 250.5 times in the effluent retentate and 425.3 and 493 times in the chlorinated effluent, respectively). These genes were upregulated 2.13 and 2.56 times in the effluent wash and 3.8 and 21.17 times in the chlorinated effluent wash.

RecB recombinase, encoded by BVL33_02390, was upregulated 5.06 and 4.66 times in the effluent retentate and wash, respectively, and 8.33 and 6.01 times in the chlorinated effluent and wash, respectively. BVL33_06310 encodes for an outer membrane protein assembly factor. This gene was upregulated 2.45 and 5.89 times in the effluent retentate and wash, respectively, and 2.96 and 3.42 times in the chlorinated effluent retentate and wash, respectively.

Another gene of note in this category is BVL33_11785, encoding for a DNA binding protein. This gene was upregulated 16.05 and 2.23 times in the effluent retentate and wash, respectively, and 9.16 and 4.32 times in the chlorinated effluent retentate and wash, respectively (Table 1).

3.4. Metabolism Genes Upregulated in Both Effluent and Chlorinated Effluent

The second most abundant upregulated category in both effluents were genes involved in metabolism (Table 1). Five of the 11 genes upregulated were ribosomal proteins (BVL33_03425, BVL33_03420, BVL33_03375, BVL33_03435 and BVL33_3500). A sixth ribosomal protein was significantly upregulated in the cellular repair category. These genes were upregulated at an average of 30.84 and 20.44 in the effluent wash and chlorinated effluent wash, respectively. This was higher than the average upregulation of genes in either sample (6.40 and 9.08, respectively). These same genes were upregulated at an average of 2.43 and 3.06 times in the effluent retentate and the chlorinated effluent retentate, respectively (Table 1).

BVL33_13885, which belonged to both the cellular repair and metabolism categories, was upregulated 117.2 and 2.01 times in the effluent retentate and wash, respectively, and 421.2 and 8.79 times in the chlorinated effluent retentate and wash, respectively (Table 1). This gene encodes for an acyltransferase. BVL33_13315, an acyl-CoA thioesterase (upregulated 2.47 and 3.04 times in effluent retentate and wash, respectively, and 2.11 and 7.39 times in chlorinated effluent retentate and wash, respectively) and BVL33_14010 an acyltransferase (upregulated 2.35 and 3.04 times in the effluent retentate and wash, respectively, and 2.19 and 4.41 times in chlorinated effluent retentate and wash, respectively), were upregulated in the same category. These genes could be involved in the same pathway. BVL33_14915 was upregulated 4.44 and 7.56 times in the effluent retentate and wash, respectively, and 2.46 and 3.19 times in the chlorinated effluent retentate and wash, respectively. This gene encodes for an acyl-CoA dehydrogenase, sharing a substrate with BVL33_13885 and BVL33_13315 (Table 1).

After these genes, the next most upregulated gene in the metabolism category was BVL33_04785, encoding for one of the two subunits comprising an integration host factor (IHF) protein complex. This gene was upregulated 12.7 and 5.81 times in the effluent retentate and wash, respectively, and 6.56 and 11.07 times in the chlorinated effluent retentate and wash, respectively.

3.5. Resistance Genes Upregulated in Both Effluent and Chlorinated Effluent

In both effluents, four different efflux pumps were upregulated among genes responsible for resistance. One of these genes, BVL33_02685, a sodium:proton antiporter, was upregulated 27.5 and 2.57 times in the effluent retentate and wash, respectively, and 144.50 and 7.71 times in the chlorinated effluent retentate and wash, respectively (Table 1).

Another mode by which resistance could be attained is by the upregulation of BVL33_10090, a Metallo-β-Lactamase (MBL) fold protein. This protein was upregulated 7.89 and 5.56 times in the effluent retentate and wash, respectively, and 14.22 and 8.98 times in the chlorinated effluent retentate and wash, respectively (Table 1).

BVL33_02100, encoding for an IS4 transposase, was upregulated 15 and 4.7 times in the effluent retentate and wash, respectively, and 77 and 6.11 times in the chlorinated effluent retentate and wash, respectively (Table 1).

3.6. Stress Response Genes Upregulated in Both Effluent and Chlorinated Effluent

Out of the four genes upregulated in both effluents from the stress response category, BVL33_12940 was highly upregulated in the retentates. This gene was associated with DNA starvation and stationary phase protection, and was upregulated 832.9 and 3.5 times in the effluent retentate and wash, respectively. The gene was also upregulated 179.5 and 2.17 times in the chlorinated effluent retentate and wash, respectively. Another gene within this category was BVL33_00220, which encodes for an NADPH-dependent FMN reductase. This gene was upregulated 2.24 and 5.24 times in the effluent retentate and wash, and 19.19 and 27.53 times in the chlorinated effluent retentate and wash, respectively (Table 1).

3.7. Virulence, Disease and Defense Genes Upregulated in Both Effluent and Chlorinated Effluent

BVL33_15450, which encodes for a HopJ type III effector protein was upregulated 146.9 and 4.97 times in the effluent retentate and wash, respectively, and 792.2 and 4.59 times in the chlorinated effluent retentate and wash, respectively. Within this category, BVL33_04950 was upregulated 17.07 and 5.88 times in the effluent retentate and wash samples, respectively, and 4.21 and 9.87 times in the chlorinated effluent retentate and wash samples, respectively. This gene encodes for Protein FilA, which is a pili related gene (Table 1).

3.8. Genes Upregulated Exclusively in the Effluent

Table S4 lists all of the genes that were upregulated in the effluent. Three genes which are involved in the translocation of proteins across the cytoplasmic membrane were observed to be upregulated in the effluent. BVL33_15770 (SecG), BVL33_00120 (Sec-C), BVL33_03480 (SecY) and BVL33_04010 (SecB) were upregulated only in the effluent (Table S4).

The effluent displayed a total of 12 metal-related genes (MRGs): five metal dependent hydrolases (BVL33_03340, BVL33_06175, BVL33_06180, BVL33_10645 and BVL33_14520), a copper oxidase (BVL_05035), a copper translocating P-type ATPase (BVL33_06435), Cu(I)-responsive transcriptional regulator (BVL33_06430), two heavy metal transport/detox protein genes (BVL33_02645 and BVL33_06425), a metal binding protein (BVL33_13790) and a zinc metalloprotease (BVL33_13370) (Table S4). Two of these metal dependent hydrolases (BVL33_03340 and BVL33_06175) were also seen in the chlorinated effluent (Table 1). Furthermore, BVL33_02700, an efflux transporter periplasmic adaptor subunit that is involved in metal efflux and antibiotic resistance, was also upregulated in effluent but not in chlorinated effluent (Table S4).

The effluent also saw an increase in motility related genes other than BVL33_04950 (Protein FilA), which was seen in both effluents. BVL33_02830, a hypothetical type IV pilus protein and BVL33_02810, encoding for a pilus assembly protein PilM, was upregulated in the effluent retentate and wash exclusively (Table S4).

3.9. Genes Upregulated Exclusively in the Chlorinated Effluent

All of the upregulated genes of the chlorinated effluent are listed in Table S5. A different set of genes involved in the translocation of proteins across the cytoplasmic membrane was upregulated in the chlorinated effluent. These include BVL33_14755 (SecF), and BVL33_16010 (SecE) (Table S5).

In the chlorinated effluent samples, seven more MRGs were seen, which are not listed in Table 1: three copper resistance genes (BVL33_06020 and BVL33_06015, BVL_07275), three metal dependent hydrolases (BVL33_08355, BVL33_08360 and BVL33_08940) and a zinc dependent oxioreductase (BVL33_08650) (Table S5).

In addition to BVL33_13770 and BVL33_15710, which were upregulated in the resistance category (Table 1), BVL33_15705, encoding for a cation/proton antiporter, was upregulated exclusively in the chlorinated effluent (Table S5). Two other efflux transporter periplasmic adaptor subunit genes were upregulated in the chlorinated effluent but not the effluent (BVL33_06415 and BVL_11495) (Table S5).

Exclusively in the chlorinated effluent, four more genes were upregulated from the virulence category: two pili assembly proteins (BVL33_08765 and BVL33_02825 (PilP)), a PilZ domain-containing protein (BVL33_06780) and a hypothetical type IV pilus biogenesis protein (PilN) (BVL33_02815) (Table S5).

3.10. Differential Upregulation of Genes in Retentate and Wash Samples

Interestingly, genes were upregulated to different extents between the retentate and wash samples ($p < 0.05$). The color scale in Table 1 shows this distinct pattern with more intense red signifying a high fold induction. On average, each gene was upregulated 60.15 and 65.65 times in the effluent retentate and the chlorinated effluent retentate while genes were only upregulated 6.34 and 8.99 times in the effluent wash and chlorinated effluent wash, respectively. Genes which were highly upregulated in the effluent retentate were upregulated to similar extents in the chlorinated effluent retentate. To exemplify, genes BVL33_14490 and BVL33_15970 were highly upregulated in the effluent retentate (329.6 and 250.5 times, respectively) and the chlorinated effluent retentate (425.3 and 493 times, respectively). These genes were not as upregulated in the wash samples with 2.13- and 2.56-fold induction in the effluent wash and 3.80- and 21.17-fold induction in the chlorinated effluent wash. Conversely, BVL33_15990 and BVL33_03500 were highly upregulated in the effluent wash samples (31.09 and 13.28 times, respectively) and the chlorinated effluent wash (80.82 and 16.66 times, respectively) but were not upregulated as extensively in the retentates samples (Table 1). No similar trend was observed in the downregulated genes (Table 2).

3.11. Validation of RNA-seq Data with RT-qPCR

To validate the RNA-seq data, extracted RNA from the 2015 wastewater samples as well as the wastewater collected in October 2017 were analyzed by RT-qPCR. Messenger RNA (mRNA) levels belonging to gene categories; cellular repair (BVL33_01915, BVL33_02390 and BVL33_15970) (Figure 2), resistance (BVL33_13780 and BVL33_05485) (Figure 3), stress response (BVL33_12425 and BVL33_12940) (Figure 4) and virulence (BVL33_09420 and BVL33_02700) (Figure 5) were quantified and normalized against *rpoB* gene expression. All of the genes selected for this analysis were shown to be upregulated in both the effluent and chlorinated effluent in the RNA-seq analysis (Table 1).

RT-qPCR analysis showed a distinct upregulation in each of the genes analysed in the effluent and chlorinated effluent (Figures 2A, 3A, 4A and 5A). This finding concurs with the RNA-seq analysis and proves increased transcriptional activity in *A. junii* post MBR treatment. A similar trend was observed for wastewater collected in 2017, except for a slight decrease between influent-wash and effluent-wash for gene BVL33_02390 (Figure 2B), a significant downregulation in both the effluent and chlorinated effluent for BVL33_12425 (Figure 4B) and similar expression values between influent and effluent of BVL33_09420 (Figure 5B). Despite sampling two years apart, a similar general upregulation in genes correlating to cellular repair, resistance, virulence and stress response was seen, indicating that *A. junii* is active post MBR treatment over a certain temporal interval.

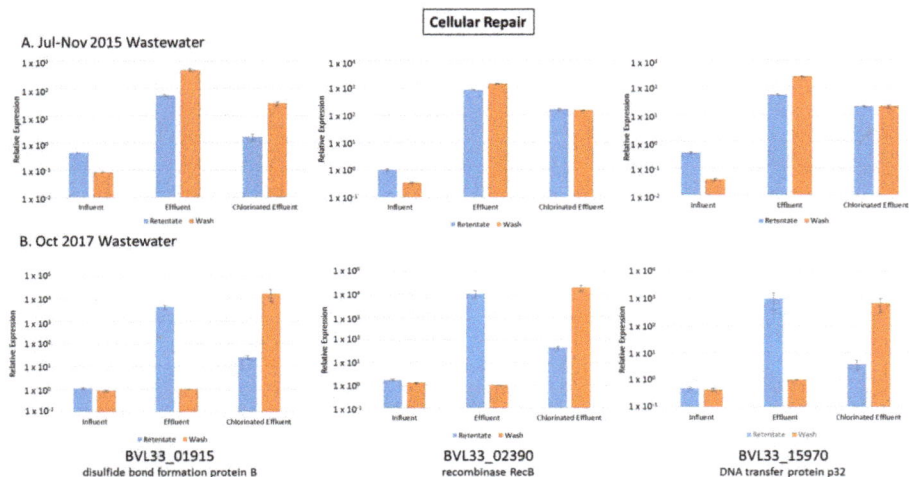

Figure 2. Relative expression profiles of *A. junii* genes involved in cellular repair. These genes displayed >2-fold change on RNA-seq analysis of wastewater sampled in July–November 2015 (Table 1). Gene expression values were normalized against a housekeeping *rpoB* gene. (**A**) Wastewater sampled in July, October and November of 2015; and (**B**) wastewater sampled in October 2017

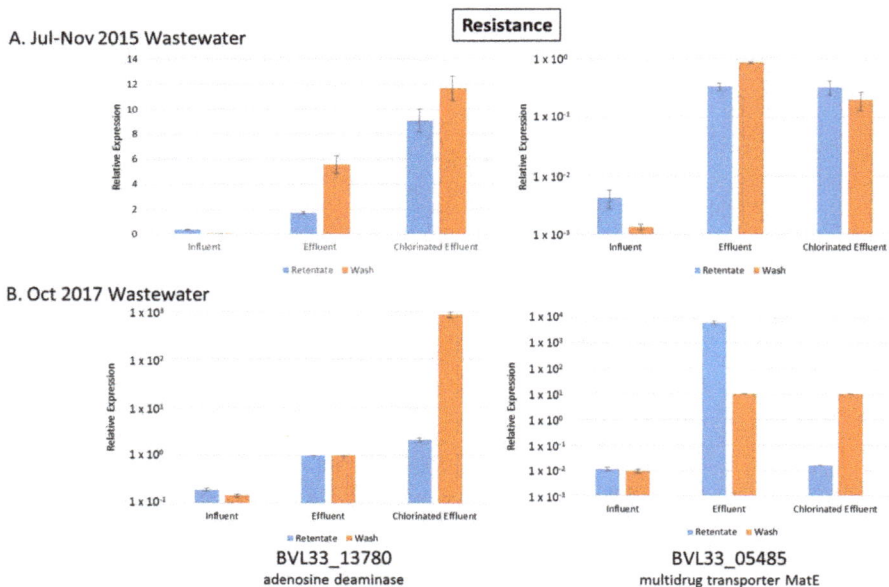

Figure 3. Relative expression profiles of *A. junii* genes involved in resistance. These genes displayed >2-fold change on RNA-seq analysis of wastewater sampled in July–November 2015 (Table 1). Gene expression values were normalized against a housekeeping *rpoB* gene. (**A**) Wastewater sampled in July, October and November of 2015; and (**B**) wastewater sampled in October 2017.

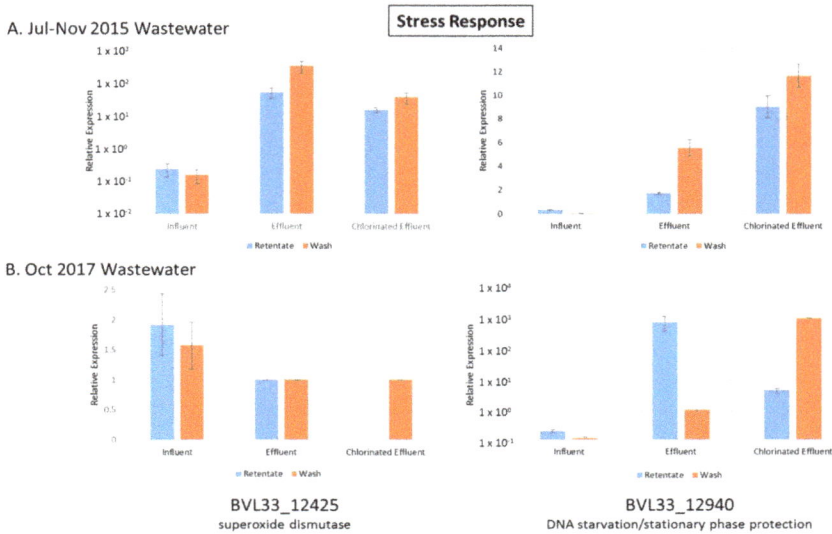

Figure 4. Relative expression profiles of *A. junii* genes involved in stress response. These genes displayed >2-fold change on RNA-seq analysis of wastewater sampled in July–November 2015 (Table 1). Gene expression values were normalized against a housekeeping *rpoB* gene. (**A**) Wastewater sampled in July, October and November of 2015; and (**B**) wastewater sampled in October 2017.

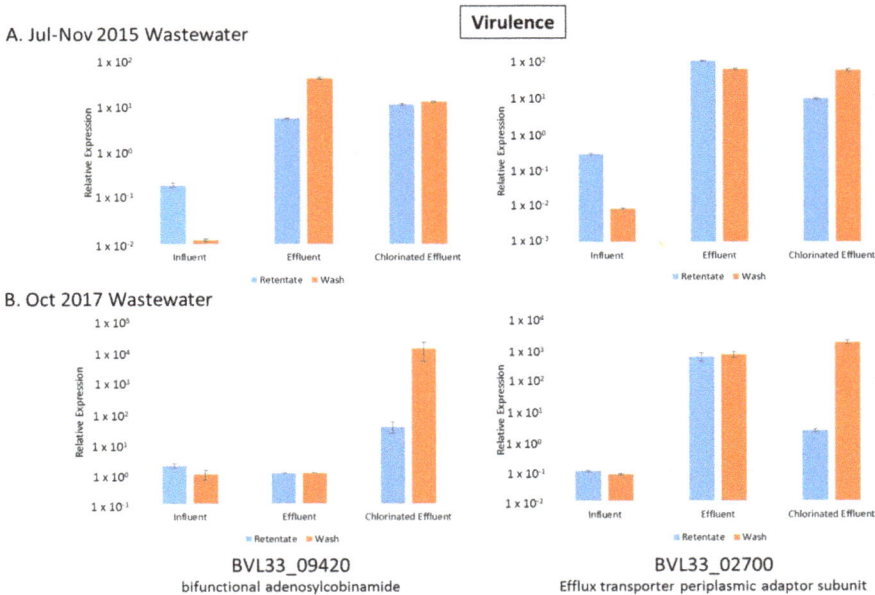

Figure 5. Relative expression profiles of *A. junii* genes involved in virulence. These genes displayed >2-fold change on RNA-seq analysis of wastewater sampled in July–November 2015 (Table 1). Gene expression values were normalized against a housekeeping *rpoB* gene. (**A**) Wastewater sampled in July, October and November of 2015; and (**B**) wastewater sampled in October 2017.

3.12. Metal ion Concentrations in March 2016 and October 2017 Wastewater Samples

Given that metal-related genes were upregulated post treatment, ICP-MS analysis was done on the wastewater samples from October 2017 to corroborate this observation. Table S6 lists the average values of each metal ion assayed. Majority of the detected metals decrease post treatment. To exemplify, ^{24}Al, ^{55}Mn, and ^{56}Fe concentrations decreased significantly in both the effluent (0.43-, 0.26-, and 0.34-fold change, respectively) and the chlorinated effluent (0.69-, 0.35-, and 0.41-fold change, respectively). In contrast, several metals increased in concentrations post treatment. ^{47}Ti showed significant increases in both the effluent (2.33-fold change) and the chlorinated effluent (2.55-fold change). ^{60}Ni showed a significant increase in the effluent (1.4-fold increase) with a slight increase in the chlorinated effluent (1.2-fold increase). ^{24}Mg showed a 1.82-fold increase in the chlorinated effluent but no significant change in the effluent. The concentration of ^{66}Zn showed a huge increase in the chlorinated effluent (4.91-fold increase) compared to influent (Figure 6 and Table S6).

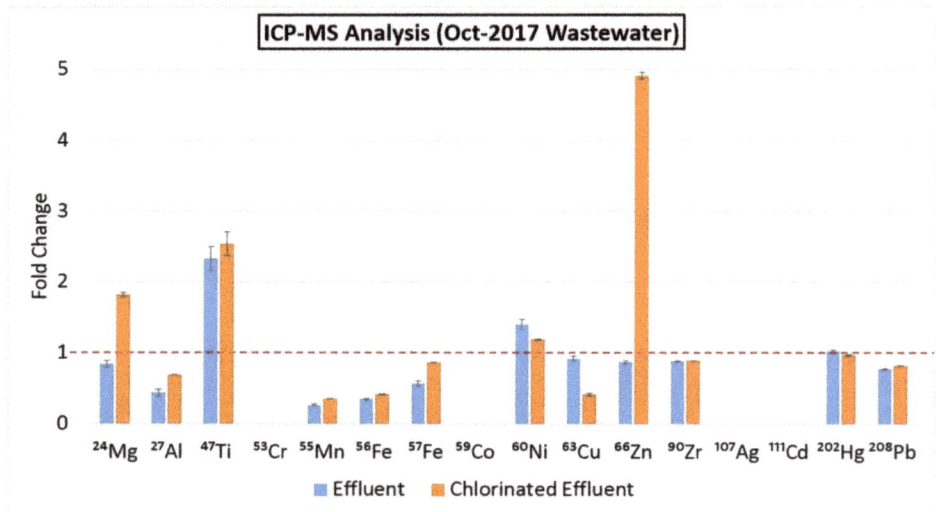

Figure 6. Fold Change of Metal Concentration in the effluent and chlorinated effluent in October 2017. Fold changes are measured with respect to the concentration in the influent. A fold change value of 1.0 implies no change (red dashed line). The numbers in superscript denote the atomic mass of the metal. Raw values of each ion as measured by ICP-MS are listed in Table S6.

4. Discussion

The WWTP in this study achieves microbial removal and inactivation by three different mechanisms. Firstly, nutrient removal is carried out by the activated sludge (reviewed in [40]). Secondly, microfiltration membranes submerged in the bioreactor separates microbes on basis of their size by filtration. Lastly, chlorination serves as a disinfection step which inactivates microbes in the effluent prior to discharge or reuse.

Despite the intended role of wastewater treatment process to remove and inactivate microorganisms, *Acinetobacter* spp. was detected in increasing abundances post MBR and chlorination (Figure 1). Four species of *Acinetobacter* (*A. lwoffi*, *A. guillouiae*, *A. junii* and *A. venetanus*) as well as unclassified species of *Acinetobacter* were detected following a similar trend, suggesting that all members of the *Acinetobacter* genus are constantly enriched compared to other bacterial populations in WWTPs. The enrichment of *Acinetobacter* spp. was also observed in another WWTP and at a different sampling period in 2012–2013 [4]. These findings suggest that *Acinetobacter* spp., particularly *A. junii*, have developed strategies to survive through the wastewater treatment train. Furthermore, since the metatranscriptomics data revealed an increase in *Acinetobacter* spp. activity down the WWTP, this suggests that members of the genus are both transcriptionally active and viable.

Metatranscriptomics was performed to elucidate the mechanisms adopted by *A. junii* to survive the treatment train. RNA-seq analysis was determined by comparing the mRNA expression values in the effluent and chlorinated effluent to the mRNA expression value of the influent. A main limitation of TFF is that the filtration membrane can foul rapidly if big particulates are not removed prior to filtration. Hence, particulates were first removed by centrifugation. Data pertaining to the influent in this study come from the supernatant fraction post centrifugation. Intact bacterial cells could be pelleted and removed, with only the supernatant fraction subjected to TFF. Therefore, the microbial load of the influent presented in this study might be an underrepresentation. However, majority of the bacterial species decreased in relative abundance in the effluent and chlorinated effluent (Figure 1 and Figures S2–S5), suggesting that the clarified influent still contains a relatively higher bacterial load despite the centrifugation step. Furthermore, gene expression comparisons are expressed as fold-expression values in this study, and are not affected by the microbial load of *A. junii* present at each stage of the WWTP.

Based on differential gene expression profiles, it is likely that *A. junii* upregulate genes related to metabolism and stress response to withstand the depleting nutrient in the sludge chambers, as well as upregulate genes related to cellular repair and resistance to counter the effects of chlorination.

To illustrate, among these gene categories, majority of the upregulated genes were related to cellular repair (*n* = 26) (Table 1). In this category, BVL33_02390, encoding for recombinase recB was significantly upregulated. recB contains a helicase and single stranded DNA dependent-ATPase, and a domain which recruits other proteins involved in the DNA repair process [41]. This protein functions with recC and recD to form recBCD, and subsequently initiates repair of double stranded breaks in DNA [42]. The recBCD complex is present in most bacterial species including *Acinetobacter* spp. [43], and was also shown to be crucial in accepting DNA in transformation processes in *A. baylyi* [43]. In response to the damages incurred during the wastewater treatment process, recBCD might be required to remedy the double stranded breaks in the DNA by recombination or through foreign DNA acquired through transformation. However, recB was the only component of this complex which was upregulated in either the effluent or the chlorinated effluent (Table 1, Tables S4 and S5). This might suggest a role for recB in *A. junii* independent of recC and recD, which might not be determined as yet. An alternative repair mechanism may be initiated to provide redundancy in cell repair. Of note is gene BVL33_11785, encoding for a DNA-binding protein, HU. HU proteins are expressed in all bacterial species and function in DNA compaction, similar to histones in eukaryotes. These binding proteins form a stable protein-chromosome, preventing accumulation of mutations [44]. This compaction has been shown to protect the genome in response to reactive oxygen species, iron and copper toxicity, thermal stress, extreme pH and irradiation [45,46]. In *E. coli*, HU proteins are involved in repair of UV-damaged DNA by a recB-dependent pathway [47]. Understanding that *recB* was also upregulated, it is possible that this same pathway along with the HU protein, are being utilized to repair damaged DNA.

Furthermore, BamE, an outer membrane protein assembly factor, was upregulated in the cellular repair category. This protein is involved in the assembly and insertion of β-barrel proteins into the outer membrane together with four other proteins, BamA, BamB, BamC, and BamD, to produce the β-barrel–assembly machinery (BAM) complex [48]. By itself, BamE has been shown to bind to phosphotidylglycerol, possibly anchoring the BAM complex to the membrane [49]. Interestingly, *E. coli* lacking the *bamE* gene were found to be susceptible to vancomycin, creating the suggestion that an increasing BamE would make *A. junii* resistant to antibiotics [50]. However, this link has yet to be confirmed.

This protein complex functions after outer membrane proteins (OMPs) are synthesized in the cell and transported across the cytoplasmic membrane by the Sec translocon [51]. Even though the genes encoding for the rest of the BamABCDE complex were not observed to be upregulated, individual genes encoding for the Sec translocon were. The effluent saw an upregulation of *secG*, *sec-C*, and *secY* while the chlorinated effluent saw increases in *secF* and *secE* (Tables S4 and S5). The main translocon channel is composed of proteins SecY, SecE and SecG [52]. While *secY*, *secE* and *secG* were not upregulated in both effluents, *secG* and *secY* gene were upregulated in the effluent and by the time *A. junii* transitions to the chlorination chamber, these proteins would already be in abundance. The subsequent upregulation of *secE* in the chlorinated effluent would complete the SecYEG channel allowing for successful translocation. SecB, also upregulated in the effluent, serves as a chaperon protein, targeting OMPs to the SecYEG channel [53]. These upregulations indicate *A. junii* are actively facilitating OMP translocation and insertion into the outer membrane, possibly to replace damaged proteins incurred in WWTP. However, the lack of upregulation of the other components of the BAM complex might suggest that the upregulated BamE binds to endogenous BamA, BamB, BamC, and BamD or the mechanism of its action in this described scenario is not as expected.

Besides repairing and protecting the cell wall structure, it was observed that *A. junii* also actively upregulated genes, for example BVL33_13885, that was associated with both the cellular repair and metabolism pathways. This gene encodes for an acyltransferase protein, which has a primary function of transporting fatty acids for β-oxidation [54]. This is the first step in the pathway which oxidizes fatty acids in place of glucose as a carbon source. *Acinetobacter* spp. have been shown to resort to synthesizing wax esters and triacylglycerols under growth limiting conditions [55,56]. These compounds serve as an alternative energy source as well as protecting the cell from desiccation, irradiation and pathogens [57]. The upregulation of other genes involved in fatty acid metabolism (BVL33_13315, BVL33_14915 and BVL33_14010) in both effluents seem to indicate the increase in reliance on alternative carbon sources (Table 1). Furthermore, within the stress response category, BVL33_00220, encoding for an NADPH-dependent FMN reductase was upregulated. This protein has also been shown previously to allow *A. baumannii* to utilize *p*-hydroxyphenylacetate as an alternative carbon source [58]. If this function is similar, it might serve useful for *A. junii* in conditions of nutrient instability such as in a WWTP. Collectively, the upregulation of the genes suggest that there exists a strong selection pressure for *A. junii* to depend on alternative carbon sources for energy in a WWTP. This would provide *Acinetobacter* spp., an additional advantage to survive the fluctuating nutrient conditions of a WWTP and outcompete other strains in a WWTP.

In addition to active cell repair and expanding its carbon sources for metabolic needs, metatranscriptomics suggest that *A. junii* also upregulate resistance genes and stress response genes to facilitate survival. To exemplify, two of the genes upregulated in both effluents (BVL33_02685 and BVL33_01605) were efflux pumps for nickel, cadmium, cobalt and zinc (Table 1). Upregulation of these genes did not seem to correlate with the heavy metal concentrations since majority of the heavy metal concentrations decreased after treatment (Figure 6 and Table S6). However, earlier studies have reported a lack of correlation between the concentration of heavy metals and related resistance genes, suggesting that acute exposure of sub-lethal concentrations of these compounds were sufficient to trigger the expression of resistance genes [59,60]. Alternatively, the upregulation of these efflux pumps may be co-selected for by chlorination so as to regulate the concentration gradient of ROS generated in the chlorination chamber [61,62] and intracellularly [17]. Efflux pumps, conferring resistance to antibiotics, were upregulated in *A. baumannii* upon chlorination. Studies have linked the upregulation of efflux pumps in bacteria to damage and mutations incurred on the genome due to the action of chlorine-induced ROS [63,64]. In this study, BVL33_02685, had a higher fold change in the chlorinated effluent than the effluent, suggesting that chlorination might impose the selective pressure for this upregulation (Table 1).

Alternatively, BVL33_00220 which encodes for a DNA starvation/stationary phase protection protein was also highly upregulated. A similar protein in *A. baylyi* was shown to bind to DNA during stationary phase, forming a condensed and stable structure. This protein also protects the genome from oxidative stress by sequestering intracellular ions. The combined effects might serve to protect

A. junii (considering common functionality) from the relative increase in metals seen in the chlorinated effluent (Figure 6 and Table S6) by entering into dormancy if the earlier approaches of active cell repair, expansion of metabolic substrates, and biofilm formation are insufficient to assist in its survival through the WWTP.

Furthermore, it was observed that the profiles of upregulated genes between the retentate and the wash samples differ significantly ($p < 0.05$) (Table 1). In an earlier study of *A. baumannii*, the RNA expression profiles was observed to differ between planktonic and biofilm associated populations [65]. It is plausible that the RNA expression profiles of the retentate and wash of *A. junii* mirror that seen in planktonic and biofilm-associated *A. baumannii*. Strains which were isolated in the wash might be hydrophobic and hence adhere to the TFF membrane stronger, as compared to strains collected in the retentate. Hydrophilic strains would remain in suspension in the WWTP and escape sedimentation. The presence of different subpopulations of the same bacteria in a single environment has been shown to facilitate persistence. In conditions of harsh or changing environments, having different phenotypes may confer the microbe the ability to persist and achieve dominance [66,67]. The lesser transcriptionally active population of *A. junii*, found in the wash samples, may be entering a state of quiescence or dormancy and will regain activity when conditions change. Bacteria entering or exiting a state of quiescence have been shown to have an increase in protein production to facilitate de novo synthesis of metabolites in extreme conditions [68,69]. This could account for significant upregulation of ribosomal proteins in the wash samples (Table 1). This upregulation in transcriptional machinery may prime the cell for the further increase in protein synthesis [70]. The five most upregulated genes in the retentate, which were not as upregulated in the wash, were a DNA transfer protein, a SPOR-domain containing protein, an acyltransferase, a DNA starvation/stationary phase protection protein and a HopJ type III protein. These genes indicate that cells in the retentate are focused on DNA preservation, cell division and fatty acid metabolism. Retentate *A. junii* seems to be replicating rapidly and at the same time protecting its genome, which would be more susceptible to mutations and strand breaks during replication. The energy required for replication may come from the β-oxidation of fatty acids.

Collectively, this study provides data to suggest that *A. junii* adopted a multi-faceted approach, ranging from utilizing alternative carbon sources, preservation of its genome integrity, repairing OMPs, promoting growth of two different subpopulations and by actively extruding metals or reactive oxygen species to survive the wastewater treatment process (Figure 7).

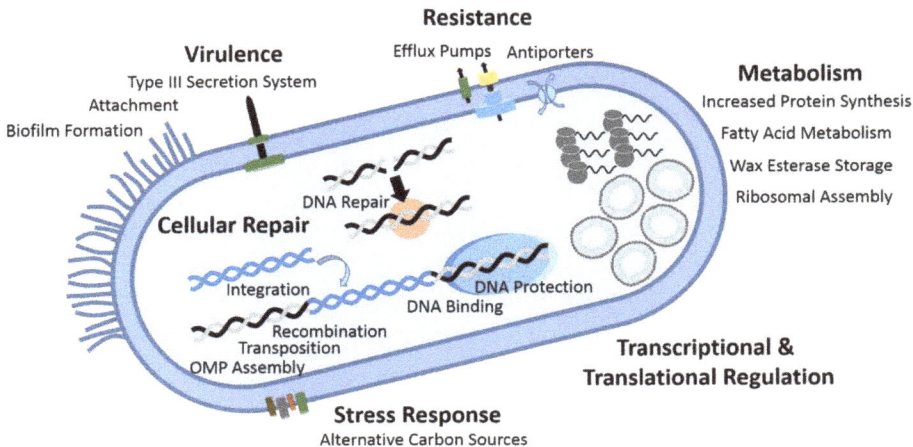

Figure 7. Classifications of upregulated genes in *Acinetobacter junii* in response to wastewater treatment.

Given that *A. junii* is able to persist well in WWTP, perhaps concerning is the observation that BVL33_02100, encoding for an IS4 transposase was significantly upregulated in the retentate. The gene encoding for this insertion factor is ubiquitous throughout all domains of life but found to be actively transposing in only a few species [71]. Three active insertion factors, IS*Aba1*, IS*Aba11* and A473, have been found in *A. baumanii* [72–74]. These genes have been shown to confer resistance to *A. baumannii* by disrupting genes which facilitate uptake of antibiotics [75] or by being powerful promoters to native resistance genes [76]. The function of this transposase may be enhanced by the upregulation of BVL33_15970, a DNA transfer protein. Both these proteins may function synergistically in lateral gene transfer, allowing for tolerance of the WWTP conditions and for *A. junii* to gain other gene functions.

In addition, the most upregulated gene in the "Virulence, Disease and Defense" category was BVL33_15450, encoding for a HopJ type III effector protein (Table 1). This protein has not been reported in *Acinetobacter* spp., but has been shown to be play a role in plant pathogenicity in *Pseudomonas* spp. [77]. While no evidence in this study links *Acinetobacter* spp. to plant pathogenicity directly, numerous studies have isolated *Acintobacter* spp., from vegetables globally, suggesting a possible relationship (reviewed in [78–81]). While the data show significant increases of *A. junii* gene expression down the WWT process, this study does not provide the actual cell density which would allow for proper assessment of risk. However, from the low number of *Acintetobacter* spp. isolated from the effluents of WWTP in Jeddah (Tables S1 and S2), it is likely that that viable *A. junii* would impose low risk. Therefore, although this study provides insights to understanding how *A. junii* persist throughout the WWTP, their persistence warrants no need for a massive revamp of existing WWTP. This is particularly the case for MBR-based WWTP, where a total eradication of *Acinetobacter* spp. would be impractical and costly.

Supplementary Materials: The following are available online at http://www.mdpi.com/2073-4441/10/4/436/s1, Table S1: Antibiotic profile of species detected in the effluent wastewater from Al-Jassim et al., 2015 [4]. Grey circles indicate susceptibility. Values > 0.5 (highlighted in red) indicate resistance. Amp: Ampicillin, Kan: Kanamycin, Erm: Erythromycin, Tet: Tetracycline, Cef: Cefuroxime, Chl: Chloramphenicol, Mer: Meropenem, Cip: Ciprofloxacin. Table S2: Antibiotic profile of species detected in the chlorinated effluent wastewater from Al-Jassim et al., 2015 [4]. Grey circles indicate susceptibility. Values > 0.5 (highlighted inet: Tetracycline, Cef: Cefuroxime, red) indicate resistance. Amp: Ampicillin, Kan: Kanamycin, Erm: Erythromycin, TChl: Chloramphenicol, Mer: Meropenem, Cip: Ciprofloxacin. Table S3: Gene names, categories and primer sequences of genes assayed by RT-qPCR. Table S4: All 288 genes of *Acinetobacter junii* that were upregulated in the effluent. The gene expression values of the effluent-retentate and effluent-wash were compared to the gene expression values of the influent-retentate and influent-wash, respectively. All upregulations were included if they were more than two-fold change and were statistically significant ($p < 0.05$) on the Baggerly proportion-based test. Table S5: All 378 genes of *Acinetobacter junii* that were upregulated in the chlorinated effluent. The gene expression values of the chlorinated effluent-retentate and chlorinated effluent-wash were compared to the gene expression values of the influent-retentate and influent-wash, respectively. All upregulations were included if they were more than two-fold change and were statistically significant ($p < 0.05$) on the Baggerly proportion-based test. Table S6: ICP-MS analysis of the influent, effluent and chlorinated effluent of October 2017 Wastewater. Numbers in superscript refer to the atomic weight of the cation. Figure S1: Venn diagrams representing the numbers of significantly upregulated (A) and downregulated (B) genes only in the effluent (orange), chlorinated effluent (yellow) and in both effluent and chlorinated effluent (purple). Only genes that displayed a two-time up/downregulation and have $p < 0.05$ on the Baggerly proportion-based test were considered as significant. Figure S2: Heat map representing relative abundances of the different kingdoms detected in the wastewater samples collected in July, October and November 2015. Colored scale bar represents the relative abundances for each heat map. IR: Influent-retentate, IW: Influent-wash, ER: Effluent-retentate, EW: Effluent-wash, CR: Chlorinated Effluent-retentate, CW: Chlorinated Effluent-wash. Figure S3: Heat map representing relative abundances of the different phyla detected in the wastewater samples collected in July, October and November 2015. Colored scale bar represents the relative abundances for each heat map. IR: Influent-retentate, IW: Influent-wash, ER: Effluent-retentate, EW: Effluent-wash, CR: Chlorinated Effluent-retentate, CW: Chlorinated Effluent-wash. Figure S4: Heat map representing relative abundances of the different classes detected in the wastewater samples collected in July, October and November 2015. Colored scale bar represents the relative abundances for each heat map. IR: Influent-retentate, IW: Influent-wash, ER: Effluent-retentate, EW: Effluent-wash, CR: Chlorinated Effluent-retentate, CW: Chlorinated Effluent-wash. Figure S5: Heat map representing relative abundances of the different families detected in the wastewater samples collected in July, October and November 2015. Colored scale bar represents the relative abundances for each heat map. IR: Influent-retentate, IW: Influent-wash, ER: Effluent-retentate, EW: Effluent-wash, CR: Chlorinated Effluent-retentate, CW: Chlorinated Effluent-wash.

Acknowledgments: The authors would like to thank George Princeton Dunsford for access to the KAUST wastewater treatment plant and Moustapha Harb for providing sampling assistance. The research reported in this publication was supported by the KAUST baseline funding BAS/1/1033-01-01 awarded to Pei-Ying Hong.

Author Contributions: P.Y.-H. and M.R.J. conceived and wrote the manuscript. M.R.J., N.A.-J., H.C. and M.F.H. carried out the experiments. M.F.H. assisted in writing the manuscript.

Conflicts of Interest: The authors declare that the research was conducted in the absence of any commercial or financial relationships that could be construed as a potential conflict of interest.

References

1. Shomar, B. Water Resources, Water Quality and Human Health in Regions of Extreme Stress: Middle East. *J. Earth Sci. Clim. Chang.* **2013**, *4*, 1–12. [CrossRef]

2. Al-Jasser, A.O. Saudi wastewater reuse standards for agricultural irrigation: Riyadh treatment plants effluent compliance. *J. King Saud Univ. Eng. Sci.* **2011**, *23*, 1–8. [CrossRef]

3. Webb, A.L.; Taboada, E.N.; Selinger, L.B.; Boras, V.F.; Inglis, G.D. Efficacy of wastewater treatment on *Arcobacter butzleri* density and strain diversity. *Water Res.* **2016**, *105*, 291–296. [CrossRef] [PubMed]

4. Al-Jassim, N.; Ansari, M.I.; Harb, M.; Hong, P.-Y. Removal of bacterial contaminants and antibiotic resistance genes by conventional wastewater treatment processes in Saudi Arabia: Is the treated wastewater safe to reuse for agricultural irrigation? *Water Res.* **2015**, *73*, 277–290. [CrossRef] [PubMed]

5. Sui, Q.; Jiang, C.; Yu, D.; Chen, M.; Zhang, J.; Wang, Y.; Wei, Y. Performance of a sequencing-batch membrane bioreactor (SMBR) with an automatic control strategy treating high-strength swine wastewater. *J. Hazard. Mater.* **2018**, *342*, 210–219. [CrossRef] [PubMed]

6. Wu, D.; Dolfing, J.; Xie, B. Bacterial perspectives on the dissemination of antibiotic resistance genes in domestic wastewater bio-treatment systems: Beneficiary to victim. *Appl. Microbiol. Biotechnol.* **2018**, *102*, 597–604. [CrossRef] [PubMed]

7. Lekunberri, I.; Balcazar, J.L.; Borrego, C.M. Metagenomic exploration reveals a marked change in the river resistome and mobilome after treated wastewater discharges. *Environ. Pollut.* **2017**, *234*, 538–542. [CrossRef] [PubMed]

8. Xiao, S.; Hu, S.; Zhang, Y.; Zhao, X.; Pan, W. Influence of sewage treatment plant effluent discharge into multipurpose river on its water quality: A quantitative health risk assessment of Cryptosporidium and Giardia. *Environ. Pollut.* **2017**, *233*, 797–805. [CrossRef] [PubMed]

9. Ajonina, C.; Buzie, C.; Rubiandini, R.H.; Otterpohl, R. Microbial pathogens in wastewater treatment plants (WWTP) in Hamburg. *J. Toxicol. Environ. Health Part A* **2015**, *78*, 381–387. [CrossRef] [PubMed]

10. Naidoo, S.; Olaniran, A.O. Treated wastewater effluent as a source of microbial pollution of surface water resources. *Int. J. Environ. Res. Public Health* **2014**, *11*, 249–270. [CrossRef] [PubMed]

11. Cai, L.; Zhang, T. Detecting human bacterial pathogens in wastewater treatment plants by a high-throughput shotgun sequencing technique. *Environ. Sci. Technol.* **2013**, *47*, 5433–5441. [CrossRef] [PubMed]

12. Judd, S. The status of membrane bioreactor technology. *Trends Biotechnol.* **2008**, *26*, 109–116. [CrossRef] [PubMed]

13. Iorhemen, O.T.; Hamza, R.A.; Tay, J.H. Membrane bioreactor (MBR) technology for wastewater treatment and reclamation: Membrane fouling. *Membranes* **2016**, *6*, 33. [CrossRef] [PubMed]

14. Pollice, A.; Laera, G.; Saturno, D.; Giordano, C.; Sandulli, R. Optimal sludge retention time for a bench scale MBR treating municipal sewage. *Water Sci. Technol.* **2008**, *57*, 319–322. [CrossRef] [PubMed]

15. Anastasi, E.M.; Wohlsen, T.D.; Stratton, H.M.; Katouli, M. Survival of *Escherichia coli* in two sewage treatment plants using UV irradiation and chlorination for disinfection. *Water Res.* **2013**, *47*, 6670–6679. [CrossRef] [PubMed]

16. Chapman, J.S. Disinfectant resistance mechanisms, cross-resistance, and co-resistance. *Int. Biodeterior. Biodegrad.* **2003**, *51*, 271–276. [CrossRef]

17. Zhang, Y.; Gu, A.Z.; He, M.; Li, D.; Chen, J. Subinhibitory concentrations of disinfectants promote the horizontal transfer of multidrug resistance genes within and across genera. *Environ. Sci. Technol.* **2017**, *51*, 570–580. [CrossRef] [PubMed]

18. Reactive Oxygen Species (ROS). 2015. Available online: http://www.biology-pages.info/R/ROS.html (accessed on 10 January 2018).

19. Jumat, M.; Hasan, N.; Subramanian, P.; Heberling, C.; Colwell, R.; Hong, P.-Y. Membrane bioreactor-based wastewater treatment plant in Saudi Arabia: Reduction of viral diversity, load, and infectious capacity. *Water* **2017**, *9*, 534. [CrossRef]

20. Harb, M.; Hong, P.-Y. Molecular-based detection of potentially pathogenic bacteria in membrane bioreactor (MBR) systems treating municipal wastewater: A case study. *Environ. Sci. Pollut. Res.* **2017**, *24*, 5370–5380. [CrossRef] [PubMed]

21. Hrenovic, J.; Goic-Barisic, I.; Kazazic, S.; Kovacic, A.; Ganjto, M.; Tonkic, M. Carbapenem-resistant isolates of *Acinetobacter baumannii* in a municipal wastewater treatment plant, Croatia, 2014. *Euro Surveill.* **2016**, *21*. [CrossRef] [PubMed]

22. Zhang, Y.; Marrs, C.F.; Simon, C.; Xi, C. Wastewater treatment contributes to selective increase of antibiotic resistance among *Acinetobacter* spp. *Sci. Total Environ.* **2009**, *407*, 3702–3706. [CrossRef] [PubMed]

23. Oerther, D.B.; Pernthaler, J.; Schramm, A.; Amann, R.; Raskin, L. Monitoring Precursor 16S rRNAs of *Acinetobacter* spp. in Activated Sludge Wastewater Treatment Systems. *Appl. Environ. Microbiol.* **2000**, *66*, 2154–2165. [CrossRef] [PubMed]

24. Joly-Guillou, M.L. Clinical impact and pathogenicity of Acinetobacter. *Clin. Microbiol. Infect.* **2005**, *11*, 868–873. [CrossRef] [PubMed]

25. Linde, H.J.; Hahn, J.; Holler, E.; Reischl, U.; Lehn, N. Septicemia due to *Acinetobacter junii*. *J. Clin. Microbiol.* **2002**, *40*, 2696–2697. [CrossRef] [PubMed]

26. Hung, Y.T.; Lee, Y.T.; Huang, L.J.; Chen, T.L.; Yu, K.W.; Fung, C.P.; Cho, W.L.; Liu, C.Y. Clinical characteristics of patients with *Acinetobacter junii* infection. *J. Microbiol. Immunol. Infect.* **2009**, *42*, 47–53. [PubMed]

27. Cayo, R.; Yanez San Segundo, L.; Perez del Molino Bernal, I.C.; Garcia de la Fuente, C.; Bermudez Rodriguez, M.A.; Calvo, J.; Martinez-Martinez, L. Bloodstream infection caused by *Acinetobacter junii* in a patient with acute lymphoblastic leukaemia after allogenic haematopoietic cell transplantation. *J. Med. Microbiol.* **2011**, *60 Pt 3*, 375–377. [CrossRef] [PubMed]

28. Cooper, R.M.; Tsimring, L.; Hasty, J. Inter-species population dynamics enhance microbial horizontal gene transfer and spread of antibiotic resistance. *eLife* **2017**, *6*, E25950. [CrossRef] [PubMed]

29. Al Atrouni, A.; Joly-Guillou, M.-L.; Hamze, M.; Kempf, M. Reservoirs of non-baumannii Acinetobacter species. *Front. Microbiol.* **2016**, *7*, 49. [CrossRef] [PubMed]

30. Wiedmann-al-Ahmad, M.; Tichy, H.V.; Schön, G. Characterization of Acinetobacter type strains and isolates obtained from wastewater treatment plants by PCR fingerprinting. *Appl. Environ. Microbiol.* **1994**, *60*, 4066–4071. [PubMed]

31. Doughari, H.J.; Ndakidemi, P.A.; Human, I.S.; Benade, S. The ecology, biology and pathogenesis of *Acinetobacter* spp.: An overview. *Microbes Environ.* **2011**, *26*, 101–112. [CrossRef] [PubMed]

32. Montaña, S.; Cittadini, R.; del Castillo, M.; Uong, S.; Lazzaro, T.; Almuzara, M.; Barberis, C.; Vay, C.; Ramírez, M.S. Presence of New Delhi metallo-β-lactamase gene (NDM-1) in a clinical isolate of *Acinetobacter junii* in Argentina. *New Microbes New Infect.* **2016**, *11*, 43–44. [CrossRef] [PubMed]

33. Guardabassi, L.; Dalsgaard, A.; Olsen, J.E. Phenotypic characterization and antibiotic resistance of *Acinetobacter* spp. isolated from aquatic sources. *J. Appl. Microbiol.* **1999**, *87*, 659–667. [CrossRef] [PubMed]

34. Molina-Romero, D.; Baez, A.; Quintero-Hernández, V.; Castañeda-Lucio, M.; Fuentes-Ramírez, L.E.; Bustillos-Cristales, M.d.R.; Rodríguez-Andrade, O.; Morales-García, Y.E.; Munive, A.; Muñoz-Rojas, J. Compatible bacterial mixture, tolerant to desiccation, improves maize plant growth. *PLoS ONE* **2017**, *12*, E0187913. [CrossRef] [PubMed]

35. Bastian, R.; Murray, D. *2012 Guidelines for Water Reuse*; EPA Office of Research and Development: Washington, DC, USA, 2012.

36. Bolger, A.M.; Lohse, M.; Usadel, B. Trimmomatic: A flexible trimmer for Illumina sequence data. *Bioinformatics* **2014**, *30*, 2114–2120. [CrossRef] [PubMed]

37. Truong, D.T.; Franzosa, E.A.; Tickle, T.L.; Scholz, M.; Weingart, G.; Pasolli, E.; Tett, A.; Huttenhower, C.; Segata, N. MetaPhlAn2 for enhanced metagenomic taxonomic profiling. *Nat. Methods* **2015**, *12*, 902. [CrossRef] [PubMed]

38. Baggerly, K.A.; Deng, L.; Morris, J.S.; Aldaz, C.M. Differential expression in SAGE: Accounting for normal between-library variation. *Bioinformatics* **2003**, *19*, 1477–1483. [CrossRef] [PubMed]

39. Broniek, G.; Langwińska-Wośko, E.; Szaflik, J.; Wróblewska, M. *Acinetobacter junii* as an aetiological agent of corneal ulcer. *Infection* **2014**, *42*, 1051–1053. [CrossRef] [PubMed]

40. Cydzik-Kwiatkowska, A.; Zielińska, M. Bacterial communities in full-scale wastewater treatment systems. *World J. Microbiol. Biotechnol.* **2016**, *32*, 66. [CrossRef] [PubMed]

41. Spies, M.; Kowalczykowski, S.C. Homologous recombination by RecBCD and RecF pathways. In *The Bacterial Chromosome 2005*; ASM Press: Washington, DC, USA, 2005; pp. 389–403.

42. Dillingham, M.S.; Kowalczykowski, S.C. RecBCD enzyme and the repair of double-stranded DNA breaks. *Microbiol. Mol. Biol. Rev. MMBR* **2008**, *72*, 642–671. [CrossRef] [PubMed]

43. Kickstein, E.; Harms, K.; Wackernagel, W. Deletions of recBCD or recD influence genetic transformation differently and are lethal together with a recJ deletion in *Acinetobacter baylyi*. *Microbiology* **2007**, *153 Pt 7*, 2259–2270. [CrossRef] [PubMed]

44. Nair, S.; Finkel, S.E. Dps protects cells against multiple stresses during stationary phase. *J. Bacteriol.* **2004**, *186*, 4192–4198. [CrossRef] [PubMed]

45. Boubrik, F.; Rouviere-Yaniv, J. Increased sensitivity to gamma irradiation in bacteria lacking protein, H.U. *Proc. Natl. Acad. Sci. USA* **1995**, *92*, 3958–3962. [CrossRef] [PubMed]

46. Li, S.; Waters, R. *Escherichia coli* strains lacking protein HU are UV sensitive due to a role for HU in homologous recombination. *J. Bacteriol.* **1998**, *180*, 3750–3756. [PubMed]

47. Miyabe, I.; Zhang, Q.M.; Kano, Y.; Yonei, S. Histone-like protein HU is required for recA gene-dependent DNA repair and SOS induction pathways in UV-irradiated *Escherichia coli*. *Int. J. Radiat. Biol.* **2000**, *76*, 43–49. [CrossRef] [PubMed]

48. Knowles, T.J.; Scott-Tucker, A.; Overduin, M.; Henderson, I.R. Membrane protein architects: The role of the BAM complex in outer membrane protein assembly. *Nat. Rev. Microbiol.* **2009**, *7*, 206–214. [CrossRef] [PubMed]

49. Knowles, T.J.; Browning, D.F.; Jeeves, M.; Maderbocus, R.; Rajesh, S.; Sridhar, P.; Manoli, E.; Emery, D.; Sommer, U.; Spencer, A.; et al. Structure and function of BamE within the outer membrane and the beta-barrel assembly machine. *EMBO Rep.* **2011**, *12*, 123–128. [CrossRef] [PubMed]

50. Ruiz, N.; Falcone, B.; Kahne, D.; Silhavy, T.J. Chemical conditionality: A genetic strategy to probe organelle assembly. *Cell* **2005**, *121*, 307–317. [CrossRef] [PubMed]

51. Denks, K.; Vogt, A.; Sachelaru, I.; Petriman, N.-A.; Kudva, R.; Koch, H.-G. The Sec translocon mediated protein transport in prokaryotes and eukaryotes. *Mol. Membr. Biol.* **2014**, *31*, 58–84. [CrossRef] [PubMed]

52. Maillard, A.P.; Chan, K.K.Y.; Duong, F. Preprotein translocation through the Sec translocon in bacteria. In *Protein Movement Across Membranes*; Springer: Boston, MA, USA, 2005; pp. 19–32.

53. Fekkes, P.; Driessen, A.J.M. Protein targeting to the bacterial cytoplasmic membrane. *Microbiol. Mol. Biol. Rev.* **1999**, *63*, 161–173. [PubMed]

54. Hunt, M.C.; Siponen, M.I.; Alexson, S.E.H. The emerging role of acyl-CoA thioesterases and acyltransferases in regulating peroxisomal lipid metabolism. *Biochim. Biophys. Acta (BBA)* **2012**, *1822*, 1397–1410. [CrossRef] [PubMed]

55. Fixter, L.M.; Nagi, M.N.; Mccormack, J.G.; Fewson, C.A. Structure, distribution and function of wax esters in *Acinetobacter calcoaceticus*. *Microbiology* **1986**, *132*, 3147–3157. [CrossRef]

56. Kalscheuer, R.; Steinbuchel, A. A novel bifunctional wax ester synthase/acyl-CoA:diacylglycerol acyltransferase mediates wax ester and triacylglycerol biosynthesis in *Acinetobacter calcoaceticus* ADP1. *J. Biol. Chem.* **2003**, *278*, 8075–8082. [CrossRef] [PubMed]

57. Ishige, T.; Tani, A.; Sakai, Y.; Kato, N. Wax ester production by bacteria. *Curr. Opin. Microbiol.* **2003**, *6*, 244–250. [CrossRef]

58. Chaiyen, P.; Suadee, C.; Wilairat, P. A novel two-protein component flavoprotein hydroxylase. *Eur. J. Biochem.* **2001**, *268*, 5550–5561. [CrossRef] [PubMed]

59. Choudhary, S.; Sar, P. Real-time PCR based analysis of metal resistance genes in metal resistant *Pseudomonas aeruginosa* strain J007. *J. Basic Microbiol.* **2016**, *56*, 688–697. [CrossRef] [PubMed]

60. Monsieurs, P.; Moors, H.; Van Houdt, R.; Janssen, P.J.; Janssen, A.; Coninx, I.; Mergeay, M.; Leys, N. Heavy metal resistance in *Cupriavidus metallidurans* CH34 is governed by an intricate transcriptional network. *Biometals* **2011**, *24*, 1133–1151. [CrossRef] [PubMed]

61. Shi, P.; Jia, S.; Zhang, X.X.; Zhang, T.; Cheng, S.; Li, A. Metagenomic insights into chlorination effects on microbial antibiotic resistance in drinking water. *Water Res.* **2013**, *47*, 111–120. [CrossRef] [PubMed]

62. Karumathil, D.P.; Yin, H.-B.; Kollanoor-Johny, A.; Venkitanarayanan, K. Effect of chlorine exposure on the survival and antibiotic gene expression of multidrug resistant *Acinetobacter baumannii* in Water. *Int. J. Environ. Res. Public Health* **2014**, *11*, 1844–1854. [CrossRef] [PubMed]

63. Kohanski, M.A.; DePristo, M.A.; Collins, J.J. Sublethal antibiotic treatment leads to multidrug resistance via radical-induced mutagenesis. *Mol. Cell* **2010**, *37*, 311–320. [CrossRef] [PubMed]

64. Bogomolnaya, L.M.; Andrews, K.D.; Talamantes, M.; Maple, A.; Ragoza, Y.; Vazquez-Torres, A.; Andrews-Polymenis, H. The ABC-type efflux pump MacAB protects *Salmonella enterica* serovar typhimurium from oxidative stress. *mBio* **2013**, *4*, e00630-13. [CrossRef] [PubMed]

65. Rumbo-Feal, S.; Gómez, M.J.; Gayoso, C.; Álvarez-Fraga, L.; Cabral, M.P.; Aransay, A.M.; Rodríguez-Ezpeleta, N.; Fullaondo, A.; Valle, J.; Tomás, M.; et al. Whole transcriptome analysis of *Acinetobacter baumannii* assessed by RNA-sequencing reveals different mRNA expression profiles in biofilm compared to planktonic cells. *PLoS ONE* **2013**, *8*, E72968. [CrossRef] [PubMed]

66. Cray, J.A.; Bell, A.N.; Bhaganna, P.; Mswaka, A.Y.; Timson, D.J.; Hallsworth, J.E. The biology of habitat dominance; can microbes behave as weeds? *Microb. Biotechnol.* **2013**, *6*, 453–492. [CrossRef] [PubMed]

67. Balaban, N.Q.; Merrin, J.; Chait, R.; Kowalik, L.; Leibler, S. Bacterial persistence as a phenotypic switch. *Science* **2004**, *305*, 1622–1625. [CrossRef] [PubMed]

68. Rittershaus, E.S.; Baek, S.H.; Sassetti, C.M. The normalcy of dormancy: Common themes in microbial quiescence. *Cell Host Microbe* **2013**, *13*, 643–651. [CrossRef] [PubMed]

69. Shaikh, A.S.; Tang, Y.J.; Mukhopadhyay, A.; Martin, H.G.; Gin, J.; Benke, P.I.; Keasling, J.D. Study of stationary phase metabolism via isotopomer analysis of amino acids from an isolated protein. *Biotechnol. Prog.* **2010**, *26*, 52–56. [CrossRef] [PubMed]

70. Klappenbach, J.A.; Dunbar, J.M.; Schmidt, T.M. rRNA operon copy number reflects ecological strategies of bacteria. *Appl. Environ. Microbiol.* **2000**, *66*, 1328–1333. [CrossRef] [PubMed]

71. Mahillon, J.; Chandler, M. Insertion sequences. *Microbiol. Mol. Biol. Rev. MMBR* **1998**, *62*, 725–774. [PubMed]

72. Mugnier, P.D.; Poirel, L.; Nordmann, P. Functional analysis of insertion sequence ISAba1, responsible for genomic plasticity of *Acinetobacter baumannii*. *J. Bacteriol.* **2009**, *191*, 2414–2418. [CrossRef] [PubMed]

73. Smith, M.G.; Gianoulis, T.A.; Pukatzki, S.; Mekalanos, J.J.; Ornston, L.N.; Gerstein, M.; Snyder, M. New insights into *Acinetobacter baumannii* pathogenesis revealed by high-density pyrosequencing and transposon mutagenesis. *Genes Dev.* **2007**, *21*, 601–614. [CrossRef] [PubMed]

74. Rose, A. TnAbaR1: A novel Tn7-related transposon in *Acinetobacter baumannii* that contributes to the accumulation and dissemination of large repertoires of resistance genes. *Biosci. Horiz.* **2010**, *3*, 40–48. [CrossRef]

75. Mussi, M.A.; Limansky, A.S.; Viale, A.M. Acquisition of resistance to carbapenems in multidrug-resistant clinical strains of *Acinetobacter baumannii*: Natural insertional inactivation of a gene encoding a member of a novel family of beta-barrel outer membrane proteins. *Antimicrob. Agents Chemother.* **2005**, *49*, 1432–1440. [CrossRef] [PubMed]

76. Heritier, C.; Poirel, L.; Nordmann, P. Cephalosporinase over-expression resulting from insertion of ISAba1 in *Acinetobacter baumannii*. *Clin. Microbiol. Infect.* **2006**, *12*, 123–130. [CrossRef] [PubMed]

77. Block, A.; Alfano, J.R. Plant targets for *Pseudomonas syringae* type III effectors: Virulence targets or guarded decoys? *Curr. Opin. Microbiol.* **2011**, *14*, 39–46. [CrossRef] [PubMed]

78. Peleg, A.Y.; Seifert, H.; Paterson, D.L. *Acinetobacter baumannii*: Emergence of a successful pathogen. *Clin. Microbiol. Rev.* **2008**, *21*, 538–582. [CrossRef] [PubMed]

79. Lee, J.S.; Lee, K.C.; Kim, K.K.; Hwang, I.C.; Jang, C.; Kim, N.G.; Yeo, W.H.; Kim, B.S.; Yu, Y.M.; Ahn, J.S. *Acinetobacter antiviralis* sp. nov., from tobacco plant roots. *J. Microbiol. Biotechnol.* **2009**, *19*, 250–256. [CrossRef] [PubMed]

80. Kang, Y.S.; Jung, J.; Jeon, C.O.; Park, W. *Acinetobacter oleivorans* sp. nov. is capable of adhering to and growing on diesel-oil. *J. Microbiol.* **2011**, *49*, 29–34. [CrossRef] [PubMed]

81. Alvarez-Perez, S.; Lievens, B.; Jacquemyn, H.; Herrera, C.M. *Acinetobacter nectaris* sp. nov. and *Acinetobacter boissieri* sp. nov., isolated from floral nectar of wild Mediterranean insect-pollinated plants. *Int. J. Syst. Evolut. Microbiol.* **2013**, *63 Pt 4*, 1532–1539. [CrossRef] [PubMed]

water | **MDPI**

Article

Distribution and Abundance of Antibiotic Resistance Genes in Sand Settling Reservoirs and Drinking Water Treatment Plants across the Yellow River, China

Junying Lu [1,2], Zhe Tian [1,2], Jianwei Yu [1,2], Min Yang [1,2] and Yu Zhang [1,2,*]

[1] State Key Laboratory of Environmental Aquatic Chemistry, Research Center for Eco-Environmental Sciences, Chinese Academy of Sciences, Beijing 100085, China; junying.lu.cn@gmail.com (J.L.); zhetian@rcees.ac.cn (Z.T.); jwyu@rcees.ac.cn (J.Y.); yangmin@rcees.ac.cn (M.Y.)

[2] University of Chinese Academy of Sciences, Beijing 100049, China

* Correspondence: zhangyu@rcees.ac.cn; Tel.: +86-010-6291-9883

Received: 17 January 2018; Accepted: 23 February 2018; Published: 28 February 2018

Abstract: Understanding how antibiotic resistance genes (ARGs) are distributed in drinking water treatment processes is important due to their potential public health risk. Little is known about the occurrence and distribution of ARGs in typical drinking water treatment processes, such as sand settling reservoirs (SSRs) and drinking water treatment plants (DWTPs), in the Yellow River, especially at the catchment scale. In this study, ARG profiling was investigated from water samples of influent (river water) and effluent (source water) of SSRs and finished water of DWTPs in six cities along the Yellow River catchment using real-time quantitative polymerase chain reaction (qPCR) and 16S rRNA gene sequencing. Seventeen ARGs and two mobile genetic elements (MGEs) were detected, among which *aadE*, *strA*, *strB*, *tetA*, *sulII*, *intI1*, and *Tn916* had high detection rates (over 80%). The absolute abundances (gene copies/mL of water) of ARGs were reduced by the SSRs and DWTPs generally, but no reductions were observed for the relative abundances (gene copies/16S rRNA gene) of ARGs. Spatial distributions of ARGs and bacteria were not observed. The distribution of bacterial genera was clustered into four dominant patterns in different water type samples. The bacterial genera *Pseudomonas*, *Massilia*, *Acinetobacter*, *Sphingomonas*, *Methylobacterium*, and *Brevundimonas* dominated the finished water, with *Brevundimonas* and *Methylobacterium* being speculated to be potential hosts for two ARGs (*strA* and *strB*) through network analysis. The enrichment of these two genera, likely caused by selection of disinfection process, may contribute to the higher relative abundance of ARGs in finished water. This study provides insight and effective assessment of the potential risk of ARGs in drinking water treatment processes at the catchment scale.

Keywords: antibiotic resistance gene; sand settling reservoirs; drinking water treatment plants; the Yellow River

1. Introduction

Antibiotic resistance genes (ARGs), as emerging environmental pollutants [1], are a major concern associated with the spread and development of antibiotic resistance. The aquatic environment is recognized as one of the most important reservoirs for antibiotic resistant bacteria (ARB) and ARGs [2,3]. With the wide use of antibiotics in human and veterinary medicine and the agricultural industry, ARGs have been detected in a variety of environments, including surface river water [1], municipal wastewater treatment plants (MWTPs) [4], drinking water treatment plants (DWTPs) [5], and water supply reservoirs [6]. Notably, growing evidence shows the existence of ARB and ARGs in drinking water systems from source water to finished water [5,7–10]. Since finished water from DWTPs is provided to local populations, the prevalence of ARGs in drinking water systems can be a potential threat to public health.

Previous studies have reported wide distribution of ARB and ARGs in source water, which gives resistance to a variety of antibiotics, including resistance to beta-lactams, aminoglycosides, fluoro-quinolone, sulfonamide, tetracycline, and cephalosporin [6,9,11–14]. Many earlier studies have focused on the occurrence and quantity of ARGs in DWTPs [5,8,15–18]. Previous quantification of ARGs (including *cat*, *cmr*, *bla*TEM, *bla*SHV, *sulI*, *sulII*, *tetW*, and *tetO*) in DWTPs and distribution systems showed that the absolute abundance of ARGs (gene copies/water volume) was lower in finished water than source water, though relative abundance (normalized to 16S rRNA gene) exhibited no obvious variation [15]. To gain insight into the mechanism of ARG variation in DWTPs, studies have examined the proceeding treatment processes. In DWTPs in the Yangtze River Delta, sulfonamide and tetracycline resistance genes were found to be more abundant after treatment with biological activated carbon (BAC) [8]. Among treatment processes, chlorine is believed to enrich the proportion of ARB and the relative abundance of ARGs due to the co-selection of resistance bacteria [5,7,8,17], indicating that ARGs may be enriched after DWTP treatment. Current studies have primarily focused on specific or scattered DWTPs, and thus further representative research on ARGs in DWTPs at the catchment scale is needed.

The Yellow River is one of the most important water sources in northern China, with many cities along its banks and watershed that are using it for drinking water. As the river water has high turbidity, sand settling reservoirs (SSRs) have been built for pretreatment in many cities. Previous studies on the Yellow River Catchment show the prevalence of a variety of antibiotics, with concentrations of antibiotics in the river (25 to 152 ng/L) and sediment (up to 184 ng/g) being greater than in rivers in Europe [19,20]. However, the pollution levels of ARGs in the Yellow River remain underreported, and little is known about the occurrence and distribution of ARGs in drinking water treatment processes such as SSRs.

This study aimed to elucidate the distribution and abundance of ARGs in SSRs and DWTPs at the catchment scale, and to reveal the relationships and effects between ARGs and their potential hosts in the treatment processes. Real-time quantitative polymerase chain reaction (qPCR) was used to document the prevalence of 17 ARGs and two mobile genetic elements (MGEs) (*intI1* and transposon *Tn916*) in river water (influent of SSRs), source water (effluent of SSRs), and finished water (effluent of DWTPs) of six cities (Lanzhou, Yinchuan, Hohhot, Zhengzhou, Jinan, and Dongying) in the Yellow River catchment in China. Target ARGs included five aminoglycoside ARGs, five macrolide ARGs, six tetracycline ARGs, and one sulfonamide ARG. The selection of ARGs were based on those identified in previous studies and our preliminary detection experiment. Among all the ARGs observed in the drinking water source and drinking water treatment plants, tetracycline and sulfonamide ARGs were most common [8,12,18,21], followed by macrolide, aminoglycoside and other ARGs [5,9]. Using Illumina MiSeq sequencing, the bacterial community structures in the drinking water treatment processes were characterized. Additionally, the co-occurrence patterns between ARGs and bacterial taxa were also analyzed by network analysis. The results of this study will provide information and insight for understanding the prevalence and propagation of ARGs in SSRs and DWTPs, as well as their correlations with bacteria, which should be useful for their management.

2. Materials and Methods

2.1. Study Sites and Sampling

Sampling was conducted in the winter from February to March 2014. Six cities along the Yellow River, including Lanzhou (LZ), Yinchuan (YC), Hohhot (HS), Zhengzhou (ZZ), Jinan (JN), and Dongying (DY), were selected for this study (Figure 1a). It should be noted that the sampling time is in winter with less rainfall and lower water flow, and there would be different results in summer, which need further studies in the future. Except for Lanzhou, each city has a SSR to reduce the high turbidity of the river water. All of the DWTPs consist of conventional drinking water treatment processes (coagulation and sedimentation, sand filtration, chloramine disinfection), except BAC filter was used in Jinan

DWTP. River water and source water samples were taken from the influent and effluent of the SSRs at 0–0.5 m depth below the water surface. The river water sample from Lanzhou was taken directly from the influent of DWTP. Finished water samples were taken from the effluent of DWTPs (Figure 1b). Triplicate samples were collected in 500-mL pre-cleaned and sterilized glass bottles, where the bottles were washed and rinsed by pure water from Millipore purification system and sampling water. A total of 15 samples were stored in thermotanks with ice bags and were delivered to the laboratory as soon as possible. Typical water quality parameters of the source water were tested, as shown in our previous study [22] and summarized in Table S1.

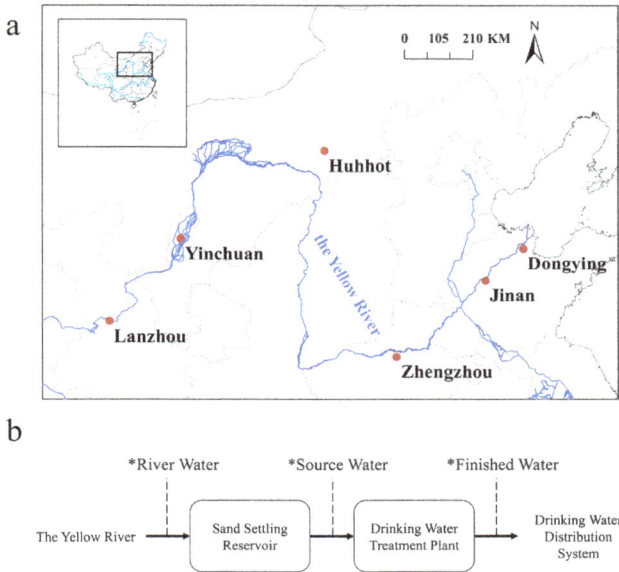

Figure 1. (**a**) Sampling locations (Lanzhou, Yinchuan, Hohhot, Zhengzhou, Jinan, Dongying); (**b**) Drinking water treatment processes, including sand settling reservoir and drinking water treatment. "*": Sampling sites.

2.2. DNA Extraction

Water samples were filtered through 0.22-μm mixed cellulose ester membrane filters (Millipore, Australia) to capture bacteria. The membrane filters were stored at −20 °C in 2-mL centrifuge tubes until DNA extraction. The DNA was extracted from the membranes using a FastDNA SPIN Kit (MP Bio, Solon, OH, USA) according to the manufacturer's instructions. The concentration of the purified DNA was quantified spectrophotometrically (NanoDrop ND-1000, Thermo, Waltham, MA, USA), with the purified DNA then stored at −20 °C until subsequent analysis and qPCR assays were performed.

2.3. Real-Time qPCR

Five aminoglycoside ARGs (*aadB*, *aadE*, *aphA1*, *strA*, and *strB*), five macrolide ARGs (*ereA*, *ermF*, *ermG*, *ermX*, and *mefA*), six tetracycline ARGs (*tetA*, *tetG*, *tetO*, *tetQ* *tetW*, *tetX*), one sulfonamide ARG (*sulII*), two MGEs (*intl1* and transposon *Tn916*), and the 16S rRNA gene were quantified using SYBR-Green real-time qPCR. The primer sequences and PCR conditions were verified in recently published papers and in this study [23] (Table S2). Positive controls contained cloned and sequenced PCR amplicons that were obtained from the sludge of wastewater treatment plants in our previous study [23]. Concentrations of the standard plasmids (ng/μL) were determined with the NanoDrop ND-1000, and their copy concentrations (copies/μL) were then calculated [24].

The 25-µL reactions of qPCR typically contained 1 × SYBR Green I, 1 × Dye (TaKaRa), 200 nM of each primer, 0.5 mg/mL BSA, and 2 µL of DNA templates. Real-time PCR was run using an ABI 7300 system (ABI, Foster City, CA, USA) with the following program: 95 °C for 30 s, 40 cycles consisting of: (i) 95 °C for 10 s; (ii) annealing temperature for 15 s; (iii) 72 °C for 15 s; and (iv) 78 °C for 26 s to collect the fluorescent signals. The melting process was automatically generated using the ABI 7300 software. Ten-fold dilution of plasmids carrying the target gene were used as calibration standards, ranging from 10^8 copies to 10^2 copies. Standard curves were constructed in each PCR run and the copy numbers of genes in each sample were interpolated using these standard curves. All of the standards, samples, and negative control (sterile water) were quantified in triplicate.

Reliable correlation coefficients ($R^2 > 0.99$) for standard curves over five orders of magnitude were obtained. Amplification efficiencies based on slopes were between 85.39% and 112.75% (Table S3). The detection limits were between 1.4×10 to 2.86×10^3 copies per µL added DNA, as shown in Table S3. Only samples in which two of the three replicates were above the limits were regarded as positive. Specificity was ensured by melting curves and gel electrophoresis. The copy number of the target gene was calculated based on the calibration curves. Relative abundances of the ARGs and MGEs were normalized to bacterial 16S rRNA genes for comparison. Absolute abundance was calculated based on number of gene copies per the water sample volume (mL). The removal of ARGs was calculated as log removal [25]. Removal of ARGs in selected treatment process = log X − log Y. X and Y are total absolute abundance of ARGs in influent and effluent of selected treatment process.

2.4. Illumina MiSeq Sequencing for 16S rRNA Gene

The DNA samples were diluted to 10 ng/µL, with 20 µL of each sample sent to a commercial laboratory (Majorbio, Beijing, China) for Illumina MiSeq sequencing. Samples were sequenced following the manufacturer's guidelines using the Illumina MiSeq sequencing instruments and reagents by paired-end strategy. The specific primers (515F: GTGCCAGCMGCCGCGGTAA, and 806R: GGACTACHVGGGTWTCTAAT) were applied for the amplification of V4 region of the 16S rRNA gene [26]. AxyPrepDNA Gel Kit was used to extract the PCR products, then the library construction and sequencing was conducted on the Illumina MiSeq platform. The raw FASTQ data were filtered using QIIME (version 1.17) with following criteria: (i) The reads were truncated that obtained an average quality score of <20 over a 50-bp sliding window, and the truncated reads shorter than 50 bp were discarded; (ii) exact barcode matching, two nucleotide mismatch in primer matching, and reads containing ambiguous characters were removed; and, (iii) only overlapping sequences longer than 10 bp were assembled according to their overlapped sequence. Operational taxonomic units (OTUs) with 97% similarity cutoff were clustered using UPARSE (version 7.1), and anomalies were eliminated using UCHIME. After filtering for quality, the sequences were aligned against the 16S rRNA sequences in the Silva database (Release119 http://www.arb-silva.de) with OTUs data to identify species at different levels. The rarefaction analysis based on Mothur v.1.21.1 was conducted to reveal the diversity indices. Raw sequencing data were deposited in the NCBI Sequence Read Archive with accession nos. SRR4253864 to SRR4253864.

2.5. Network Analysis

Network analysis was used to explore the underlying associations among genes and microbial taxa [23,27]. To visualize the correlations between ARGs and bacterial taxa, we constructed a co-occurrence network among the 17 ARGs and two MGEs quantified by qPCR and the 464 bacterial genera that were identified by MiSeq sequencing with the random matrix theory-based network inference method of the 17 water samples [28]. Briefly, all of the possible pairwise Pearson correlations among the 483 items, including the 17 ARGs, two MGEs, and 464 bacterial genera, that occurred in at least nine samples were calculated to construct a correlation matrix [28,29]. Network analyses were performed using an online analysis pipeline at http://ieg2.ou.edu/MENA. Network visualization was conducted on the interactive platform of Gephi (version 0.9.1).

2.6. Statistical Analysis

Correlations among ARGs and MGEs were evaluated by Pearson's bivariate correlation analysis (SPSS 20.0, IBM, Armonk, NY, USA). One-way ANOVA were performed to evaluate the significance of differences among the samples. Heatmap and principal coordinate analysis (PCoA) were performed in R environment using pheatmap [30] and VEGAN [31].

3. Results and Discussion

3.1. Distribution of ARGs in Water Treatment Processes

The concentrations of the 17 ARGs, two MGEs, and 16S rRNA gene quantified by qPCR are listed in Tables S4 and S5 and summarized in Figure 2. One-way ANOVA were performed to evaluate the significance of differences among sample sites, however no significant difference was found. Among the targeted ARGs and MGEs in this study, *aadE*, *intI1*, *Tn916*, *strA*, *strB*, *tetA*, and *sulII* exhibited high detection rates (over 80%). For ARGs, *strA* and *strB* were the most abundant. Total absolute abundance of ARGs ranged from $1.51 \times 10^4 \pm 1.49 \times 10^4$ copies/mL in river water, $7.78 \times 10^3 \pm 7.13 \times 10^3$ copies/mL in source water, and $6.91 \times 10^2 \pm 6.79 \times 10^2$ copies/mL in finished water. Total relative abundance of ARGs ranged from $3.96 \times 10^{-3} \pm 3.65 \times 10^{-3}$ copies/16S rRNA gene in river water, $4.25 \times 10^{-3} \pm 3.84 \times 10^{-3}$ copies/16S rRNA gene in source water, and $1.82 \times 10^{-2} \pm 1.81 \times 10^{-2}$ copies/16S rRNA gene in finished water. In terms of total absolute abundance, the highest ARG level was found in Hohhot river water, whereas in terms of total relative abundance, the highest ARG level was found in Zhengzhou finished water. When compared with the previous study, the absolute abundances of *sulII*, *tetA*, *tetG*, *tetO*, *tetW*, and *tetX* in river water were, on average, lower than those in Huangpu River [9]. These results suggest that ARGs are prevalent in the river, source, and finished water of the drinking water treatment processes in the Yellow River catchment. Spatial distribution of ARGs and MGEs was not found, and the Bray-Curtis-based principal coordinate analysis (PCoA) of the ARGs did not show geographical clustering (Figure S2).

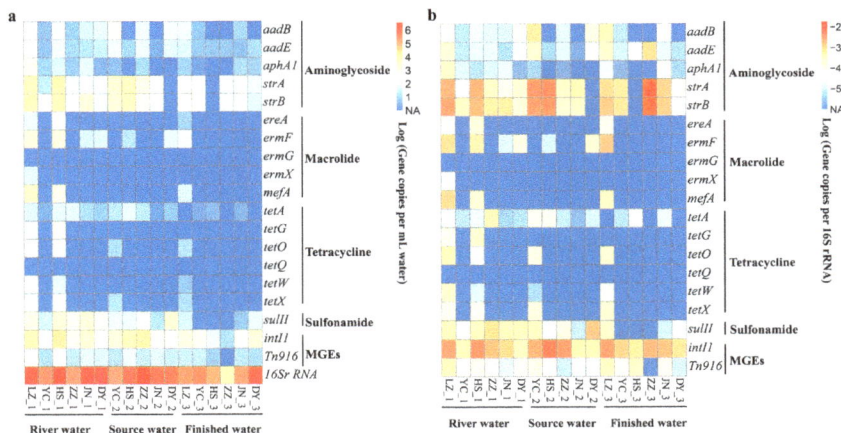

Figure 2. (**a**) Absolute and (**b**) relative abundances of the 17 ARGs, two MGEs, and 16s rRNA in water samples. Label under each column denotes sample site name followed by water type (1 is river water, 2 is source water, 3 is finished water).

To clarify the variation of ARGs and MGEs in the drinking water treatment processes, comparisons among river water, source water, and finished water were performed (Figure 3 and Figure S3). Variation in the total absolute abundance of ARGs showed that they decreased with treatment processes.

On average, the total absolute abundance of ARGs decreased from 1.07×10^4 copies/mL in river water to 5.39×10^3 copies/mL in source water and 5.99×10^2 copies/mL in finished water. In addition, total relative abundance of ARGs increased, on average, from 2.7×10^{-3} copies/16S rRNA gene in river water to 3.68×10^{-3} copies/16S rRNA gene in source water and 7.43×10^{-3} copies/16S rRNA gene in finished water. For MGEs, total absolute abundance decreased, on average, from 1.51×10^6 copies/mL in river water to 1.15×10^6 copies/mL in source water and 3.19×10^5 copies/mL in finished water, whereas total relative abundance remained relatively stable at 1.58×10^{-3} copies/16S rRNA gene in river water to 2.15×10^{-3} copies/16S rRNA gene in source water and 1.50×10^{-3} copies/16S rRNA gene in finished water. The decrease in total absolute abundance of ARGs and MGEs could be explained by the elimination of bacteria during the water treatment processes, especially treated by drinking water treatment plants that have disinfection processes to eliminate them. As for total relative abundance of ARGs and MGEs, no variation trend was observed in this study. This result suggested the possibility that ARGs and MGEs are carried by bacteria that can survive from the water treatment processes, while bacteria did not carry those genes were eliminated more effectively, which caused the proportions of ARGs and MGEs to remain unchanged or even increase. The results are consistent with a previous study which showed *bla*TEM, *bla*SHV, *sulII* and *cat* genes' absolute abundance decreased while the relative abundance increased [15]. To better compare removal effect of two major treatment processes (SSRs and DWTPs), the levels of ARGs removal in the treatment processes are provided in Table 1.

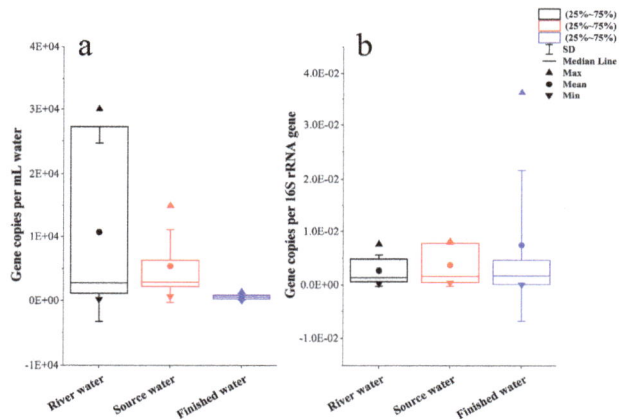

Figure 3. Total absolute (**a**) and relative (**b**) abundances of ARGs (copies/mL water) in river, source, and finished water.

SSRs were observed to have different removal of ARGs in different cities. In Yinchuan, only *aphA1* was removed at a level of 0.32-log, other ARGs include *aadB*, *aadE*, *strA*, *strB*, *tetA*, *tetO*, *tetX*, and *sulII* increased at the levels range from −0.45-log to −2.16-log. Similarly, in Zhengzhou, only *tetA* and *sulII* were removed at a level of <1-log, while all aminoglycoside resistance genes that were detected increased. Better SSRs removal level of ARGs was observed in Donying, where *aphA1*, *strA*, and *strB* have removal level range from 0.46-log to 2.14-log. In Hohhot and Jinan, SSRs have the removal of all detected ARGs at the levels range from 0.22-log to 3.07-log except *tetA* increased in Hohhot and *strA* increased in Jinan. SSRs were special treatment process in the Yellow River catchment. With its relatively slow flow rate and sedimentation function, it could be partly compared with sedimentation tank in DWTPs. However, SSRs did not show high removal level like of sedimentation in DWTPs reported in recent study [25]. In Yinchuan, Zhengzhou, and Dongying, total absolute abundance of ARGs increased at levels range from −0.24-log to −1.46-log after SSRs treatment, while 16S rRNA genes were removed at levels range from 0.29-log to 0.67-log in all SSRs except Zhengzhou. Therefore, the results concerning dissemination of ARGs in SSRs, and whether there were transfer and selection of ARGs in SSRs remain further study.

Table 1. The log removal of ARGs in the treatment processes of six sample cities.

Gene	SSRs						DWTPs				
	YC	HS	ZZ	JN	DY	LZ	YC	HS	ZZ	JN	DY
16S rRNA gene	0.29	0.67	−1.28	0.41	0.36	1.05	−0.35	0.82	2.88	0.62	−0.29
aadB	−2.09	2.17	−1.06	1.67	−0.32	0.24	1.99	ND	1.98	−1.09	2.13
aadE	−1.38	1.06	−1.00	0.85	−0.37	0.87	1.19	0.16	1.10	−0.31	0.67
aphA1	0.32	1.67	−0.28	1.77	0.46	1.89	0.35	ND	1.10	−1.19	−1.05
strA	−2.16	0.22	−0.93	−0.29	2.14	1.37	0.91	3.89	0.74	0.11	−2.05
strB	−1.23	0.19	−0.34	0.94	2.07	1.33	0.72	3.84	0.58	−0.24	−1.81
ereA	ND	2.55	ND	ND	ND	0.90	ND	ND	ND	ND	ND
ermF	ND	1.77	ND	1.76	−0.17	0.85	ND	1.40	ND	ND	2.07
ermG	ND	ND	ND	ND	ND	ND	ND	ND	ND	ND	ND
ermX	ND	ND	ND	ND	ND	1.80	ND	ND	ND	ND	ND
mefA	ND	3.07	ND	ND	ND	1.78	ND	ND	ND	ND	ND
tetA	−1.02	−0.14	0.37	0.84	0.00	1.74	0.92	1.31	1.48	−0.21	0.97
tetG	ND	2.77	ND	ND	ND	2.14	ND	ND	ND	ND	ND
tetO	−1.28	2.31	ND	ND	ND	1.36	1.28	ND	ND	ND	ND
tetQ	ND	ND	ND	ND	ND	ND	ND	ND	ND	ND	ND
tetW	ND	1.91	ND	ND	ND	1.54	ND	ND	ND	ND	ND
tetX	−1.34	2.38	ND	ND	ND	0.86	1.34	ND	ND	ND	ND
sulII	−0.45	1.35	0.25	2.03	−0.54	1.40	1.97	1.94	2.65	0.37	1.48
intl1	−0.96	0.40	−1.29	0.85	−0.28	0.94	0.40	1.92	2.98	−0.05	−0.57
Tn916	−0.10	0.59	−0.09	−0.01	0.22	1.16	−0.20	0.90	1.44	0.38	0.34
Total	−1.46	0.30	−0.24	0.82	−0.39	1.30	0.87	3.08	0.81	−0.06	1.00

Note: ND means no detected.

DWTPs are important constructions designed to enhance drinking water quality. In this study, total absolute abundance of ARGs decreased in all DWTPs at levels range from 0.81-log to 3.08-log, except Jinan, where ARGs concentration increased 0.06-log after DWTPs. Generally, DWTPs showed removal effects of ARGs, which also been provided in previous studies [5,8,25]. In Lanzhou, Yinchuan, Hohhot and Zhengzhou, all detected ARGs decreased at level range from 0.16-log to 3.89-log after DWTPs. However, in Donying, *aphA1*, *strA*, *strB* increased partly because the increasing of 16S rRNA. Total absolute and relative abundances increased in the Jinan finished water. This is noteworthy because the DWTP in Jinan uses BAC as advanced treatment. Previous study has shown that BAC can increase the relative abundance of ARGs and ARB as it aggregates microbes [5,7]. Therefore, our results suggest that BAC may play a key role in affecting ARGs, and its influence on ARGs in drinking water needs further assessment.

The correlations among the relative abundances of ARGs and MGEs are shown in Table S6. Results showed that *ereA* was significantly correlated with *intl1*, which also has been shown in a recent study of greenhouse soil [32]. In addition, some of the detected ARGs had significant correlations with transposon *Tn916*, including one aminoglycoside ARG (*aphA1*), four macrolide ARGs (*ereA*, *ermF*, *ermX*, and *mefA*), and three tetracycline ARGs (*tetO*, *tetW*, and *tetX*). *Intl1* and the *Tn916*/Tn1545 transposon family contain a variety ARGs, including tetracycline, macrolide, and aminoglycoside resistance genes [33,34]. The correlations between *intl1* and *Tn916*/Tn1545 in this study indicate the potential association with the additional horizontal transfer of ARGs. As MGEs were not eliminated efficiently in the drinking water treatment processes, as shown in Figure S3, the risk of ARG horizontal transfer is worth the attention.

3.2. Diversity and Distribution Patterns of Bacteria along Water Treatment

Based on PCoA analysis, the overall bacterial community structure showed more commonalities in the finished water from all of the cities that were analysed (Figure S4). The Shannon and Simpson diversity indices (Table S7) showed that bacterial community diversity declined after treatment via DWTPs in all of the sample sites, except for Hohhot, but no significant differences were observed between river and source water. The differences in the diversity of the bacterial community between the river and source water was varied, with two showing a slight rise and three demonstrating a decline. In this study, all of the DWTPs applied conventional processes, which include coagulation and sedimentation, sand filtration, and chloramine disinfection except for Jinan which has additional BAC treatment. One of the major function of these processes is to eliminate the microorganisms from water. Among them, chloramine disinfection has strong selection to *Proteobacteria*. This result is consistent with previous studies that have shown the reduction of bacterial diversity due to selection of chloramine resistance [17,25].

At the phylum level, as shown in Figure S5 and Table S8, *Proteobacteria* increased considerably in the finished water and became dominant. At the genus level, however, the distribution of bacteria could be divided into four patterns (Figure 4). Bacteria showing A and C pattern distributions, including *Mycobacterium*, *Synechococcus*, *Planctomycetaceae*, *Actinobacteria*, and *Rhodobacter*, were dominant both in river and source water, but their proportions decreased in finished water. In contrast, the proportion of bacteria showing B pattern distribution, including *Pseudomonas*, *Massilia*, *Acinetobacter*, *Sphingomonas*, *Methylobacterium*, *Brevundimonas*, and *Deinococcus*, increased markedly in the finished water and became dominant. The genera exhibiting these different patterns are listed Table S9. Previous metagenomic research has indicated that *Proteobacteria* are the main antibiotic resistant bacteria in drinking water [17], which may relate to their chlorination resistance [35]. Among the genera that were dominant in the finished water, all were *Proteobacteria*, except for *Deinococcus*. In previous drinking water studies, *Pseudomonas* has been regarded as an opportunistic bacterial pathogen that can spread acquired antibiotic resistance preferentially via vertical transmission [36]. Further studies have also reported that *Massilia* contains sulfonamide resistance genes [37], *Acinetobacter* is associated with multi-drug resistance [38], *Sphingomonas* is positively correlated to ARGs encoding the RND (resistance-nodulation-cell division protein family) transportation system [39], *Methylobacterium* is resistant to disinfection [40], and *Brevundimonas* possesses innate resistance to fluoroquinolones [41]. By comparing the distribution of genera (Figure 4) with the distribution of ARGs (Figure 2), a correlation between predominate ARGs (*strA* and *strB*) and bacterial genera in finished water was observed. As these bacteria are associated with antibiotic resistance, they are of public health concern.

To further prove the correlation between ARGs and bacteria, the co-occurrence patterns between ARGs and bacterial taxa were investigated using network analysis, as shown in Figure 5. Two bacterial genera, *Brevundimonas* and *Methylobacterium*, were suggested as the possible hosts of two ARGs (*strA* and *strB*) (Table S10). Notably, these two genera were enriched in the finished water and their proportions were highly increased in the finished water of Zhengzhou. This may explain the increase in the total relative abundance of ARGs in the finished water sample as the *strA* and *strB* ARG subtypes in the finished water of Zhengzhou were dominant. Although *Brevundimonas* species has rarely been reported as a pathogen causing human infection, it has been isolated in some infection cases [41]. *Methylobacterium* species has been considered as a serious concern in hospitals due to its contamination of tap water [42]. There has been no report of *strA-strB* streptomycin-resistance of *Brevundimonas* and *Methylobacterium* yet, aminoglycosides were recommended as important treatment of *Methylobacterium* species infection [43]. However, since strA-strB streptomycin-resistance genes widely distribute in bacteria, and strA-strB are often encoded on transposon borne on conjugative plasmid [44], potential aminoglycosides resistance in *Brevundimonas* and *Methylobacterium* that is caused by horizontal transfer need to be further studied.

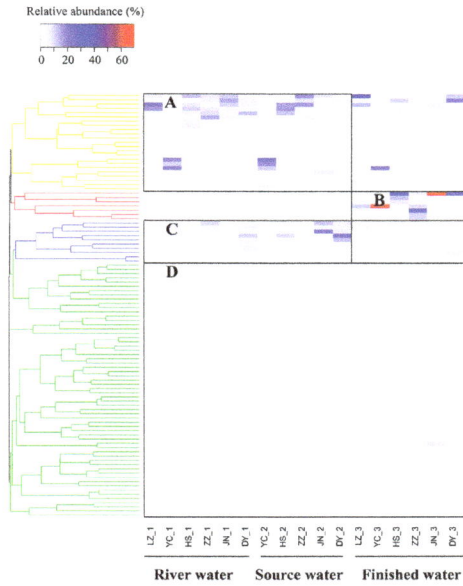

Figure 4. Heatmap of distribution profiles of bacterial genera across different water samples. Each row represents the results of genus percentage by sequencing. Dendrograms for the rows were constructed from the Bray-Curtis dissimilarity distance. Black boxes represent different patterns of genera: (**A**) widely detected in all water types, dominant in river and source water; (**B**) highly enriched in finished water, but rarely detected in river and source water; (**C**) mainly detected in river and source water, but not in finished water; and, (**D**) relatively low in all water types.

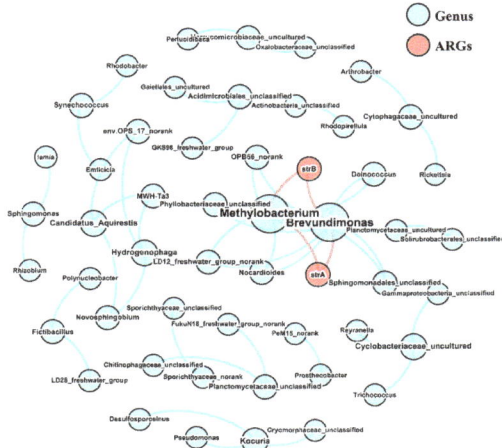

Figure 5. Network analysis revealing the co-occurrence patterns among ARG subtypes and bacteria genera. Nodes are colored according to ARG subtypes. Connection represents correlation. Size of each node is proportional to the number of connections.

4. Conclusions

In this study, variations in 17 ARGs and two MGEs in drinking water treatment processes in the Yellow River catchment were documented by qPCR. At the catchment scale, the absolute abundances of ARGs and MGEs decreased after drinking water treatment, whereas the relative abundances of ARGs and MGEs did not, suggesting the potential risk of ARGs in drinking water. The bacterial community in the drinking water treatment processes was analyzed by sequencing. The distribution of bacterial genera was characterized into four patterns, with two dominant bacterial genera (*Brevundimonas and Methylobacterium*) found to be associated with two enriched ARGs (*strA* and *strB*) in the finished water. As these two genera are reported to be resistant to disinfection, their high proportion in finished water in the present study confirms the impact of disinfection to antibiotic resistance in drinking water at the catchment scale.

Supplementary Materials: The following are available online at http://www.mdpi.com/2073-4441/10/3/246/s1, Figure S1. Number of ARGs subtypes in 17 water samples; Figure S2. Bray-Curtis-based Principal coordinates analysis of (a) relative and (b) absolute abundance of ARGs in 17 water samples; Figure S3. (A) Total absolute abundance of MGEs (copies/mL water) in river water, source water and finished water. (B) Total relative abundance of MGEs (copies/16S rRNA) in river water, source water and finished water; Figure S4. Bray-Curtis-based Principal coordinates analysis of microbial communities in 17 water samples; Figure S5. Bacteria proportion on phylum level of 17 water samples; Table S1. River water quality parameters; Table S2. PCR primers for the investigated ARGs, MGEs and bacterial 16s rRNA gene; Table S3. Quality control of the real-time qPCR methods for the all target genes; Table S4. Absolute abundances of the 17 ARGs and 2 MGEs (copies/mL water) in water samples; Table S5. Relative abundances of the 18 ARGs and 2 MGEs (gene copies/16S rRNA gene copies) in water samples; Table S6. Correlation of ARGs with MGEs; Table S7. Coverage and diversity indices of bacterial communities by Miseq sequencing; Table S8. Relative abundances of bacterial community compositions grouped by phylum in different water samples. The abundance is presented in terms of the percentage of the targeted phylum in the total sequences of a sample (%); Table S9. Genera distribution patterns in 17 water samples; Table S10. Genera percentage of bacteria co-occurrence with ARGs.

Acknowledgments: This study was supported by National Natural Science Foundation of China (No. 21437005).

Author Contributions: Yu Zhang conceived and designed the experiments; Jianwei Yu collected water samples; Junying Lu performed the experiments; Junying Lu analyzed the data and wrote the paper. Yu Zhang, Min Yang and Zhe Tian revised the manuscript.

Conflicts of Interest: The authors declare no conflict of interest.

References

1. Pruden, A.; Pei, R.; Storteboom, H.; Carlson, K.H. Antibiotic resistance genes as emerging contaminants: Studies in northern colorado. *Environ. Sci. Technol.* **2006**, *40*, 7445–7450. [CrossRef] [PubMed]
2. Zhang, X.-X.; Zhang, T.; Fang, H.H. Antibiotic resistance genes in water environment. *Appl. Microbiol. Biotechnol.* **2009**, *82*, 397–414. [CrossRef] [PubMed]
3. Baquero, F.; Martínez, J.-L.; Cantón, R. Antibiotics and antibiotic resistance in water environments. *Curr. Opin. Biotechnol.* **2008**, *19*, 260–265. [CrossRef] [PubMed]
4. Li, D.; Yu, T.; Zhang, Y.; Yang, M.; Li, Z.; Liu, M.; Qi, R. Antibiotic resistance characteristics of environmental bacteria from an oxytetracycline production wastewater treatment plant and the receiving river. *Appl. Environ. Microbiol.* **2010**, *76*, 3444–3451. [CrossRef] [PubMed]
5. Xu, L.; Ouyang, W.; Qian, Y.; Su, C.; Su, J.; Chen, H. High-throughput profiling of antibiotic resistance genes in drinking water treatment plants and distribution systems. *Environ. Pollut.* **2016**, *213*, 119–126. [CrossRef] [PubMed]
6. Huerta, B.; Marti, E.; Gros, M.; López, P.; Pompêo, M.; Armengol, J.; Barceló, D.; Balcázar, J.L.; Rodríguez-Mozaz, S.; Marcé, R. Exploring the links between antibiotic occurrence, antibiotic resistance, and bacterial communities in water supply reservoirs. *Sci. Total Environ.* **2013**, *456*, 161–170. [CrossRef] [PubMed]
7. Bai, X.; Ma, X.; Xu, F.; Li, J.; Zhang, H.; Xiao, X. The drinking water treatment process as a potential source of affecting the bacterial antibiotic resistance. *Sci. Total Environ.* **2015**, *533*, 24–31. [CrossRef] [PubMed]

8. Guo, X.; Li, J.; Yang, F.; Yang, J.; Yin, D. Prevalence of sulfonamide and tetracycline resistance genes in drinking water treatment plants in the yangtze river delta, china. *Sci. Total Environ.* **2014**, *493*, 626–631. [CrossRef] [PubMed]

9. Jiang, L.; Hu, X.; Xu, T.; Zhang, H.; Sheng, D.; Yin, D. Prevalence of antibiotic resistance genes and their relationship with antibiotics in the huangpu river and the drinking water sources, shanghai, china. *Sci. Total Environ.* **2013**, *458–460*, 267–272. [CrossRef] [PubMed]

10. Schwartz, T.; Kohnen, W.; Jansen, B.; Obst, U. Detection of antibiotic-resistant bacteria and their resistance genes in wastewater, surface water, and drinking water biofilms. *FEMS Microbiol. Ecol.* **2003**, *43*, 325–335. [CrossRef] [PubMed]

11. Coleman, B.L.; Louie, M.; Salvadori, M.I.; McEwen, S.A.; Neumann, N.; Sibley, K.; Irwin, R.J.; Jamieson, F.B.; Daignault, D.; Majury, A.; et al. Contamination of canadian private drinking water sources with antimicrobial resistant *Escherichia coli*. *Water Res.* **2013**, *47*, 3026–3036. [CrossRef] [PubMed]

12. Bergeron, S.; Boopathy, R.; Nathaniel, R.; Corbin, A.; LaFleur, G. Presence of antibiotic resistant bacteria and antibiotic resistance genes in raw source water and treated drinking water. *Int. Biodeterior. Biodegrad.* **2015**, *102*, 370–374. [CrossRef]

13. Fernando, D.M.; Tun, H.M.; Poole, J.; Patidar, R.; Li, R.; Mi, R.; Amarawansha, G.E.; Fernando, W.G.; Khafipour, E.; Farenhorst, A.; et al. Detection of antibiotic resistance genes in source and drinking water samples from a first nations community in canada. *Appl. Environ. Microbiol.* **2016**, *82*, 4767–4775. [CrossRef] [PubMed]

14. Egervarn, M.; Englund, S.; Ljunge, M.; Wiberg, C.; Finn, M.; Lindblad, M.; Borjesson, S. Unexpected common occurrence of transferable extended spectrum cephalosporinase-producing *Escherichia coil* in swedish surface waters used for drinking water supply. *Sci. Total Environ.* **2017**, *587*, 466–472. [CrossRef] [PubMed]

15. Xi, C.; Zhang, Y.; Marrs, C.F.; Ye, W.; Simon, C.; Foxman, B.; Nriagu, J. Prevalence of antibiotic resistance in drinking water treatment and distribution systems. *Appl. Environ. Microbiol.* **2009**, *75*, 5714–5718. [CrossRef] [PubMed]

16. Farkas, A.; Butiuc-Keul, A.; Ciataras, D.; Neamtu, C.; Craciunas, C.; Podar, D.; Dragan-Bularda, M. Microbiological contamination and resistance genes in biofilms occurring during the drinking water treatment process. *Sci. Total Environ.* **2013**, *443*, 932–938. [CrossRef] [PubMed]

17. Shi, P.; Jia, S.; Zhang, X.-X.; Zhang, T.; Cheng, S.; Li, A. Metagenomic insights into chlorination effects on microbial antibiotic resistance in drinking water. *Water Res.* **2013**, *47*, 111–120. [CrossRef] [PubMed]

18. Zhang, S.; Lin, W.; Yu, X. Effects of full-scale advanced water treatment on antibiotic resistance genes in the yangtze delta area in china. *FEMS Microbiol. Ecol.* **2016**, *92*. [CrossRef] [PubMed]

19. Xu, W.; Zhang, G.; Zou, S.; Ling, Z.; Wang, G.; Yan, W. A preliminary investigation on the occurrence and distribution of antibiotics in the Yellow River and its tributaries, china. *Water Environ. Res.* **2009**, *81*, 248–254. [CrossRef] [PubMed]

20. Zhou, L.J.; Ying, G.G.; Zhao, J.L.; Yang, J.F.; Wang, L.; Yang, B.; Liu, S. Trends in the occurrence of human and veterinary antibiotics in the sediments of the Yellow River, Hai river and liao river in northern china. *Environ. Pollut.* **2011**, *159*, 1877–1885. [CrossRef] [PubMed]

21. Xu, Y.; Guo, C.; Luo, Y.; Lv, J.; Zhang, Y.; Lin, H.; Wang, L.; Xu, J. Occurrence and distribution of antibiotics, antibiotic resistance genes in the urban rivers in beijing, china. *Environ. Pollut.* **2016**, *213*, 833–840. [CrossRef] [PubMed]

22. Li, X.; Yu, J.; Guo, Q.; Su, M.; Liu, T.; Yang, M.; Zhao, Y. Source-water odor during winter in the yellow river area of china: Occurrence and diagnosis. *Environ. Pollut.* **2016**, *218*, 252–258. [CrossRef] [PubMed]

23. Tian, Z.; Zhang, Y.; Yu, B.; Yang, M. Changes of resistome, mobilome and potential hosts of antibiotic resistance genes during the transformation of anaerobic digestion from mesophilic to thermophilic. *Water Res.* **2016**, *98*, 261–269. [CrossRef] [PubMed]

24. Pei, R.; Kim, S.-C.; Carlson, K.H.; Pruden, A. Effect of river landscape on the sediment concentrations of antibiotics and corresponding antibiotic resistance genes (arg). *Water Res.* **2006**, *40*, 2427–2435. [CrossRef] [PubMed]

25. Su, H.C.; Liu, Y.S.; Pan, C.G.; Chen, J.; He, L.Y.; Ying, G.G. Persistence of antibiotic resistance genes and bacterial community changes in drinking water treatment system: From drinking water source to tap water. *Sci. Total Environ.* **2017**, *616–617*, 453–461. [CrossRef] [PubMed]

26. Caporaso, J.G.; Lauber, C.L.; Walters, W.A.; Berg-Lyons, D.; Lozupone, C.A.; Turnbaugh, P.J.; Fierer, N.; Knight, R. Global patterns of 16s rrna diversity at a depth of millions of sequences per sample. *Proc. Natl. Acad. Sci. USA* **2011**, *108*, 4516–4522. [CrossRef] [PubMed]

27. Li, B.; Yang, Y.; Ma, L.; Ju, F.; Guo, F.; Tiedje, J.M.; Zhang, T. Metagenomic and network analysis reveal wide distribution and co-occurrence of environmental antibiotic resistance genes. *ISME J.* **2015**, *9*, 2490–2502. [CrossRef] [PubMed]

28. Deng, Y.; Jiang, Y.-H.; Yang, Y.; He, Z.; Luo, F.; Zhou, J. Molecular ecological network analyses. *BMC Bioinform.* **2012**, *13*, 113. [CrossRef] [PubMed]

29. Deng, Y.; Zhang, P.; Qin, Y.; Tu, Q.; Yang, Y.; He, Z.; Schadt, C.W.; Zhou, J. Network succession reveals the importance of competition in response to emulsified vegetable oil amendment for uranium bioremediation. *Environ. Microbiol.* **2016**, *18*, 205–218. [CrossRef] [PubMed]

30. Kolde, R. Pheatmap: Pretty Heatmaps. R Package Version 1.0.8. 2015. Available online: http://CRAN.R-project.org/package=pheatmap (accessed on 27 February 2018).

31. Oksanen, J.; Kindt, R.; Legendre, P.; O'Hara, B.; Stevens, M.H.H.; Oksanen, M.J.; Suggests, M. The vegan package. *Community Ecol. Package* **2007**, *10*, 631–637.

32. Li, J.; Xin, Z.; Zhang, Y.; Chen, J.; Yan, J.; Li, H.; Hu, H. Long-term manure application increased the levels of antibiotics and antibiotic resistance genes in a greenhouse soil. *Appl. Soil Ecol.* **2017**, *121*, 193–200. [CrossRef]

33. Santoro, F.; Vianna, M.E.; Roberts, A.P. Variation on a theme; an overview of the tn916/tn1545 family of mobile genetic elements in the oral and nasopharyngeal streptococci. *Front. Microbiol.* **2014**, *5*, 535. [CrossRef] [PubMed]

34. Roberts, A.P.; Mullany, P. Tn916-like genetic elements: A diverse group of modular mobile elements conferring antibiotic resistance. *FEMS Microbiol. Rev.* **2011**, *35*, 856–871. [CrossRef] [PubMed]

35. Mathieu, L.; Bouteleux, C.; Fass, S.; Angel, E.; Block, J. Reversible shift in the α-, β-and γ-proteobacteria populations of drinking water biofilms during discontinuous chlorination. *Water Res.* **2009**, *43*, 3375–3386. [CrossRef] [PubMed]

36. Vaz-Moreira, I.; Nunes, O.C.; Manaia, C.M. Diversity and antibiotic resistance in *Pseudomonas* spp. from drinking water. *Sci. Total Environ.* **2012**, *426*, 366–374. [CrossRef] [PubMed]

37. Wang, N.; Yang, X.; Jiao, S.; Zhang, J.; Ye, B.; Gao, S. Sulfonamide-resistant bacteria and their resistance genes in soils fertilized with manures from jiangsu province, southeastern china. *PLoS ONE* **2014**, *9*, e112626. [CrossRef] [PubMed]

38. Zhang, Y.; Marrs, C.F.; Simon, C.; Xi, C. Wastewater treatment contributes to selective increase of antibiotic resistance among *Acinetobacter* spp. *Sci. Total Environ.* **2009**, *407*, 3702–3706. [CrossRef] [PubMed]

39. Jia, S.; Shi, P.; Hu, Q.; Li, B.; Zhang, T.; Zhang, X.-X. Bacterial community shift drives antibiotic resistance promotion during drinking water chlorination. *Environ. Sci. Technol.* **2015**, *49*, 12271–12279. [CrossRef] [PubMed]

40. Simões, L.C.; Simoes, M.; Vieira, M.J. Influence of the diversity of bacterial isolates from drinking water on resistance of biofilms to disinfection. *Appl. Environ. Microbiol.* **2010**, *76*, 6673–6679. [CrossRef] [PubMed]

41. Han, X.Y.; Andrade, R.A. Brevundimonas diminuta infections and its resistance to fluoroquinolones. *J. Antimicrob. Chemother.* **2005**, *55*, 853–859. [CrossRef] [PubMed]

42. Furuhata, K.; Kato, Y.; Goto, K.; Hara, M.; Yoshida, S.-I.; Fukuyama, M. Isolation and identification of methylobacterium species from the tap water in hospitals in japan and their antibiotic susceptibility. *Microbiol. Immunol.* **2006**, *50*, 11–17. [CrossRef] [PubMed]

43. Lee, C.-H.; Tang, Y.-F.; Liu, J.-W. Underdiagnosis of urinary tract infection caused by methylobacterium species with current standard processing of urine culture and its clinical implications. *J. Med. Microbiol.* **2004**, *53*, 755–759. [CrossRef] [PubMed]

44. Sundin, G.W.; Bender, C.L. Dissemination of the stra-strb streptomycin-resistance genes among commensal and pathogenic bacteria from humans, animals, and plants. *Mol. Ecol.* **1996**, *5*, 133–143. [CrossRef] [PubMed]

water   MDPI

Article

The Effects of Antibiotics on Microbial Community Composition in an Estuary Reservoir during Spring and Summer Seasons

Zheng Xu [1], Yue Jiang [1], Shu Harn Te [2], Yiliang He [1,*] and Karina Yew-Hoong Gin [2,3]

[1] School of Environmental Science and Engineering, Shanghai Jiao Tong University, Shanghai 200240, China; xuzheng-2004@163.com (Z.X.); jiangyue0524@sjtu.edu.cn (Y.J.)
[2] NUS Environmental Research Institute (NERI), National University of Singapore, Singapore 138602, Singapore; eritsh@nus.edu.sg (S.H.T.); ceeginyh@nus.edu.sg (K.Y.-H.G.)
[3] Department of Civil and Environmental Engineering, National University of Singapore, Singapore 138602, Singapore
* Correspondence: ylhe@sjtu.edu.cn; Tel.: +86-21-54744008

Received: 23 November 2017; Accepted: 17 January 2018; Published: 6 February 2018

Abstract: The increased antibiotic pollutants in aquatic environments pose severe threats on microbial ecology due to their extensive distribution and antibacterial properties. A total of 16 antibiotics including fluoroquinolones (FQs) (ofloxacin (OFX), ciprofloxacin (CFX), norfloxacin (NFX)), Sulfonamides (SAs) (sulfamonomethoxine (SMM), sulfadiazine (SDZ), sulfaquinoxaline (SQX)), Tetracyclines (TCs) (tetracycline (TC), doxycycline (DC)), β-lactams (penicillin G (PEN G), penicillin V (PEN V), cefalexin (LEX)), Macrolides (MLs) (erythromycin-H_2O (ETM), tylosin (TYL)) and other antibiotics (Polymix-B (POL), Vancomycin (VAN), Lincomycin (LIN)) were detected in the surface water of the Qingcaosha Reservoir. Multivariate statistical analysis indicated that both water quality and physicochemical indexes have less contributions on variations of these antibiotics, suggesting the concentrations of antibiotics inside the reservoir are mainly affected by upstream runoff and anthropic activity along the river. Antibiotics including TYL, PEN G and ETM showed significant correlations with variations of bacterial community composition, and closely connected with various gram-negative bacteria in co-occurrence/exclusion patterns of the network, suggesting these bacterial taxa play important roles in the course of migration and transformation of related antibiotics. In conclusion, further research is required to evaluate the potential risk of genetic transfer of resistance to related bacteria induced by long-term exposure to low levels of antibiotics in the environment.

Keywords: bacterial community; co-occurrence pattern; antibiotics; estuary reservoir; surface water

1. Introduction

Estuary reservoirs as important water sources for cities in estuarine areas of the world, are strongly influenced both by estuarine ecologic changes and anthropic activities in surrounding areas [1–3]. In recent years, due to the extensive use of antibiotics in the world, the ecological environments in estuarine areas are facing serious ecological threats. Research showed that almost 80% to 90% of these compounds were excreted into the environments after initial metabolisms in organisms [4,5]. Finally, high levels of antibiotics were discharged into estuarine ecosystems via river runoff and sewage outfalls from land-based multiple sources, which have negative effects on human health and the safety of estuarine ecosystems [6,7].

In estuarine aquatic ecosystems, bacterial communities play important roles during the microbial food web that recycles and consumes organic matters [8]. Additionally, studies indicate that the bacterial community composition in aquatic ecosystems was strongly affected by environmental factors such as changes of hydrological conditions and different water trophic statuses [9,10]. Although antibiotics in aquatic ecosystems have been widely studied in recent years, and much has been achieved in the fields of toxic effects of antibiotics on aquatic plants and animals [11–13], still less is known about how the antibiotics affect the bacterial community composition in aquatic ecosystems.

The Qingcaosha (QCS) Reservoir is the largest estuary reservoir of China, located at the Yangtze estuary near Shanghai. The reservoir covers a total area of 66.27 km^2, with a depth ranging from 2.5 m to 13.5 m. The main function of the reservoir is compensating for drinking water shortages in Shanghai city, inputting water with high turbidity from the Yangtze River estuary and outputting clean water to the water plant after the self-purification inside the reservoir [14–16]. Recently, it has been reported that multi-antibiotics were detected in the Yangtze estuary, even in the tap water of Shanghai [17–19]. However, still only few studies have discussed the presence of antibiotics inside the QCS Reservoir, especially the influences of these compounds on bacterial community changes, as well as the effect on co-occurrence patterns of cyanobacterial and non-cyanobacterial taxa within the aquatic ecosystem.

This study aims to utilise a systematic method that encompasses the measurement of water quality, antibiotic concentrations and bacterial community composition data sets to evaluate the relationships between water environmental parameters, concentration levels of antibiotics and bacterial community variations in the QCS Reservoir. Especially, it aims to determine how these antibiotics affect bacterial community composition (including co-occurrence patterns between cyanobacteria and non-cyanobacterial taxa) in the estuary aquatic ecosystem during spring and summer season.

2. Materials and Methods

2.1. Sampling Sites and In Situ Measurements

Three locations, sites 1 to 3, were selected as sampling points in the reservoir representing inlet, internal and exit conditions, respectively (Figure S1). To better understand the spatial (sites) and temporal (months) variations of antibiotic concentrations, as well as their co-occurrence pattern with and impact on bacteria and cyanobacteria in surface water of this estuary reservoir during the warm season (from spring to summer) [20,21], we collected water samples from April to September 2014 at a depth of 0.5 m below the surface. All water samples were collected in triplicate, stored in sterile polypropylene bottles and processed immediately upon arrival at the laboratory. For physical and chemical detections (water quality indices, physiochemical parameters and antibiotics), we processed and measured two technical replicates for each of the biological replicates; while for high-throughput sequencing, triplicate samples for each site were combined by equal volume and processed for sequencing.

2.2. Physic-Chemical Parameters and Environmental Factors

Water quality and chemical parameters including water temperature, dissolved oxygen (DO) and pH were measured in situ by using a multi-parameter water quality analyzer (Multi3410, WTW Company, Weilheim, Germany). Other physic-chemical parameters include electrical conductivity (EC), turbidity, total phosphorus (TP), total nitrogen (TN), total carbon (TC), total organic carbon (TOC), inorganic carbon (IC) and ammonium nitrogen (NH$_4$$^+$-N) were analyzed according to standard methods for water and wastewater monitoring analysis [22]. The concentrations of chlorophyll-a (chl-a), which represented phytoplankton biomass, were measured using a Phytoplankton Analyzer (PHYTO-PAM) (Heinz Walz GMBH, Effeltrich, Germany) [23]. After filtering samples through 0.45 μm cellulose acetate membranes, the concentrations of K$^+$, Na$^+$, Ca^{2+}, Mg^{2+}, Al^{3+} and Si^{4+} ions were detected by inductively coupled plasma (ICP) spectroscopy [15]. The F$^-$, Cl$^-$ and SO$_4$$^{2-}$ ions contents were detected using Metrohm 830 Ion Chromatography (Metrohm AG, Herisau, Switzerland) [15].

2.3. Antibiotics Detection and Analysis

Based on the results of our preliminary experiment within QCS Reservoir in an earlier stage, we referred to related researches about antibiotics in the Yangtze estuary [17–19], and antibiotics including the fluoroquinolones (FQs) (such as ofloxacin (OFX), ciprofloxacin (CFX), norfloxacin (NFX)), Sulfonamides (SAs) (such as sulfamonomethoxine (SMM), sulfadiazine (SDZ), sulfaquinoxaline (SQX)), Tetracyclines (TCs) (such as tetracycline (TC), doxycycline (DC)), β-lactams (such as penicillin G (ETM), penicillin V (PEN V), cefalexin (LEX)), Macrolides (MLs) (such as erythromycin-H_2O (ETM), tylosin (TYL)) and other common antibiotics (such as Polymix-B (POL), Vancomycin (VAN), Lincomycin (LIN)) were selected as target compounds in our research. The antibiotic standards and internal standards (purities of all the chemicals are >98%) were purchased from Dr. Ehrenstorfer GmbH (Augsburg, Germany), and the detection methods of antibiotics were according to our colleague Yue Jiang's research, and the further information about preparation, extraction, sample detection and analysis can be found in the Supplementary Methods.

2.4. Genome DNA Extraction

For genomic Deoxyribonucleic acid (DNA) extraction, a total of 500 mL water sample was filtrated through 0.22 μm cellulose acetate filter membranes, and DNA was extracted directly from the same amount of membranes using a E.Z.N.A. ™ Water DNA Kit (OMEGA bio-tek, Houston, TX, USA) according to the manufacturer's instruction.

2.5. The 16S rRNA Gene Sequencing via (Polymerase Chain Reaction) PCR Amplification

To determine the diversity and variability of the bacterial community at different sites of the reservoir, we ran PCR for 18 water samples using a primer set targeting the V4 region of the 16S rRNA gene. This primer set (515F (5′-GTGCCAGCMGCCGCGGTAA-3′) and 806R (5′-GGACTACVSGGGTATCTAAT-3′)) exhibited fewer biases and more accurate taxonomic and phylogenetic information for individual bacterial taxa [24]. The details of PCR amplification procedures can be found in Supplementary Methods.

After purification, the PCR product were sent to Personal Biotechnology Co., Ltd. (Shanghai, China) for paired-end amplicon sequencing (2×150 bp) using Illumina MiSeq (Illumina, San Diego, CA, USA). Raw data were processed according to procedures as described previously [25], using the Quantitative Insights into Microbial Ecology (QIIME) pipeline (version 1.7.0, http://qiime.org/) for quality control (Supplementary Methods). The raw sequencing dataset is available for download from NCBI Sequence Read Achieve under BioProject PRJNA397386.

2.6. Statistical Analysis

The average value and standard deviation of each biotic and abiotic parameter were calculated using Microsoft Excel 2016 (Microsoft, Redmond, WA, USA). Before the multivariate statistical analysis, the relative abundance of bacterial OTUs (Operational Taxonomic Units) were square-root transformed to reduce the disturbance of highly abundant OTUs in analysis progress, while the environmental data sets (water quality indices, physiochemical parameters and antibiotics) were normalized using means and standard deviations of the variables.

The similarity matrices of biological and physicochemical characteristics in samples were constructed based on Bray−Curtis similarity and Euclidean distance, respectively. RELATE-BEST was used to evaluate the relationships between environmental factors and bacterial OTUs, and distance-based linear models (DistLM) were created to model and assess the contribution of each of the environmental variables and antibiotics to variations of microbial community composition by using PRIMER v6 and PERMANOVA+ (PRIMER-E Ltd., Plymouth, UK) [26]. Furthermore, the nonmetric multidimensional scaling (NMDS) was conducted to investigate the temporal and spatial variations of bacterial community composition. Alpha-diversity including Chao1, Ace, Chao/Ace, Simpson and Shannon indices between samples were calculated using the observed number of OTUs [27].

Highly abundant OTUs, which contributed to >20% reads for at least four samples and ≥1% reads in all the samples, were selected for further correlation analysis. It is noted that among these selected bacterial OTUs, no alternation was made to the original abundance within any sample. The degree of correlations between environmental variables, antibiotics and highly abundant OTUs across the whole sampling period were calculated using Pearson's correlation coefficient (r) and p-value with the rcor.test algorithm provided in the ltm package in R (version 3.4.2, https://www.r-project.org/) [28]. The correlations between OTU pairs were also determined similarly. During the analysis process, the false discovery rate was constant kept below 5% based on the Benjamini–Hochberg procedure [29]. Finally, significant correlations (positive/negative) were visualized using network generated by the Frucherman Reingold algorithm on the Gephi package (version 0.9.1, https://gephi.org/). Within the network, relevant topological and node/edge matrices including betweenness centrality, closeness centrality, degrees and hub values were also enumerated through the network analysis plug-in [30,31].

3. Results

3.1. Physic-Chemical Parameters and Environmental Factor in QCS Reservoir

During the sampling period, water temperature varied from 15.3 to 29.1 °C, which increased rapidly from April to July, and decreased gradually from July to September (Figure 1A). The pH value showed repeated down and uptrends, where highest level was detected at the internal point from May to July, and at the exit point for other periods (Figure 1B). The concentration of chl-a presented a rapid decrease from April to May, then ascended with increased water temperature. The internal point showed the highest chl-a most of the sampling time except April and August (Figure 1C). Similar to the pH value, chl-a kept at the lowest level at the inlet site across the whole period compared with other two sites. The turbidity at internal and exit were relatively stable, ranging from 6.83 to 13.7, but higher readings were observed from August to September (34.9 to 125) at the inlet (Figure 1D). The concentrations of NH^+_4-N, IC, DO and TN exhibited decreased trends with increased water temperature (Figure 1E–I). Among these, levels of NH^+_4-N, IC, DO were higher at internal and exit sites, especially, for DO at internal point from April to July. Nutrients (TN and TP) concentrations of the inflow reduced significantly after flowing through the reservoir, except for TP, which increased and maintained at a higher level from July to August at internal point (Figure 1H,I). The variations of other environmental factors can be seen in Supplementary Materials (Table S1).

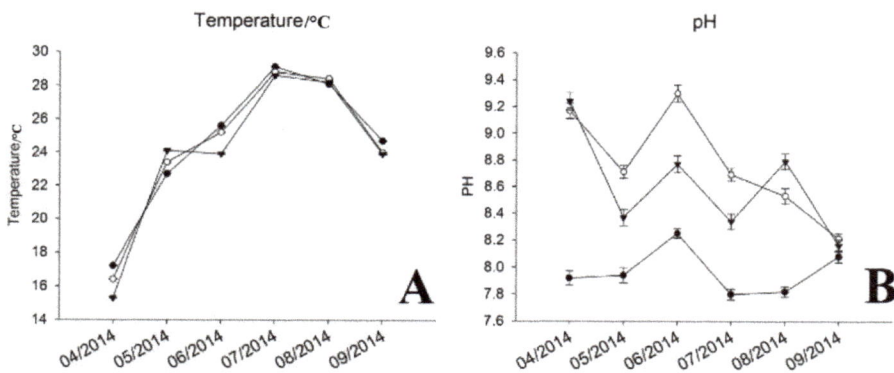

Figure 1. *Cont.*

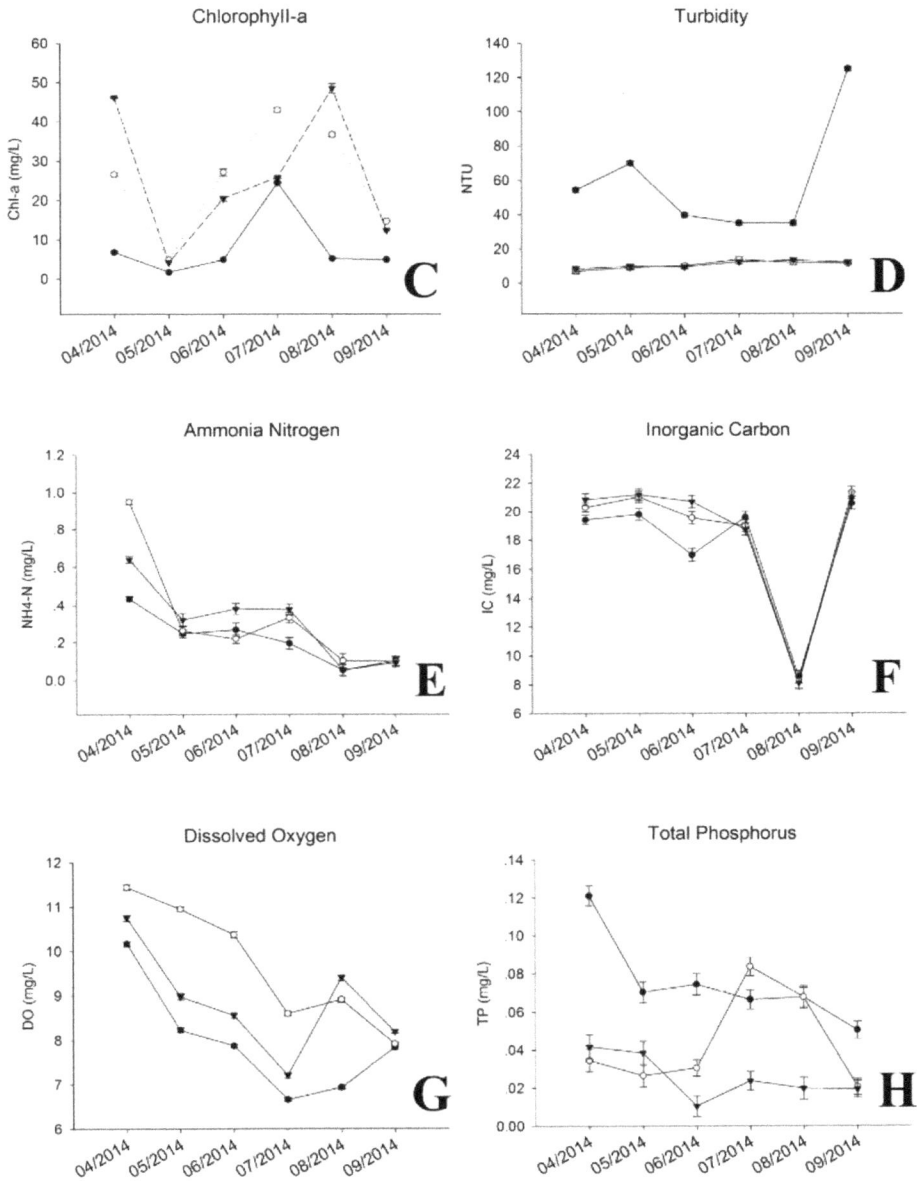

Figure 1. *Cont.*

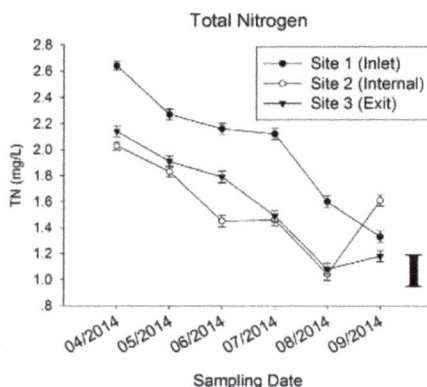

Figure 1. Water chemistry and environmental parameters. (**A**) Temperature; (**B**) pH; (**C**) Chlorophyll-a; (**D**) Turbidity; (**E**) Ammonia Nitrogen (NH_4-N^+); (**F**) Inorganic Carbon (IC); (**G**) Dissolved Oxygen (DO); (**H**) Total Phosphorus (TP); (**I**) Total Nitrogen (TN).

3.2. Antibiotics in QCS Reservoir

All 16 antibiotics including sulfonamides (SAs), fluoroquinolones (FQs), β-lactmas, macrolides (MLs), tetracyclines (TCs) and other common antibiotics were detected at three sampling sites throughout the sampling period (Figure S2). The concentrations of SAs exhibited an increasing trend from July to September at all three sites (Figure S2A–C), SDZ ranged from 0 to 48.90 ng/L, SMM ranged from 0 to 97.03 ng/L, and SQX ranged from 4.1 to 30.05 ng/L. The FQs revealed lower concentrations from May to July at all three sites (Figure S2D–F). Among these, NFX ranged from 48.01 to 193.84 ng/L, CFX ranged from 0 to 18.09 ng/L, and OFX ranged from 0 to 7.88 ng/L. In the group of β-lactmas, the concentration of LEX was very stable (average 14.43 ng/L) except at one time-point (July) (Figure S2G). In contrast, PEN G and PEN V revealed a large range of variation, which ranged from 0 to 133.50 ng/L and 21.7 to 157.76 ng/L, respectively (Figure S2H,I). The MLs including TYL and ETM exhibited lower concentrations across the whole sampling period, which ranged from 0 to 3.42 ng/L and 0 to 20.71 ng/L, respectively (Figure S2J,K). The concentrations of tetracyclines exhibited irregular variations in QCS Reservoir, TC and DC ranged from 4.11 to 26.69 ng/L and 20.67 to 170.97 ng/L, respectively (Figure S2L,M). The other common antibiotics such as POL ranged from 0 to 43.98 ng/L, VAN ranged from 0 to 9.50 ng/L, and LIN ranged from 0 to 9.30 ng/L (Figure S2N–P).

3.3. Dynamic Analysis of Bacterial Community Composition Based on the 16S rRNA Sequencing Data

Bacterial community composition analysis, as assessed by sequencing of V4 region of the 16S rRNA gene, identified a total of 5132 bacterial OTUs at genus level based on 97% similarity. The rarefaction curves indicated that the number of sequences were sufficient to cover the majority of species in the bacterial community within each sample (Figure S3). In QCS Reservoir, the dominant bacterial phyla identified were Proteobacteria (including α-, β- and γ-Proteobacteria, 31.3%), Actinobacteria (24.8%), Cyanobacteria (10.8%), Bacteroidetes (10.4%), Planctomycetes (8.2%), Verrucomicrobia (5.4%) and Chlorobi (2.2%) (Figure 2A). The other minor bacterial phyla (average abundance <2%) including Acidobacteria, Chloroflexi, Gemmatimonadetes, Firmicutes and Nitrospirae were also detected, in sum contributing less than six percent of total observed sequences. Classification at level of class showed that eight major non-cyanobacterial taxa including α-, β-, γ-Proteobacteria, Acidimicrobiia, Actinobacteria, Flavobacteriia, Sphingobacteriia and OPB56 contributed more than 60% of total sequences on average in each site (Figure 2B). Thereinto, α- and β-Peoteobacteria had higher abundance at inlet site (16.5% and 16.4%, respectively) than other two sites (11.2% and 9.5% at internal site, 10.9% and 8.9% at exit site). But the γ-Proteobacteria appeared to represent an opposite trend with slightly higher abundance at internal and exit sites (5.9% and 4.8%, respectively), compared to the inlet site (3.7%). Different from Proteobacteria, the actinobacterial abundance was relatively stable at all three sites during the sampling period (average abundance ~15%). In contrast, a waving trend of Acidimicrobiia abundance was observed at internal and exit sites (internal: 4.3% to 12%; exit: 4.4% to 13%). In addition, the abundance of Flavobacteriia was obviously higher from April to June at both internal and exit sites (10.6% and 5.6%, respectively) compared to the inlet site (1.7%). The Synechococcophycideae, as the dominant taxa within cyanobacterial populations, varied from 1% to 30.5% of total sequences at internal and exit sites from July to September.

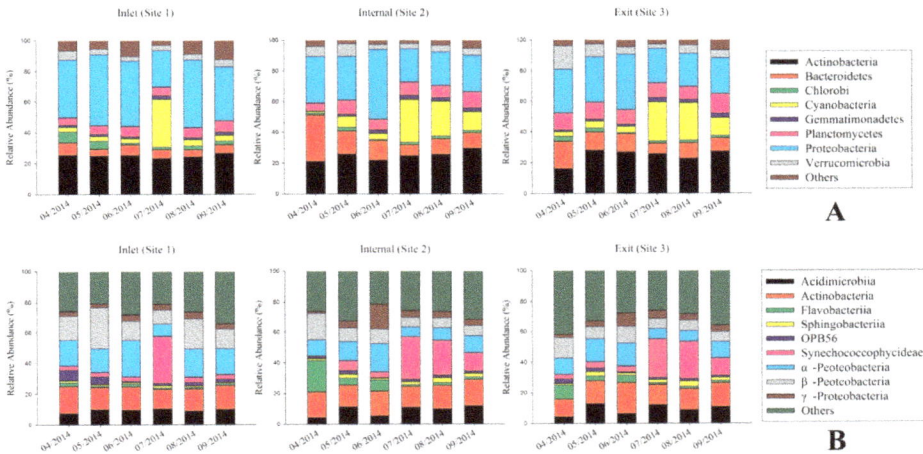

Figure 2. Relative abundance of 16S rRNA bacterial OTUs across the whole sampling period (**A**) Phylum level; (**B**) Class level.

3.4. Covariance Analysis of Bacterial Community Composition and Environmental Variables

A marginal test of biotic and abiotic factors based on distance-based linear modeling (DistLM) identified 14 environmental factors (TN, NH_4^+-N, pH, DO, EC, turbidity, temperature, chl-a, K^+, Na^+, Ca^{2+}, Mg^{2+}, Cl^- and F^-) and four antibiotics (PEN G, TYL and ETM) which were significantly correlated with the variations of bacterial community composition ($p < 0.05$) (Tables S2 and S3). In addition, DistLM (best procedure and AICc criterion) indicated that most environmental variables had no significant effect on the concentrations of antibiotics except for the turbidity ($p = 0.038$) (Table S4).

Permutational MANOVA (9999 permutations) was calculated based on a reduced model across all 18 samplings in different sampling sites with time series, both significant temporal and spatial effects ($p < 0.005$) of environmental variables, antibiotic concentrations and bacterial community compositions were observed respectively between these three sites within QCS Reservoir (Table S5).

The richness and diversity (represented by Chao1/Ace and inverse Simpson index, respectively) of the bacterial community in each sample was calculated based normalized OTU abundance (Figure 3). Our findings showed that the inlet site had higher richness and diversity indices, while the internal and exit sites shared similar diversity but different from those found in the inlet especially in April. However, the diversity indices significantly decreased at all three sites in July, coinciding with the period of high cyanobacterial abundance. In contrast, the richness indices of samples were relatively stable over the time with a declining trend observed from inlet to exit.

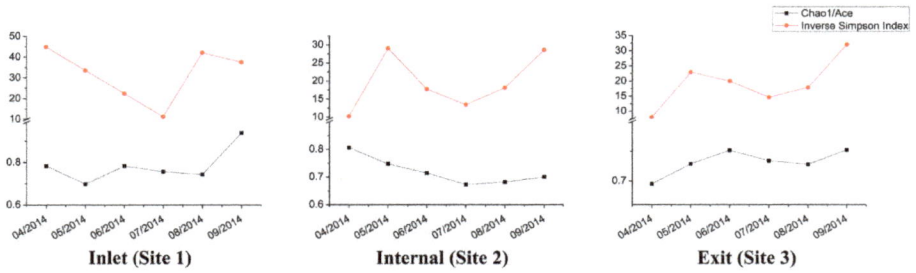

Figure 3. Bacterial OTU Richness (Chao/Ace, Black) and diversity (Inverse Simpson Index, Red).

The NMDS plots based on Bray Curtis similarity showed significant distinction of bacterial community composition between inlet and the other two sites (Figure 4A,B). Most environmental factors exhibited close correlations with bacterial community variation (Figure 4A). Most samples (except samples in April at internal and exit sites) were significantly associated with high water temperature and TOC concentration. Samples of internal and exit sites collected from July to September gathered closely together and correlated with chl-a. In contrast, samples of inlet site showed scattered distribution, and associated with high turbidity, TP and Al^{3+}. The distributions of antibiotics were different from the environmental factors, where PEN G and ETM were positively associated with the inlet samples while TYL exhibited positive correlation with internal and exit samples in April. Other antibiotics were not significantly correlated with changes of bacterial community (Figure 4B).

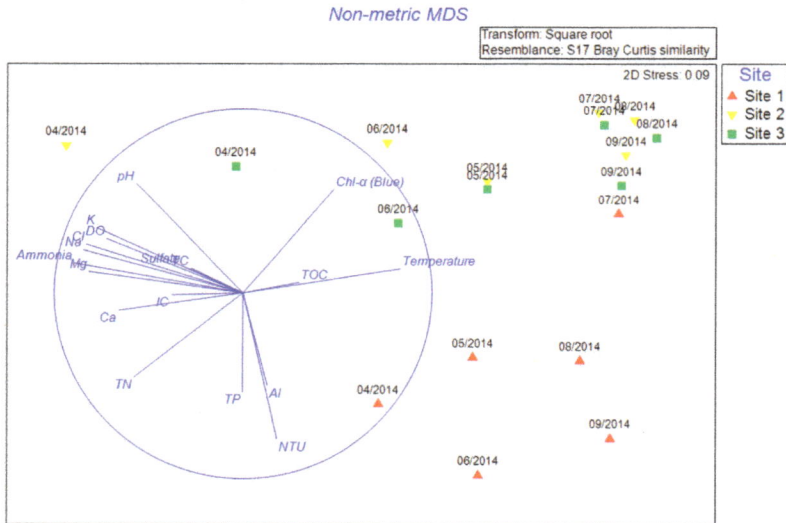

Figure 4. *Cont.*

Figure 4. The nonmetric multidimensional scaling (NMDS) reflecting the distribution of bacterial communities with environmental variables (**A**) and antibiotics (**B**) in estuary reservoir.

3.5. Multivariate Analysis of Biotic and Abiotic Factors in QCS Reservoir

A total of 99 variables, including 65 major OTUs (contributing more than one percent to any samples), 18 environmental variables and 16 antibiotics were shown in single interconnected network. A total of 4851 tested correlations were calculated by using rcor.test in the ltm package, with 474 ultimately considered significant (Table S6). Among them, 366 were positive correlations and 108 were negative. A visual correlation network was constructed with r score as the edge-weight, and values of betweenness centrality represented the size of nodes in the network (Figure 5A,B).

To illustrate the interactions among environmental variables, antibiotics and bacterial OTUs in QCS Reservoir, we explored the co-occurrence and co-exclusion patterns of these biotic/abiotic factors based on Pearson's correlation coefficient (*r*). Totally, 366 pairs of positive correlations were identified from 44 genera, 15 environmental variables and 9 antibiotics. Meanwhile, 108 pairs of negative correlations were identified from 27 genera, 15 environmental variables and 2 antibiotics (Figure 5A,B). By modularity analysis, all nodes were separated into different modules in each network. Within each module, the nodes were connected closely by co-occurrence or co-exclusion patterns. Between different modules, each module was linked with other modules through the key nodes, which exhibited high betweenness centrality (>20%) in the network.

In the co-occurrence pattern network (Figure 5A), all nodes were separated into six modules including modules I (28.09%), II (21.35%), III (21.35%), IV (12.36%), V (11.24%) and VI (5.62%). Module I to III accounted for almost 71% of total network, which included 48 bacterial OTUs from 23 genera, 3 antibiotics (TYL, PEN G and ETM) and 12 environmental variables (NH_4-N^+, temperature, DO, TN, TP, turbidity, Chl-a, Mg^{2+}, Na^+, K^+, Al^{3+} and Cl^-). Module VI was only composed of five antibiotics (including SQX, NFX, PEN V, LIN and POL) without bacterial OTUs and other environmental variables. In module I, TYL exhibited significant and positive correlations with five environmental factors (including Mg^{2+}, Na^+, K^+, Cl^- and NH_4-N^+) and four bacterial OTUs (*Comamonadaceae* (β-Proteobacteria), *calciphila* (Bacteroidetes), *Chitinophagaceae* (Bacteroidetes) and

Candidatus Xiphinematobacter (Verrucomicrobia)). In module II, PEN G showed positive correlations with two environmental variables (TP and turbidity), one antibiotic (ETM) and four bacterial OTUs (*Holophagaceae* (Acidobacteria), *curvus* (β-Proteobacteria), *OPB56* (Chlorobi) and *Nitrospira* (Nitrospirae)). Additionally, ETM revealed positive correlations with PEN G and two bacterial OTUs belonged to *curvus* (β-Proteobacteria) and *Rhodospirillaceae* (α-Peoteobacteria). Within minor modules, antibiotic (TC) exhibited positive correlations with [*Cerasicoccaceae*] (Verrucomicrobia) and *Fluviicola* (Bacteroidetes) in module V, and SQX in module VI positively correlated with *Cytophagaceae* (Bacteroidetes) in module III.

In the co-exclusion pattern network (Figure 5B), all nodes were also separated into six modules—I (37.04%), II (35.19%), III (16.67%), IV (3.7%), V (3.7%) and VI (3.7%). Module I to III accounted for almost 89% of the total network, which included 34 bacterial OTUs from 26 genera, 1 antibiotic (PEN G) and 13 environmental variables (including temperature, DO, pH, turbidity, TN, TP, NH_4-N^+, Chl-a, K^+, Na^+, Ca^{2+}, Mg^{2+} and Cl^-). In module III, PEN G exhibited significant negative correlations with pH and *Pirellulaceae* (Planctomycetes). In addition, the antibiotic LEX revealed significant negative correlations with *PHOS-HD29* (δ-Proteobacteria).

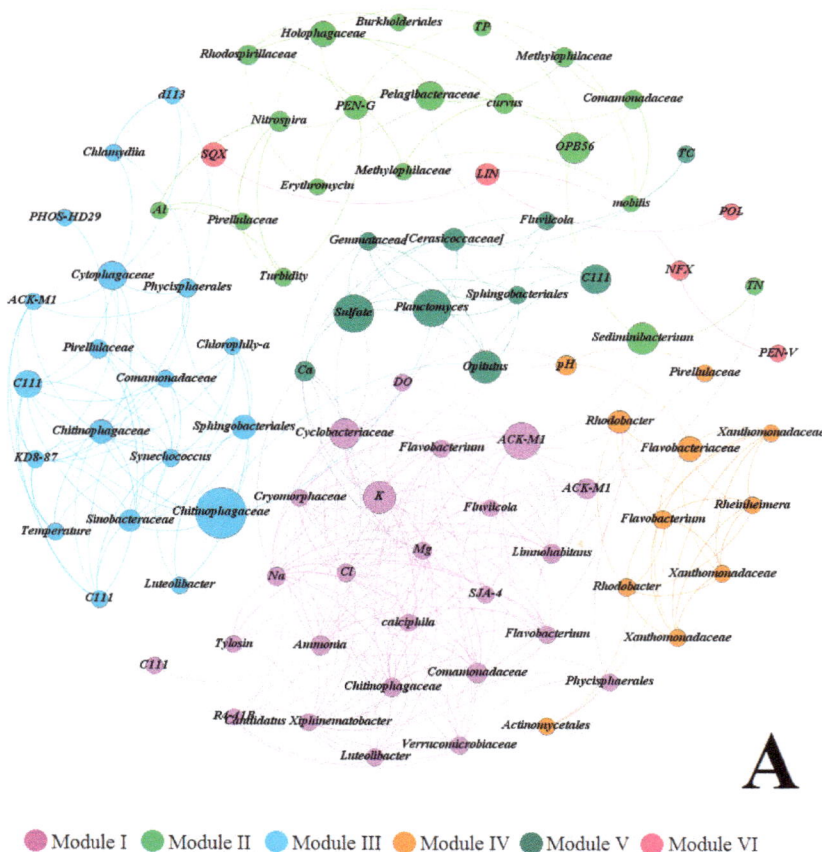

Figure 5. *Cont.*

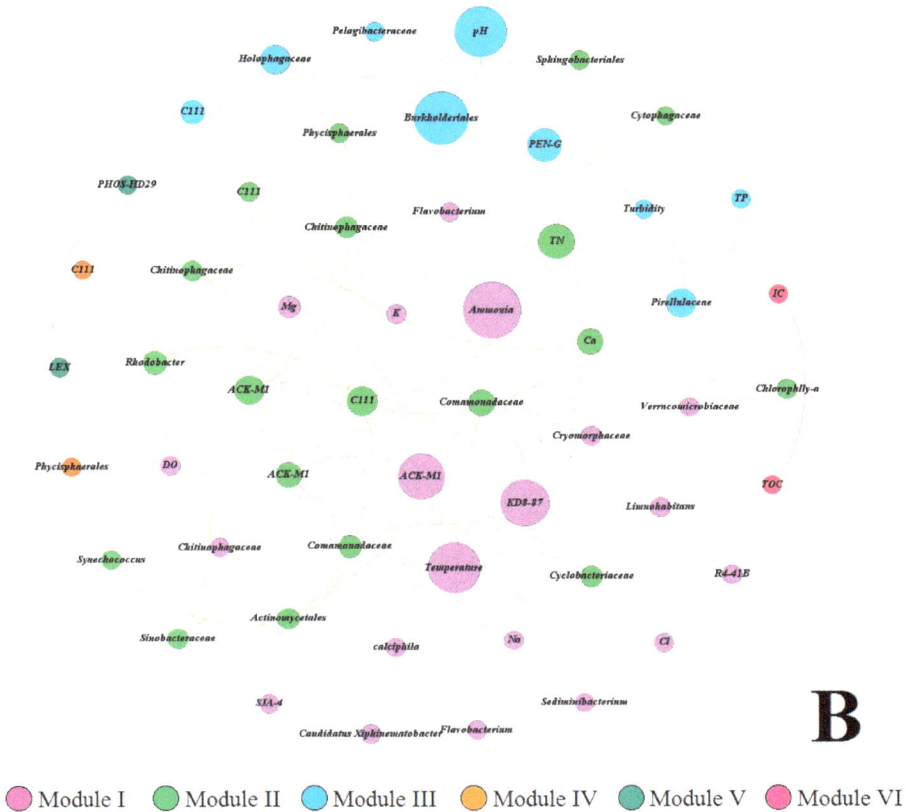

Figure 5. The network analysis showing the co-occurrence patterns between environmental variables, antibiotics and bacterial OTUs. A connection stands for a significant ($p < 0.05$) and strong positive (**A**) or negative correlation (**B**). The nodes were colored according to modularity class. The size of each node represents the value of betweenness centrality.

We further explored the co-occurrence and co-exclusion patterns of cyanobacterial and non-cyanobacterial OTUs within an organic correlation sub-network (Figure 6). Our results showed that temperature was the dominant environmental factor accelerating the proliferation of *Synechococcus* (dominant cyanobacterial taxa in QCS Reservoir). In addition, the concentration of chl-a, as well as several bacterial OTUs (such as *C111* (Actinobacteria), *Sinobacteraceae* (γ-Proteobacteria), *Comamonadaceae* (β-Proteobacteria), *Pirellulaceae* (Planctomycetes), *Luteolibacter* (Verrucomicrobia) and *KD8-87* (Gemmatimonadetes)) displayed significant and positive correlations with *Synechococcus*. In contrast, two bacterial OTUs belonged to *Actinomycetales* (Actinobacteria) and *Rhodobacter* (α-Peoteobacteria) revealed negative correlations with *Synechococcus*.

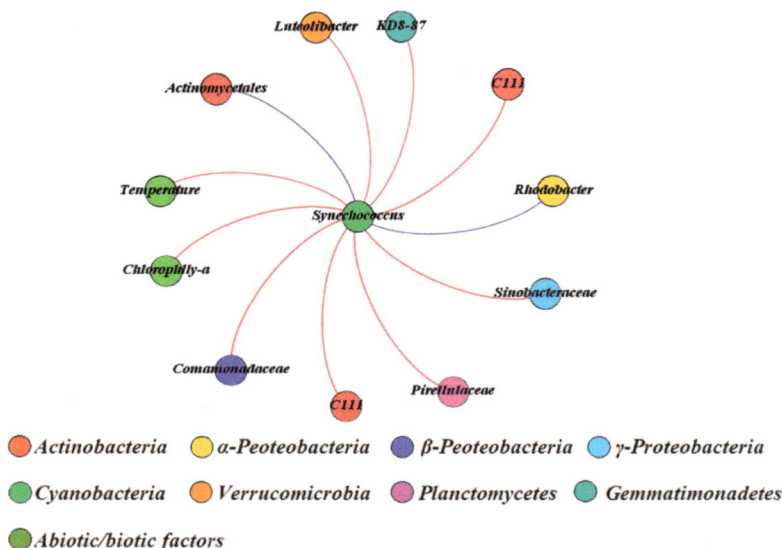

Figure 6. The organic correlation sub-network reflecting the pairwise correlations between *Synechococcus* and other bacterial OTUs, environmental factors and antibiotics. The nodes were colored according to different bacterial phylum. A red edge represents a significant positive correlation. A blue edge represents a significant negative correlation.

4. Discussion

Antibiotics as human and veterinary medicines are widely used in the prevention and treatment of diseases, and also as additives in livestock and breeding fields [17,18,32]. Research has indicated that the annual global consumption of antibiotics is about 150,000 tons, and almost 17% of the consumption is used in China [33]. With increasing use of antibiotics as medicines and animal growth promoters, plenty of these compounds have been released into aquatic environments, and pose direct/indirect threats on aquatic ecosystems [34]. In this study, we provided a systematic view on the relationships between environmental factors, antibiotics and bacterial community in an estuary aquatic ecosystem, and further explored the influences of antibiotics on co-occurrence/exclusion patterns of bacterial community.

4.1. Correlations between Environmental Parameters and Antibiotic Concentrations

In the present study, all sixteen antibiotics were detected at inlet, internal and exit of the QCS Reservoir from April to September (Figure S2). The distributions of these antibiotics varied temporally and spatially (Table S5). Apart from the LEX and TYL, all antibiotics exhibited increasing trend of different degrees from July to September (Figure S2), indicating a seasonal pattern of these compounds in the upstream area strongly polluted by agricultural and industrial activities and dense population along the Yangtze River [17]. We speculated that the inflow from the Yangtze River and WWTPs effluents were main sources of these antibiotics. This is supported by related research showing that the concentrations of antibiotics were mainly effected by upstream runoff and precipitation of Yangtze estuary [17,18,35].

Within these antibiotics, PEN G and TYL showed relative strong correlations with a variety of environmental variables (Figure 5A,B). This revealed that a huge concentration variation of PEN G and TYL were expected under different environmental conditions. Research also indicated that the process of water body self-purification further promoted the sedimentation effects of some antibiotics from water phase inside the reservoir [36], which is contributing to the decline of antibiotic concentrations in the surface water of the reservoir. Several antibiotics including SQX, NFX, PEN V, LIN and POL exhibited co-occurrence pattern and did not correlated to environmental factors (Figure 5A,B), which implied that these compounds might have similar changing patterns but were less affected by water quality and trophic status. Interestingly, although earlier research has indicated that the TOC have positively correlated with FQs concentrations in the sediment of the Yangtze estuary [18], we did not find any significant correlation between TOC and FQs concentrations in the surface water of the reservoir (Table S6) in our study. We speculate that the contradictory results could be attributed to the different metabolic transformation mechanisms of FQs between the surface water and sediment phases.

4.2. Correlations between Environmental Parameters and Bacterial Community Composition

In this study, high-throughput sequencing (HTS) technology was used to evaluate the microbial community diversity and composition in the surface water spanning from spring to summer at different sites of the reservoir. Combining information on the changes of water quality and physicochemical parameters, we further explored the co-occurrence/exclusion patterns of the bacterial community with biotic/abiotic factors, also including the influence on co-occurrence patterns between cyanobacterial and non-cyanobacterial taxa during summer.

Significant spatial and temporal effects of bacterial community composition were found in QCS Reservoir (Figure 2A,B). This is mainly because the inlet site was so close to the Yangtze River, serving as the source water into the reservoir. That means the aquatic ecological environment at inlet site was very similar with conditions in the Yangtze River characterized for higher concentrations of nutrients (N, P) and turbidity, which was affected by seriously non-point pollution and soil erosion in upstream [37–39]. Therefore, to some extent, the microbial community composition at inlet site was very similar to the structure of microorganisms in the Yangtze estuary. By contrast, the internal and exit sites are located in the midstream and downstream of the reservoir, respectively, with characteristics of obvious lower concentrations of TN, TP and turbidity (Figure 1). Besides, the lower water flow velocity and longer retention times might facilitate the water purification and further increased the transparency of water column in these areas [15], which might partly decrease the abundance of particle-associated bacteria (such as lineages of β-Proteobacteria) in surface water at these sites. Our study also indicated that both species richness and diversity indices obviously decreased from inlet to exit site (Figure 3). Although the water quality was improved obviously at internal and exit sites, the aquatic environment at these sites provided suitable conditions for the proliferation of cyanobacterial, and resulted in rapid increasing in abundance of *Synechococcus* at these sites in summer season (July to September) (Figure 2A,B). Results of sub-network further indicated that water temperature was the major environmental driving factor accelerated the *Synechococcus* growth, and chl-a concentration also significantly increased during this period (Figure 6). In addition, several heterotrophic bacterial OTUs belonging to Actinobacteria, Gemmatimonadetes, Planctomycetes, Verrucomicrobia, β- and γ-Proteobacteria exhibited the co-occurrence patterns with *Synechococcus*, which indicated that mutualism mechanisms might exist between these bacterial taxa in aspect of carbon source utilization [40–44]. Further research is underway to confirm whether these related bacterial taxa can be bio-indicators to predict for the proliferation of *Synechococcus*.

The modularity analysis on correlation network indicated that the individual bacterial taxon did not exist independently in aquatic ecosystem, but closely correlated with other bacterial taxa and responded to the changes of surrounding environment together, which was further represented in co-occurrence/exclusion patterns (Figure 5A,B). The collective behavior of bacterial community is called "quorum-sensing (QS)" [45–47], which can strengthen the cooperations between different bacterial taxa, and also the adaptability of bacterial community to environment changes [48]. In our research, distinct co-occurrence/exclusion patterns between bacterial community were represented by different modules within networks. The different environmental variables closely correlated with co-occurrence/exclusion patterns were also included in each module. Within each module, the bacterial taxa widely connected with others, but only few bacterial taxa with high betweeness centrality connected other bacteria in different modules (Figure 5A,B). This implied that the bacterial taxa with high betweeness centrality as bridges between different modules played critical roles in the whole network [31]. In module I of the co-occurrence network, the dominant bacterial OTUs mainly classified as Bacteroidetes and Verrucomicrobia were co-occurred with environmental variables including NH_4-N^+, DO, Mg^{2+}, Na^+, K^+ and Cl^-, which indicated that these bacterial taxa mainly existed in samples at internal and exit sites in April (Figure 4A). While in module II, major environmental variables including TN, TP, turbidity and Al^{3+}, were co-occurred with high abundant β-Proteobacterial OTUs, which illustrated that the β-Proteobacterial taxa were mainly existed in samples from inlet site. In addition, temperature, chl-a and dominant cynaobacterial OTUs represented the composing characteristics of module III, which also implied bacterial OTUs in this module were mainly distributed at internal and exit sites in summer. Moreover, the other modules also indicated the characteristics of bacterial community under different environmental driving factors. Between different modules, *ACK-M1* (Actinobacteria) in module I, *Sediminibacterium* (Bacteroidetes) in module II and *Chitinophagaceae* (Bacteroidetes) in module III as bridges linked with other modules, also exhibited high betweeness centrality. This implied the important roles of these key bacterial taxa in aquatic ecosystems, although their ecological functions were still unknown. Additionally, no connections were found between module I, II and III, but all these three modules were connected with module V, which indicated the bacterial community in module V have stronger functional heterogeneities compared with other modules. On the other hand, the co-exclusion network showed the co-exclusion patterns between environmental variables and bacterial OTUs (Figure 5B). The results indicated that the co-exclusion patterns of environmental variables and bacterial OTUs have distinct spatial (inlet and the other two sites) and temporal (spring and summer) effects. In addition, the environmental factors including temperature, NH_4-N^+ and pH, also bacterial OTUs including *ACK-M1* (Actinobacteria), *KD8-87* (Gemmatimonadetes) and *Burkholderiales* (β-Proteobacteria) as bridges linked with other modules, which implied the potential important roles of these biotic/abiotic factors in the network.

4.3. Correlations between Antibiotic Concentrations and Bacterial Community Composition

Most of the former correlative researches have been mainly focused on the conditions of relative higher antibiotic concentrations, and how these compounds affected the bacterial community in laboratory experiments, which strongly limited our understanding about the potential ecological impact of these antibiotics on actual aquatic ecosystems [49,50]. Hence in our study, high-throughput sequencing (HTS) combined with correlation network analysis to explore the co-occurrence/exclusion patterns and effects between antibiotics and the bacterial community in the actual estuary ecosystem, as well as the influences of these antibiotics on co-occurrence pattern of cyanobacterial and non-cyanobacterial taxa.

Total of 11 antibiotics from all 16 antibiotics, were found in the co-occurrence/exclusion pattern networks (Figure 5A,B). Among these antibiotics, SQX (module VI) exhibited a positive correlation with bacterial OTU belonging to *Cytophagaceae* (Bacteroidetes) in a different module (module III) (Figure 5A). Although the strains of *Cytophagaceae* revealed sensitivity to kinds of antibiotics [51,52], we still unknown the association between *Cytophagaceae* and SQX. In addition, three dominant antibiotics including TYL, PEN G and ETM were found to have significant effects ($p < 0.05$) on variations of bacterial community composition in QCS Reservoir (Table S3). Also, these three antibiotics distributed within different modules (including module I and II of co-occurrence patterns and module III of co-exclusion pattern) of each network (Figure 5A,B). The modular classifications of these antibiotics within the co-occurrence pattern network further reflected the different distribution characteristics between these compounds in QCS Reservoir (Figure 4B). Among these, TYL was closely associated with samples in April at internal and exit sites, while ETM and PEN G were more relevant with samples at inlet site.

Each antibiotic (TYL, ETM and PEN G) exhibited extensive co-occurrence patterns with environmental variables and bacterial OTUs within modules, and we further found that these co-occurred bacterial OTUs all belonged to gram-negative bacteria (including Acidobacteria, Bacteroidetes, Nitrospirae, Verrucomicrobia, α- and β-Proteobacteria) in QCS Reservoir. Although previous research showed that some gram-negative bacteria were effective to resist and biodegrade kinds of antibiotic including macrolides and β-Lactams in a variety of ways [53], the specific interaction mechanisms between these bacterial taxa and antibiotics were still not clear. It is noted that the PEN G exhibited a co-exclusion pattern with bacterial OTUs from Planctomycetes in our research (Figure 5B). Related research has indicated that the β-lactams mainly restrained bacterial growth by impeding the synthesis of peptidoglycan in the cell walls [54]. However, the cell structure of Planctomycetes was characterized by the absence of peptidoglycan in their cell walls [55]. These results implied that PEN G may have other potential inhibition mechanisms on bacteria such as Planctomycetes. Different to widely co-occurrence patterns of these antibiotics (TYL, ETM and PEN G) within modules, no connections of these compounds with other biotic/abiotic factors were found across different modules in the network. This indicated that the variations of these antibiotics might directly affect the bacterial taxa within co-occurrence patterns first, and then further influence the whole bacterial community structure through the changes of these co-occurred bacterial taxa in aquatic ecosystem.

In a word, no significant inhibiting effects of most detected antibiotics were found on bacterial community composition in this estuary reservoir, also no influences of antibiotics were found on the co-occurrence pattern between cyanobacteria and non-cyanobacterial taxa, which were mainly because of the lower concentrations of these antibiotics. However, there might exist a potential risk of genetic transfer of resistance to related bacteria induced by exposure to long-term exposure to low levels of antibiotics in the environment. Especially, TYL, ETM and PEN G exhibited co-occurrence patterns with multiple gram-negative bacterial taxa in the network, which indicated that these bacterial taxa played important roles during the migrating and transforming course of these antibiotics in aquatic ecosystem.

5. Conclusions

In this study, we evaluated the relationships between environmental factors, antibiotics and bacterial community composition in surface water of a large reservoir located in Yangtze estuary from spring to summer (April to September). Both significant spatial and temporal effects were found in bacterial community composition between inlet and the other two sites of the reservoir. The environmental factors showed significant influence on bacterial community composition, while having little effect on distributions of most antibiotics. No significant inhibitory effects of most antibiotics on bacterial community were found in our research. Among these antibiotics, PEN G, TYL and ETM closely correlated with variations of bacterial community composition, and exhibited co-occurrence patterns with some gram-negative bacterial taxa, which implied important functions of these bacterial taxa in the course of migration and transformation of antibiotics. Further study is required to explore the interaction mechanisms between these antibiotics and bacterial taxa. In addition, the antibiotics in low concentrations revealed no significant effect on the co-occurrence pattern between cyanobacteria and non-cyanobacterial taxa. Based on above results, continued research is necessary to evaluate the potential risk of genetic transfer of resistance to related bacteria induced by long-term exposure to low levels of antibiotics in the environment.

Supplementary Materials: The following are available online at http://www.mdpi.com/2073-4441/10/2/154/s1, Figure S1: Aerial schematic of Qingcaosha (QCS) Reservoir and annotated sampling locations (Site 1 (Inlet), Site 2 (Internal), Site 3 (Exit)); Figure S2: Concentrations of antibiotics at the three sites. (A) SDZ, (B) SMM, (C) SQX, (D) NFX, (E) CFX, (F) OFX, (G) LEX, (H) PEN G, (I) PEN V, (J) TYL, (K) ETM, (L) TC, (M) DC, (N) POL, (O) VAN, (P) LIN; Figure S3: Rarefaction curves of sequences in each sample; Table S1: Supplemental water chemistry and environmental parameters; Table S2: DistLM results of abundant bacterial community data against environmental variables (999 permutations); Table S3: DistLM results of abundant bacterial community data against antibiotics (999 permutations); Table S4: DistLM results of environmental variables against antibiotics (999 permutations); Table S5: Permutational MANOVA results of environmental variables, antibiotic concentrations and bacterial community compositions (9999 permutations); Table S6: Multiple linear correlations by rcor.test in ltm package (R language).

Acknowledgments: This research grant is supported by the joint program between Shanghai Jiaotong University (SJTU) and National University of Singapore (NUS), and we are also grateful to the Campus for Research Excellence And Technological Enterprise (CREATE) programme under the joint program Energy and Environmental Sustainability Solutions for Megacities (E2S2) between Shanghai Jiaotong University (SJTU) and the National University of Singapore (NUS), also Singapore National Research Foundation (NRF) under its Environmental & Water Technologies Strategic Research Programme and administered by PUB, the Singapore's National Water Agency (Grant number: 1102-IRIS-14-02). All sources of funding of the study should be disclosed. Please clearly indicate grants that you have received in support of your research work. Clearly state if you received funds for covering the costs to publish in open access.

Author Contributions: Zheng Xu conceived and designed the experiments, and mainly performed the HTS experiment, also analyzed the sequencing data and wrote the paper. Yue Jiang mainly performed the antibiotics detection experiment; Shu Harn Te helped polish the language. Yiliang He and Karina Yew-Hoong Gin reviewed and edited the manuscript.

Conflicts of Interest: We declare that we have no financial and personal relationships with other people or organizations that can inappropriately influence our work, there is no professional or other personal interest of any nature or kind in any product, service and/or company that could be construed as influencing the position presented in, or the review of, the manuscript entitled.

References

1. Yong, H.J.; Yang, J.S.; Park, K. Changes in Water Quality After the Construction of an Estuary Dam in the Geum River Estuary Dam System, Korea. *J. Coast. Res.* **2014**, *30*, 1278–1286.
2. Chen, W.; Chen, K.; Kuang, C.; Zhu, D.Z.; He, L.; Mao, X.; Liang, H.; Song, H. Influence of sea level rise on saline water intrusion in the Yangtze River Estuary, China. *Appl. Ocean Res.* **2016**, *54*, 12–25. [CrossRef]
3. Cloern, J.E.; Abreu, P.C.; Carstensen, J.; Chauvaud, L.; Elmgren, R.; Grall, J.; Greening, H.; Johansson, J.O.; Kahru, M.; Sherwood, E.T. Human Activities and Climate Variability Drive Fast-Paced Change across the World's Estuarine-Coastal Ecosystems. *Glob. Chang. Biol.* **2015**, *22*, 513–529. [CrossRef] [PubMed]
4. Bound, J.P.; Voulvoulis, N. Pharmaceuticals in the aquatic environment—A comparison of risk assessment strategies. *Chemosphere* **2004**, *56*, 1143–1155. [CrossRef] [PubMed]
5. Kümmerer, K. Antibiotics in the aquatic environment—A review—Part II. *Chemosphere* **2009**, *75*, 435–441. [CrossRef] [PubMed]
6. Rizzo, L.; Manaia, C.; Merlin, C.; Schwartz, T.; Dagot, C.; Ploy, M.C.; Michael, I.; Fatta-Kassinos, D. Urban wastewater treatment plants as hotspots for the release of antibiotics in the environment: A review. *Water Res.* **2013**, *47*, 957–995.
7. Manzetti, S.; Ghisi, R. The environmental release and fate of antibiotics. *Mar. Pollut. Bull.* **2014**, *79*, 7–15. [CrossRef] [PubMed]
8. Crump, B.C.; Baross, J.A.; Simenstad, C.A. Dominance of particle-attached bacteria in the Columbia River estuary, USA. *Aquat. Microb. Ecol.* **1998**, *14*, 7–18. [CrossRef]
9. Rieck, A.; Herlemann, D.P.R.; Jürgens, K.; Grossart, H.P. Particle-Associated Differ from Free-Living Bacteria in Surface Waters of the Baltic Sea. *Front. Microbiol.* **2015**, *6*. [CrossRef] [PubMed]
10. Yung, C.M.; Ward, C.S.; Davis, K.M.; Johnson, Z.I.; Hunt, D.E. Insensitivity of Diverse and Temporally Variable Particle-Associated Microbial Communities to Bulk Seawater Environmental Parameters. *Appl. Environ. Microbiol.* **2016**, *82*, 3431–3437. [CrossRef] [PubMed]
11. Li, W.; Shi, Y.; Gao, L.; Liu, J.; Cai, Y. Occurrence of antibiotics in water, sediments, aquatic plants, and animals from Baiyangdian Lake in North China. *Chemosphere* **2012**, *89*, 1307–1315. [CrossRef] [PubMed]
12. Liu, J.; Lu, G.; Wang, Y.; Yan, Z.; Yang, X.; Ding, J.; Jiang, Z. Bioconcentration, metabolism, and biomarker responses in freshwater fish Carassius auratus exposed to roxithromycin. *Chemosphere* **2014**, *99*, 102–108. [CrossRef] [PubMed]
13. Huang, D.J.; Hou, J.H.; Kuo, T.F.; Lai, H.T. Toxicity of the veterinary sulfonamide antibiotic sulfamonomethoxine to five aquatic organisms. *Environ. Toxicol. Pharmacol.* **2014**, *38*, 874–880. [CrossRef] [PubMed]
14. Jin, X.; He, Y.; Kirumba, G.; Hassan, Y.; Li, J. Phosphorus fractions and phosphate sorption-release characteristics of the sediment in the Yangtze River estuary reservoir. *Ecol. Eng.* **2013**, *55*, 62–66. [CrossRef]

15. Jin, X.; He, Y.; Zhang, B.; Hassan, Y.; George, K. Impact of sulfate and chloride on sediment phosphorus release in the Yangtze Estuary Reservoir, China. *Water Sci. Technol.* **2013**, *67*, 1748–1756. [CrossRef] [PubMed]

16. Ou, H.S.; Wei, C.H.; Deng, Y.; Gao, N.Y. Principal component analysis to assess the composition and fate of impurities in a large river-embedded reservoir: Qingcaosha Reservoir. *Environ. Sci. Process. Impacts* **2013**, *15*, 1613–1621. [CrossRef] [PubMed]

17. Yan, C.; Yang, Y.; Zhou, J.; Liu, M.; Nie, M.; Shi, H.; Gu, L. Antibiotics in the surface water of the Yangtze Estuary: Occurrence, distribution and risk assessment. *Environ. Pollut.* **2013**, *175*, 22–29. [CrossRef] [PubMed]

18. Shi, H.; Yang, Y.; Liu, M.; Yan, C.; Yue, H.; Zhou, J. Occurrence and distribution of antibiotics in the surface sediments of the Yangtze Estuary and nearby coastal areas. *Mar. Pollut. Bull.* **2014**, *83*, 317–323. [CrossRef] [PubMed]

19. Chen, K.; Zhou, J.L. Occurrence and behavior of antibiotics in water and sediments from the Huangpu River, Shanghai, China. *Chemosphere* **2014**, *95*, 604–612. [CrossRef] [PubMed]

20. Huang, Z.; Xie, B.; Yuan, Q.; Xu, W.; Lu, J. Microbial community study in newly established Qingcaosha Reservoir of Shanghai, China. *Appl. Microbiol. Biotechnol.* **2014**, *98*, 9849–9858. [CrossRef] [PubMed]

21. Sun, Z.; Li, G.; Wang, C.; Jing, Y.; Zhu, Y.; Zhang, S.; Liu, Y. Community dynamics of prokaryotic and eukaryotic microbes in an estuary reservoir. *Sci. Rep.* **2014**, *4*, 6966. [CrossRef] [PubMed]

22. Wei, F. *Monitoring and Analysis Methods of Water and Wastewater*; China Environmental Science Press: Beijing, China, 2002.

23. Gera, A.; Alcoverro, T.; Mascaró, O.; Pérez, M.; Romero, J. Exploring the utility of Posidonia oceanica chlorophyll fluorescence as an indicator of water quality within the European Water Framework Directive. *Environ. Monit. Assess.* **2012**, *184*, 3675–3686. [CrossRef] [PubMed]

24. Caporaso, J.G.; Lauber, C.L.; Walters, W.A.; Berglyons, D.; Huntley, J.; Fierer, N.; Owens, S.M.; Betley, J.; Fraser, L.; Bauer, M. Ultra-high-throughput microbial community analysis on the Illumina HiSeq and MiSeq platforms. *ISME J.* **2012**, *6*, 1621–1624. [CrossRef] [PubMed]

25. Caporaso, J.G.; Kuczynski, J.; Stombaugh, J.; Bittinger, K.; Bushman, F.D.; Costello, E.K.; Fierer, N.; Peña, A.G.; Goodrich, J.K.; Gordon, J.I. QIIME allows analysis of high-throughput community sequencing data. *Nat. Methods* **2010**, *7*, 335–336. [CrossRef] [PubMed]

26. Shu, H.T.; Tan, B.F.; Thompson, J.R.; Gin, Y.H. Relationship of microbiota and cyanobacterial secondary metabolites in Planktothricoides-dominated bloom. *Environ. Sci. Technol.* **2017**, *51*, 4199–4209.

27. Ng, C.; Le, T.H.; Goh, S.G.; Liang, L.; Kim, Y.; Rose, J.B.; Yewhoong, K.G. A Comparison of Microbial Water Quality and Diversity for Ballast and Tropical Harbor Waters. *PLoS ONE* **2016**, *11*, e0154652. [CrossRef] [PubMed]

28. Woodhouse, J.N.; Kinsela, A.S.; Collins, R.N.; Bowling, L.C.; Honeyman, G.L.; Holliday, J.K.; Neilan, B.A. Microbial communities reflect temporal changes in cyanobacterial composition in a shallow ephemeral freshwater lake. *ISME J.* **2016**, *10*, 1337–1351. [CrossRef] [PubMed]

29. Benjamini, Y.; Hochberg, Y. Controlling The False Discovery Rate—A Practical And Powerful Approach To Multiple Testing. *J. R. Stat. Soc. B* **1995**, *57*, 289–300.

30. Bastian, M.; Heymann, S.; Jacomy, M. Gephi: An Open Source Software for Exploring and Manipulating Networks. In Proceedings of the Third International Aaai Conference on Weblogs and Social Media, San Jose, CA, USA, 17–20 May 2009; Available online: https://gephi.org/publications/gephi-bastian-feb09.pdf (accessed on 23 November 2017).

31. Yu, H.; Pm, K.; Sprecher, E.; Trifonov, V.; Gerstein, M. The Importance of Bottlenecks in Protein Networks: Correlation with Gene Essentiality and Expression Dynamics. *PLoS Comput. Biol.* **2007**, *3*, e59. [CrossRef] [PubMed]

32. Gothwal, R.; Shashidhar, T. Antibiotic Pollution in the Environment: A Review. *CLEAN Soil Air Water* **2015**, *43*, 479–489. [CrossRef]

33. Xu, W.H.; Zhang, G.; Zou, S.C.; Li, X.D.; Liu, Y.C. Determination of selected antibiotics in the Victoria Harbour and the Pearl River, South China using high-performance liquid chromatography-electrospray ionization tandem mass spectrometry. *Environ. Pollut.* **2007**, *145*, 672–679. [CrossRef] [PubMed]

34. Waksman, S.A. *Microbial Antagonisms and Antibiotic Substances*; The Commonwealth Fund: New York, NY, USA, 1947; pp. 1033–1034.

35. Bao, L.J.; Maruya, K.A.; Snyder, S.A.; Zeng, E.Y. China's water pollution by persistent organic pollutants. *Environ. Pollut.* **2012**, *163*, 100–108. [CrossRef] [PubMed]

36. Kümmerer, K. Antibiotics in the aquatic environment—A review—Part I. *Chemosphere* **2009**, *75*, 417–434. [CrossRef] [PubMed]

37. Zhang, L.; Liang, D.; Ren, L.; Shi, S.; Li, Z.; Zhang, T.; Huang, Y. Concentration and source identification of polycyclic aromatic hydrocarbons and phthalic acid esters in the surface water of the Yangtze River Delta, China. *J. Environ. Sci.* **2012**, *24*, 335–342. [CrossRef]

38. Floehr, T.; Xiao, H.; Scholz-Starke, B.; Ottermanns, R.; Ross-Nickoll, M. Solution by dilution?—A review on the pollution status of the Yangtze River. *Environ. Sci. Pollut. Res. Int.* **2013**, *20*, 6934–6971. [CrossRef] [PubMed]

39. Huang, P.B.; Jiao, N.Z.; Feng, J.; Shu, Q.L. Research progress on Planctomycetes' diversity and ecological function in marine environments. *Microbiol. China* **2014**, *41*, 1891–1902.

40. Eiler, A.; Olsson, J.A.; Bertilsson, S. Diurnal variations in the auto- and heterotrophic activity of cyanobacterial phycospheres (*Gloeotrichia echinulata*) and the identity of attached bacteria. *Freshw. Biol.* **2010**, *51*, 298–311. [CrossRef]

41. Li, J.; Zhang, J.; Liu, L.; Fan, Y.; Li, L.; Yang, Y.; Lu, Z.; Zhang, X. Annual periodicity in planktonic bacterial and archaeal community composition of eutrophic Lake Taihu. *Sci. Rep.* **2015**, *5*, 15488. [CrossRef] [PubMed]

42. Buck, U.; Grossart, H.P.; Amann, R.; Pernthaler, J. Substrate incorporation patterns of bacterioplankton populations in stratified and mixed waters of a humic lake. *Environ. Microbiol.* **2010**, *11*, 1854–1865. [CrossRef] [PubMed]

43. Jones, S.E.; Newton, R.J.; Mcmahon, K.D. Evidence for structuring of bacterial community composition by organic carbon source in temperate lakes. *Environ. Microbiol.* **2010**, *11*, 2463–2472. [CrossRef] [PubMed]

44. Agostini, V.O.; Macedo, A.J.; Muxagata, E. Evaluation of antibiotics as a methodological procedure to inhibit free-living and biofilm bacteria in marine zooplankton culture. *Anais da Academia Brasileira de Ciencias* **2016**, *88*, 733–746. [CrossRef] [PubMed]

45. Cornforth, D.M.; Popat, R.; Mcnally, L.; Gurney, J.; Scottphillips, T.C.; Ivens, A.; Diggle, S.P.; Brown, S.P. Combinatorial quorum sensing allows bacteria to resolve their social and physical environment. *Proc. Natl. Acad. Sci. USA* **2014**, *111*, 4280–4284. [CrossRef] [PubMed]

46. Zhang, W.; Li, C. Exploiting Quorum Sensing Interfering Strategies in Gram-Negative Bacteria for the Enhancement of Environmental Applications. *Front. Microbiol.* **2016**, *6*, 1535. [CrossRef] [PubMed]

47. Ayaz, E.; Gothalwal, R. Effect of environmental factors on bacterial quorum sensing. *Cell. Mol. Biol.* **2014**, *60*, 46–60. [PubMed]

48. Kievit, T.R.D.; Iglewski, B.H. Bacterial Quorum Sensing in Pathogenic Relationships. *Infect. Immun.* **2000**, *68*, 4839–4849. [CrossRef] [PubMed]

49. Mcknight, U.S.; Rasmussen, J.J.; Kronvang, B.; Binning, P.J.; Bjerg, P.L. Sources, occurrence and predicted aquatic impact of legacy and contemporary pesticides in streams. *Environ. Pollut.* **2015**, *200*, 64–76. [CrossRef] [PubMed]

50. Li, D.; Yang, M.; Hu, J.; Zhang, Y.; Chang, H.; Jin, F. Determination of penicillin G and its degradation products in a penicillin production wastewater treatment plant and the receiving river. *Water Res.* **2008**, *42*, 307–317. [CrossRef] [PubMed]

51. Filippini, M.; Svercel, M.; Laczko, E.; Kaech, A.; Ziegler, U.; Bagheri, H.C. Fibrella aestuarina gen. nov., sp. nov., a filamentous bacterium of the family Cytophagaceae isolated from a tidal flat, and emended description of the genus Rudanella Weon et al. 2008. *Int. J. Syst. Evol. Microbiol.* **2011**, *61*, 184–189. [CrossRef] [PubMed]

52. Joung, Y.; Kim, H.; Kang, H.; Lee, B.I.; Ahn, T.S.; Joh, K. Lacihabitans soyangensis gen. nov., sp. nov., a new member of the family Cytophagaceae, isolated from a freshwater reservoir. *Int. J. Syst. Evol. Microbiol.* **2014**, *64*, 3188–3194. [CrossRef] [PubMed]

53. Wright, G.D. Bacterial resistance to antibiotics: Enzymatic degradation and modification. *Adv. Drug Deliv. Rev.* **2005**, *57*, 1451–1470. [CrossRef] [PubMed]

54. Heijenoort, J.V.; Gutmann, L. Correlation between the structure of the bacterial peptidoglycan monomer unit, the specificity of transpeptidation, and susceptibility to β-lactams. *Proc. Natl. Acad. Sci. USA* **2000**, *97*, 5028–5030. [CrossRef] [PubMed]

55. Fuerst, J.A.; Sagulenko, E. Beyond the bacterium: Planctomycetes challenge our concepts of microbial structure and function. *Nat. Rev. Microbiol.* **2011**, *9*, 403–413. [CrossRef] [PubMed]

water

MDPI

Article

Occurrence, Seasonal Variation and Risk Assessment of Antibiotics in Qingcaosha Reservoir

Yue Jiang [1], Cong Xu [1], Xiaoyu Wu [1], Yihan Chen [1], Wei Han [2], Karina Yew-Hoong Gin [3] and Yiliang He [2,*]

[1]   School of Environmental Science and Engineering, Shanghai Jiao Tong University, Shanghai 200240, China; jiangyue0524@sjtu.edu.cn (Y.J.); xucong90@sjtu.edu.cn (C.X.); wxysal@yeah.net (X.W.); chenyhok1987@sjtu.edu.cn (Y.C.)
[2]   Sino-Japan Friendship Centre for Environmental Protection, Beijing 100029, China; hanwei_2002@126.com
[3]   Department of Civil and Environmental Engineering, Faculty of Engineering, National University of Singapore, Singapore 117576, Singapore; ceeginyh@nus.edu.sg
*   Correspondence: ylhe@sjtu.edu.cn; Tel.: +86-21-5474-4008

Received: 28 November 2017; Accepted: 19 January 2018; Published: 29 January 2018

Abstract: Qingcaosha Reservoir is an important drinking water source in Shanghai. The occurrence of five groups of antibiotics was investigated in the surface water of this reservoir over a one-year period. Seventeen antibiotics were selected in this study based on their significant usage in China. Of these antibiotics, 16 were detected, while oxytetracycline was not detected in any sampling site. The detected frequency of tylosin was only 47.92% while the other 15 antibiotics were above 81.25%. The dominant antibiotic was different in four seasons: norfloxacin was dominant in spring, and penicillinV was dominant in summer, autumn and winter, with medium concentrations of 124.10 ng/L, 89.91 ng/L, 180.28 ng/L, and 216.43 ng/L, respectively. The concentrations and detection frequencies of antibiotics were notably higher in winter than in other seasons, demonstrating that low temperature and low flow may result in the persistence of antibiotics in the aquatic environment. Risk assessment suggested that norfloxacin, ciprofloxacin, penicillinV, and doxycycline in the surface water presented high ecological risks.

Keywords: antibiotics; Qingcaosha reservoir; risk assessment

---

1. Introduction

Since their discovery, antibiotics have been an effective means of treatment or prevention of bacterial infections. Currently, antibiotics are not only widely used as human medicine, but also as veterinary medicine and animal growth promoters [1]. In fact, the annual consumption of antibiotics for both human use and for veterinary use is substantial. The total usage of 36 frequently detected antibiotics in China was estimated to be 92,700 tons in 2013; 48% human use and 52% veterinary use [2]. However, the excessive use of antibiotics inevitably leads to the resistance of bacteria [3], which means that the microbes once susceptible to antibiotics are increasingly difficult to treat [4].

With increasing attention being paid to antibiotics as contaminants, antibiotics are found to be ubiquitous in wastewater treatment plants [5,6], surface water and sediments [7,8]. It has been demonstrated that the estuary zone may act as a reservoir for antibiotics coming from multiple sources because antibiotics are transported from terrestrial sources into the estuary region through river runoff. There are various sources of antibiotics in the aquatic environment. Wastewater from residential facilitates, hospitals, animal husbandry and the pharmaceutical industry is considered as the main source of antibiotics [9–11]. The antibiotics taken by humans or animals cannot be fully metabolized, and consequently enter sewage or manure via excreted urine or feces [12]. Since antibiotics can only be partially removed in wastewater treatment plants [13], the antibiotic residues will be discharged into the aquatic environment. Other important sources of antibiotics in the environment include the manure and sludge used in agricultural sites [7], as well as discarded pharmaceuticals, sludge and solid waste of the pharmaceutical industry in landfills [14].

It is notable that the antibiotics were found in surface water and groundwater which serve as drinking sources [15,16]. This gives rise to the concern that antibiotics may occur in drinking water and threaten human health. After all, the conventional drinking water treatment processes were proved not to be effective in removing all antibiotics [17]. As a matter of fact, trace-level antibiotics have been detected in tap water and drinking water samples [18,19].

Qingcaosha Reservoir, located at the estuary of the Yangtze River, is one of the major drinking water sources in Shanghai. The daily water supply is about 7.19 million m^3, which contributes to more than 50% of the total water supply of the city. The good water quality of Qingcaosha is essential to human health. Florfenicol and thiamphenicol were detected in the tap water of Shanghai by screening 21 antibiotics in the water samples [19]. The contamination of antibiotics in the Huangpu River, another drinking water source in Shanghai, has already been investigated, which showed its moderate contamination level of antibiotics [20,21]. The occurrence of antibiotics in the Yangtze Estuary and the coastal areas nearby was also reported [22,23]. In Qingcaosha Reservoir, the sedimentation and sorption release characteristics of phosphorus fractions [24], the effects of sudden salinity changes on physiological parameters and related gene transcription in M. aeruginosa [25] have been reported. However, the presence of antibiotics has still not been discussed.

The objective of this study was to determine the occurrence of 17 selected antibiotics in the surface water of Qingcaosha Reservoir (sulfonamides, tetracyclines, fluoroquinolones) or seldom reported antibiotics in this region (β-lactams, macrolides) and evaluate the ecological risk of these antibiotics in the reservoir. The detection of the water samples in Qingcaosha Reservoir was over a one-year period to understand the seasonal variation of the antibiotics.

2. Materials and Methods

2.1. Chemicals and Standards

Antibiotics standards of sulfonamides (SAs) including sulfadiazine (SDZ), sulfamonomethoxine (SMM), and sulfaquinoxaline (SQX); fluoroquinolones (FQs) including norfloxacin (NFX), ciprofloxacin (CFX), and ofloxacin (OFX); β-lactams including cefalexin (LEX), penicillinG (PENG), and penicillinV (PENV); macrolides (MLs) including tylosin (TYL), and erythromycin-H$_2$O (ETM-H$_2$O); tetracyclines (TCs) including oxytetracycline (OTC), tetracycline (TC), and doxycycline (DC); others including polymix-B (POL), vancomycin (VAN), lincomycin (LIN), and labeled compounds including ciprofloxacin-D$_8$ (CFX-D$_8$), norfloxacin-D$_5$ (NFX-D$_5$), amoxicillin-D$_4$ (AMX-D$_4$), sulfadiazine-D$_4$ (SDZ-D$_4$), and Doxycycline-D$_3$ (DOX-D$_3$) were purchased from Dr. Ehrenstorfer (GmbH, Augsburg, Germany). The purities of all the chemicals are >98%. The detailed properties of the target compounds are shown in Table A1. Each compound was prepared by diluting the stock solution with methanol at 1000 mg/L and mixture of working standards containing each compound at 10 mg/L. Methanol and acetonitrile were of High-Performance Liquid Chromatography (HPLC) grade and purchased from Sinopharm Chemical Reagent Co., Ltd. (Ourchem, Guangzhou, China). Ultra-pure water (MQ) was obtained from a Milli-Q water purification system (Millipore, Billerica, MA, USA). All standard solutions were stored in the refrigerator at −20 °C. Formic acid, hydrochloric acid, and disodium ethylenediamine tetracetate (Na$_2$EDTA) were of analytical grade and purchased from Sinopharm Chemical Reagent Co., Ltd. (Tianjin, China).

2.2. Sample Collection

QCS reservoir is located at Yangtze River Estuary and supplies Shanghai with more than 7.19 million m$^3 \cdot$d^{-1} of drinking water. The geographical coordinates of the reservoir are 31°48′ N, 121°57′ E. In order to improve the channel flow conditions and avoid damaging the flood control of

Yangtze Estuary, the reservoir has been constructed in a narrow and long shape. The inlet and outlet sluices and the output pipe station of the reservoir are managed by operators. When the inlet sluice was open, the flux of inflow water was set from 700 to 900 $m^3 \cdot s^{-1}$. When the outlet sluice was open, the flux of outflow water was set from 100 to 300 $m^3 \cdot s^{-1}$. The output pipe was located at the water pump station; it transported water from the reservoir to a drinking water treatment plant and the flux was about 60–83 $m^3 \cdot s^{-1}$ [26].

Twelve sampling campaigns were conducted monthly from May 2016 to April 2017. The water quality parameters are shown in Figure A1. The location of sampling sites is illustrated in Figure 1. Sampling point S1 was located in the Yangtze Estuary, in front of the inflow sluice and outside the reservoir, where the velocity and turbidity of water are different from the site inside the reservoir. S2 was located downstream of the outflow sluice in the reservoir, which is used to control the water level together with the inflow sluice. S3 was located at the water pump station, which transported the water to the water plant. S4 was located in the middle of the reservoir, which is an important inspection location. Before sampling, the bottles were sequentially cleaned by methanol and Milli-Q water. All of the water samples were collected 0.5 m below the surface using a water grab sampler, then stored in bottles. The samples were immediately transported to the laboratory and filtered through 0.45 μm filters (Anpel, Shanghai, China) and stored in the refrigerator at 4 °C.

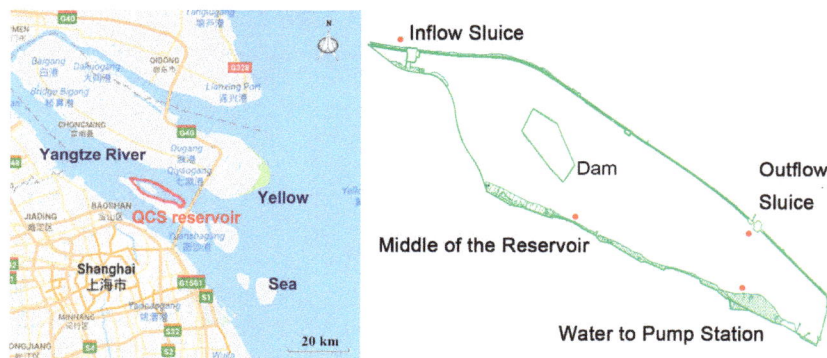

Figure 1. Sampling sites in Qingcaosha Reservoir.

2.3. Sample Preparation and Extraction

The filtered water samples (1 L) were extracted using Poly-Sery Hydrophile-Lipophile Balance Solid-Phase Extraction (HLB SPE) (Anpel, Shanghai, China) cartridges (200 mg, 6 mL). All cartridges were preconditioned with 10 mL methanol and 10 mL Milli-Q water at a flow rate of 1 mL/min. Before extraction, the samples were acidified to pH = 3 with hydrochloric acid, followed by the addition of 0.5 g Na_2EDTA as the chelating agent, then spiked with 50 ng labeled compound (1 mg/L). The extraction rate by SPE was 3–5 mL/min. After being extracted, antibiotics were eluted from SPE cartridges with 10 mL methanol, and the eluent was condensed to dryness under a gentle stream of N_2. Finally, 0.5 mL of the external standard (NFX-D_5, 80 μg/L) was added.

The target antibiotics were performed by a TSQ Quantum Access Max ultra-performance liquid chromatograph-tandem mass spectrometry system (UPLC-MS/MS) (Thermo Fisher Scientific, Waltham, MA, USA). The separation was performed with an Agilent C18 column (50 mm × 2.1 mm, 1.8 μm) (Agilent, Santa Clara, CA, USA). The temperature of the column was maintained at 30 °C, the flow rate was 0.3 mL/min, and the injection volume was 10 μL. Mobile phase: eluent A is ultrapure water containing 0.1% formic acid (*v/v*) and eluent B is acetonitrile. The gradient program was as follows: 95% A (0–1.3 min), 95–60% A (1.3–8 min), 60% A (8–10 min), 95% A (10–12 min). The mass

spectrometric analysis was operated with a positive electrospray ionization (ESI⁺) source in multiple
reaction monitoring (MRM) modes. Tandem mass spectrometric parameters for the target antibiotics
are shown in Table A2.

2.4. Quality Assurance/Quality Control

Standard solutions (from 5 to 400 µg/L in seven points), spiking with the same internal standards,
showed strong linearity; correlation coefficients (r^2) of the standard curves were higher than 0.99.
The limits of detection (LODs) for antibiotics were defined as the concentrations corresponding to the
signal-to-noise (S/N) ratio of 3 and ranged from 0.21–2.41 ng/L. The limits of quantification (LOQs)
for antibiotics were defined as the concentrations corresponding to the signal-to-noise (S/N) ratio of
10 and ranged from 0.46–8.32 ng/L. The mean recoveries of 16 antibiotics, spiked to the filtered surface
water (*n* = 3), were between 69–125% [27,28]. Moreover, triplicate Milli-Q water samples, used as field
blank samples, were all detected below the LOD.

2.5. Risk Assessment

Risk quotient (RQ) has been widely used to evaluate the ecological risk of individual antibiotics
in the aquatic environment. RQ was calculated from the ratio between the measured environmental
concentration (MEC) and the predicted no effect concentration (PNEC). The value of the PNEC was
obtained from acute toxicity data (EC50 or LC50) in Environmental Protection Agency (EPA) and
divided by a safety factor of 1000. The equations are shown as follows:

$$RQ = \frac{MEC}{PNEC} \tag{1}$$

$$PNEC = \frac{EC_{50}\ or\ LC_{50}}{1000} \tag{2}$$

In Equation (1), the highest MEC of all sampling sites and lowest PNEC of each trophic level
was used.

3. Results

3.1. Occurrence of Antibiotics in Qingcaosha Reservoir

Of the 17 antibiotics compounds from six groups, only one antibiotic (OTC) was not detected at
any sampling site; TYL was detected with the lowest detection rate of 47.92% while four antibiotics
(SQX, NFX, TC, DC) were the most frequently detected compounds with a detection rate of 100%,
and the detection rates of the other twelve antibiotics were all above 80%. The concentration and
detection frequencies of antibiotics in Qingcaosha Reservoir are shown in Figure 2.

In the group of sulfonamides (SAs), SQX showed 100% detection frequency, while SDZ and
SMM showed detection frequency over 89%. With the properties of recalcitrance and hydrophilicity,
SAs were prevalent in the surface water [6,29]. In the group of fluoroquinolones (FQs), three antibiotics
(NFX, CFX, OFX) still frequently showed detection frequency over 81.25%. Among FQs, CFX showed
the highest concentration (283.5 ng/L) in November. The mean concentrations of antibiotics in this
group followed the rank order: NFX (129.63 ng/L) > CFX (25.12 ng/L) > OFX (7.64 ng/L). In the group
of β-lactams, the detection frequencies of LEX, PENG, PENV were more than 89.58%. The highest
concentration was observed for PENV, reaching 404.9 ng/L. In the group of macrolides (MLs), TYL and
ETM-H₂O were detected at a low concentration level, in the range of not detedted (n.d)–6.2 ng/L and
n.d–37.4 ng/L, respectively, which were notably lower than Australian urban water (0–60 ng/L) [29]
but higher than the South Yellow Sea (n.d–1.7 ng/L) [30]. The detected frequency of TYL (47.92%) is
lowest among 16 antibiotics. In the group of tetracyclines (TCs), TC and DC were 100% detected with
concentrations up to 35.9 ng/L and 266.7 ng/L, respectively. In the group of others, the concentrations
of the antibiotics were relatively low compared with the other five groups of antibiotics. LIN has been

reported in ranges from 4–171 ng/L in Urban Water [31] and 1–50 ng/L in Surface water in Spain [29] compared to a range from n.d–11.5 ng/L in Qingcaosha Reservoir.

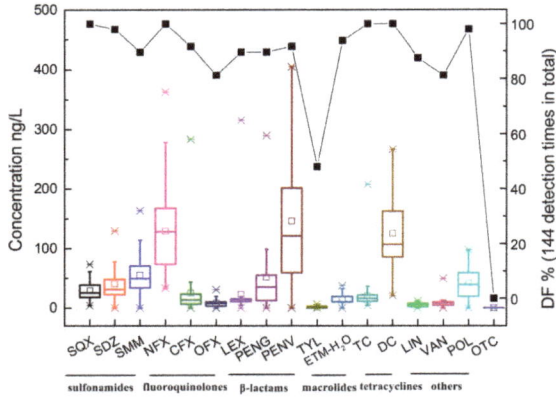

Figure 2. The concentration and detection frequencies of antibiotics in Qingcaosha Reservoir. Color "×" belongs to the box chart and the Y-axis of the box diagram is on the left, it represents the maximum and minimum concentration for each antibiotic. Color "□" belongs to the box chart and the Y-axis of the box diagram is on the left, it represents the mean value for each antibiotic. "■" represents the detection frequency (DF) of every antibiotic and the Y-axis of the DF is on the right. The line connecting with the box is up to 99% and down to 1% which lefts the maximum and minimum value of each antibiotic as Color "×":

3.2. Seasonal Variation of Antibiotics

The seasonal variations of six groups of antibiotics investigated at four sampling sites are shown in Figure 3. The concentrations of antibiotics in summer were notably lower than in winter, and slightly lower than in spring and autumn.

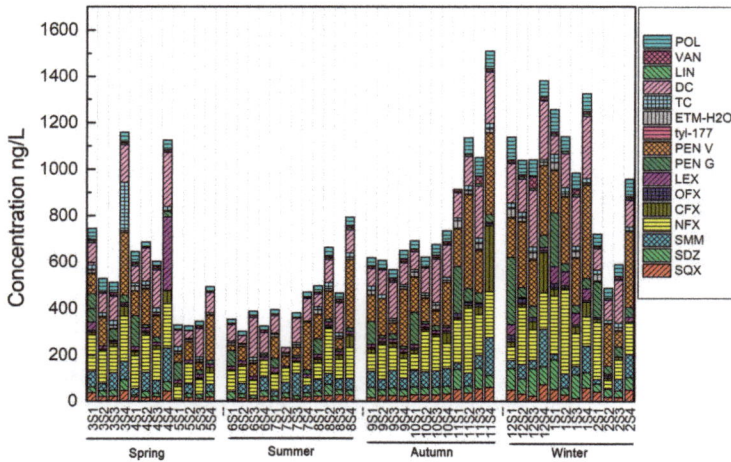

Figure 3. Total concentration of six groups of antibiotics in Qingcaosha Reservoir over four seasons. S1 to S4 represent the every sampling sites and the number before S1 to S4 represent the months.

Among all the antibiotics, the concentrations were highest in winter. The median concentrations of antibiotics in winter were approximately 1.10–5.32 times higher than in the other three seasons. Concentrations of SAs ranged from 0 to 143.8 ng/L in spring, from 0 to 97.0 ng/L in summer, from 17.2 to 163.2 ng/L in autumn, and from 12.92 to 161.3 ng/L in winter. Moreover, the result of this study showed that dominating antibiotics in different seasons were different. For example, NFX was found to be the main compound in spring while PENV was significantly high in the other three seasons.

3.3. Spatial Variation of Antibiotics

Among all sampling sites, the middle of the reservoir (S4) was most contaminated, with the total concentration of 1510.86 ng/L in October, followed by inflow sluice (S1) (1256.71 ng/L) in January, outflow sluice (S2) (1140.58 ng/L) in January and water pump station (S3) (1050.86 ng/L) in November. For each group of antibiotics, the highest concentration of macrolides was found at S1 (41.07 ng/L) in December, while the highest concentration of other groups was found at S4 (Figure A2).

4. Discussion

4.1. Occurrence, Seasonal Variation

Compared with previous studies (Table 1), the concentrations of SDZ and SQX are relatively higher than those in the Yangtze Estuary [22], Huangpu River [21], and Urban water [29]. The concentration of SMM (n.d–163.2 ng/L) in this study is lower than those in the Huangpu River (ranged from 2.05–623.27 ng/L) [32] and the Pearl River (ranged from n.d–1080 ng/L) [33].

CFX and NFX were significantly higher than those in Yangtze Estuary (n.d–14.2 ng/L and n.d–2.27 ng/L, respectively) [15] and Pearl River Estuary (n.d–34.2 ng/L and n.d–33.6 ng/L, respectively) [33]. OFX in Qingcaosha Reservoir was much lower than the South Yellow Sea (n.d–497.6 ng/L) [30] and Chaohu (n.d–182.7 ng/L) [34]. It is reported that FQs showed relatively low persistence in water and had strong sorption to the solid phase [35], so in many studies, the detection frequencies of FQs were relatively low, which is different to this study. Compared with other studies, the concentration of OFX was found in municipal sewage and animal wastewater in the area of the South Yellow Sea, so the concentration was relatively high, but the pollution level of OFX in Qingcaosha Reservoir was lower while NFX and CFX were higher than other water bodies.

PENG has been reported in ranges from n.d–250 ng/L [29] compared to a range from n.d–289.9 ng/L in this study. In many types of research, β-lactams were thought not to be a concern as environmental pollutants due to the characteristic of fast hydrolysis. However, the result showed that β-lactams account for a large proportion of the total. This phenomenon elucidates that although these antibiotics are, generally, considered to degrade easily, a pseudo-persistence may be occurring as a result of the continuous discharge.

ETM-H_2O is used not only in livestock as treatment and food additions, but also in the treatment of humans. Because of the strong sorption to sediments and high hydrophobicity of the MLs, the concentrations of the MLs in the aquatic environment were found to be very low. In previous investigations, TC and DC have also been detected in the Huangpu River [20] and Yangtze Estuary [21]. The detection frequencies of TCs in the Huangpu River were high; this might be due to the large usage and discharge in the river [20]. Normally, TCs were seldom reported in the natural water due to their strong degradation ability as well as absorption to particles or soil. However, these antibiotics were found to be ubiquitous in rivers in Shanghai; this may reflect the large usage and discharge of the TCs in this area.

It can be summarized that tetracyclines and fluoroquinolones are widely used as both human and veterinary medicines to treat diseases or to promote growth in livestock [22]. Because both human and livestock excreta, with metabolized or un-metabolized drugs, pass into sewage systems, the detection frequencies and concentrations of these antibiotics are high. SQX and TYL are only used in veterinary applications so they are less prevalent than other antibiotics in the aquatic environment.

Table 1. Comparison of antibiotics concentrations in surface water from different sites (ng/L).

Compounds	Sampling Locations							
	Qingcaosha	Yangtze Estuary	Huangpu River	Chaohu	Pearl River Estuary	South Yellow Sea	Urban Water	Surface Water
	China	China	China	China	China	China	Australia	Spanish
SDZ	n.d–129.8	0.28–71.8	1.39–112.5	n.d–45.6	n.d–726	-	n.d–30	-
SMM	n.d–163.2	0.53–89.1	2.05–623.27	n.d–8.8	n.d–1080	n.d–9.3	-	8–43
SQX	4.1–73.68	n.d–23.5	n.d–64.2	-	-	-	-	-
NFX	32.8–278.2	n.d–14.2	n.d–0.2	n.d–70.2	n.d–174	n.d–21.1	30–1150	-
CFX	n.d–283.5	n.d–2.27	n.d–34.2	n.d–23.3	n.d–33.6	-	n.d–1300	-
OFX	n.d–30.6	n.d–12.4	n.d–28.5	n.d–182.7	2.5–108	n.d–497.6	-	-
LEX	n.d–71.7	-	-	-	-	-	n.d–100	-
PENG	n.d–289.9	-	-	-	-	n.d–11.8	n.d–250	-
PENV	n.d–404.9	-	-	-	-	-	n.d–100	-
TYL	n.d–6.2	-	-	-	-	-	0–60	0.5–16
ETM-H$_2$O	n.d–37.4	-	-	-	-	n.d–1.7	n.d	-
OTC	n.d	n.d–22.5	n.d–219.8	-	-	-	n.d–100	-
TC	4.1–35.9	n.d–2.37	n.d–113.89	-	-	-	n.d–80	-
DC	32.4–266.7	n.d–5.63	n.d–112.3	n.d–42.3	-	-	n.d–40	-
POL	n.d–97.5	-	-	-	-	-	-	-
VAN	n.d–49.2	-	-	-	-	-	-	-
LIN	n.d–11.5	-	-	-	-	-	1–50	4–171
References	This study	[22]	[20,21]	[35]	[33]	[30]	[29]	[31]

n.d: not detected; - : not analyzed.

The varying presence of antibiotics between four seasons may be due to the usage and prosperities of antibiotics, flow conditions and water temperature. It is worth noting that the concentrations of detected antibiotics in high flow and warm conditions were lower than those in low flow and cold conditions [20,36]. From May to September, in order to prevent the eutrophication of water in the reservoir, the flow condition of Qingcaosha Reservoir was high, and from November to April, to restrain the invasion of the salt tide, the flow condition was low. The great dilution by the large flux of the Yangtze River in summer (normal above 50,000 m^3/s) led to the low concentration of antibiotics. Moreover, due to the higher microbial activity and stronger sunlight in summer, the bio-degradation and photo-degradation of antibiotics might be higher in summer than other seasons [37]. Therefore, lower concentrations were observed in summer than in the other three seasons for most antibiotics.

It is worth noting that S4 was located in the middle of the reservoir and was close to the suburb area and S3 was located at the water pump station, which transported the water to the water plant. The above data shows that the contamination level in S3 was less serious than in the other three sampling sites. The total concentration of antibiotics at S4 was very high; this result might be explained by the settled particles releasing antibiotics into the water. Doretto indicated that settled particles with low organic carbon contents had high antibiotics desorption capacity [38]. Sedimentation of the particles was remarkable in S4 while it was not significant downstream of the reservoir (S2/S3) (Figure A1), suggesting that there was a high potential for surface water contamination. Furthermore, the key factor affecting the sedimentation was flow rate; the rank order of the flow rate was as follows: S1 (0.2 m/s) > S4 (0.01~0.19 m/s) > S3, S2 (0.01~0.03 m/s) [39]. Particles were fully settled in S4 because of the relatively high velocity upstream of the reservoir. The suspended particles downstream of the reservoir were phytoplankton which would not release contaminants into the water. Qingcaosha Reservoir is located at the estuary of the Yangtze River and most domestic sewage effluents are now continuously discharged downstream of the Yangtze River since the completion of the wastewater control project. Yan claimed that wastewater treatment plants located in the upper reaches of the Yangtze River would be the primary reason for the higher concentration of antibiotics downstream of the Yangtze River [40]. Furthermore, several drain outlets were found in the Yuxi River which was located downstream of the Yangtze River; effluents from the pond and village area were discharged into the estuary without any purification treatment. Therefore, the point source of the antibiotics also exists [34].

Thus, although the tetracyclines had strong absorption to sediment, they were still detected with high frequency. The concentration of DC was found to be much higher than that of TC; the input contamination from upstream of the reservoir might be the main reason. The dominant contamination at S1 was PENV while at S2 and S4 it was PENG. This result might be due to the wide range of applications of PENG and PENV in clinical applications, considering their high effectiveness and low toxicity [29,30].

4.2. Environmental Risk Assessment of Antibiotics

In this study, the potential ecological risks of antibiotics were assessed by using the risk quotients (RQs) approach, according to the European technical guidance document (TGD) on risk assessment. The value of RQ was defined as the ratio of the measured environmental concentration (MEC) and the predicted no-effect concentration (PNEC). The value of PNEC was assessed based on the toxicity data which were obtained from the Ecological Structure Activity Relationships (ECOSAR) and shown in Table A3. In order to better distinguish the ecological risk levels, according to the individual RQ value, three risk levels were classified (0.01–0.1: low risk; 0.1–1: medium risk; >1: high risk) [22,41].

The risk quotients (RQs) of antibiotics in the reservoir are shown in Figure 4. According to the RQs, seven antibiotics (SDZ, SMM, SQX, PENG, TYL, OTC and POL) posed a low risk to the relevant sensitive aquatic organisms (*S. capricornutum, S. vacuolatus, P. subcapitata, M. aeruginosa*) in four seasons; four antibiotics (NFX, CFX, PENV, DC) caused high risk. The RQ values of NFX, CFX, PENV, and DC suggested that these antibiotics might present a significant risk to the algae in Qingcaosha Reservoir.

Normally, RQs in winter, spring and autumn are remarkably higher than in summer. For example, OFX, ETM-H$_2$O and LIN caused high risk in winter, while they caused medium risk in summer, and the proportions of samples classified as high risk during the entire sampling period were 16.7%, 58.3%, 33.3%, respectively. However, LEX, may cause medium risk in the aquatic environment in April and low risk in other months.

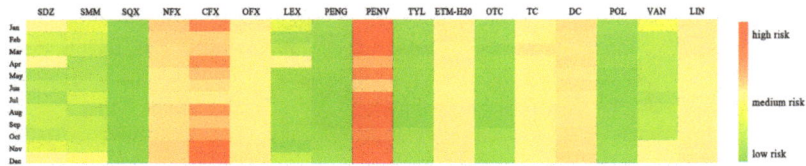

Figure 4. Risk assessment of antibiotics in Qingcaosha Reservoir from May 2016 to April 2017.

Studies have demonstrated that the residual trace antibiotics in the aquatic environment may impose selective stress on the microbe communities and accelerate the spread of antibiotic resistance genes (ARGs) [42]. ARGs such as *sul I*, *sul II*, *tet (C)*, *tet (G)* were the most prevalent resistance genes in raw water in Yangtze River Delta, and the absolute abundances of the *sul* and *tet* class genes ranged from 10^{10} to 10^{12} copies/L [43]. In another drinking water source in Shanghai, 11 ARGs were detected with high concentrations, and *sul II* was present at the highest concentration (4.19×10^8 copies/L). This phenomenon might reflect the widespread use of sulfonamides in this region [19]. Furthermore, ARGs could be transferred between bacteria through transposons, plasmids and integrons [44]. Xu observed that the abundances of ARGs were significantly correlated to the levels of mobile genetic elements, indicating that *intI-1* and transposons may contribute to the abundances of ARGs in drinking water [45]. Hence, the prevalence of the antibiotics in Qingcaosha Reservoir exhibited not only ecological risk in the water phase but also the risk of spread of the antibiotic resistance genes, which should be researched further in the future studies.

5. Conclusions

The occurrence and seasonal variations of six groups of antibiotics in Qingcaosha Reservoir were detected by using SPE and UPLC-MS/MS; antibiotics in all of the sampling sites were at the ng/L level. All antibiotics were frequently detected during the year-long period except for OTC and TYL. β-lactams showed the highest concentrations of antibiotics compared to other groups of antibiotics, suggesting its extensive use in this region, while macrolides exhibited relatively low concentrations. The seasonal variation of the total antibiotics indicated that residues of antibiotics in the surface water in winter were higher than in the other three seasons. The concentrations of the antibiotics at S4 were highest, suggesting that the antibiotics released by settled particles may be the major source. Since the completion of the wastewater control project, most domestic sewage effluents are discharged downstream of the Yangtze River; this might be the major antibiotics source of the reservoir. The risk assessment results based on the RQs clearly revealed that six antibiotics, which included NFX, CFX, PENV, DC, OFX, ETM-H$_2$O and LIN, posed a significant ecological risk to the relevant algae in the surface water of Qingcaosha Reservoir. Moreover, more attention should be paid to the fate of these antibiotics, considering the spread of antibiotic resistance genes in this region.

Acknowledgments: This study was funded by the National Major Science and Technology Program for Water Pollution Control and Treatment (No. 2014ZX07206001) and the Campus for Research Excellence And Technological Enterprise (CREATE) program.

Author Contributions: Yue Jiang, Cong Xu and Xiaoyu Wu conducted the experiments and wrote the paper; Yihan Chen designed the experiments; Wei Han analyzed the data; Karina Yew-Hoong Gin and Yilang He were in charge of the paper.

Conflicts of Interest: The authors declare no conflict of interest.

Appendix A

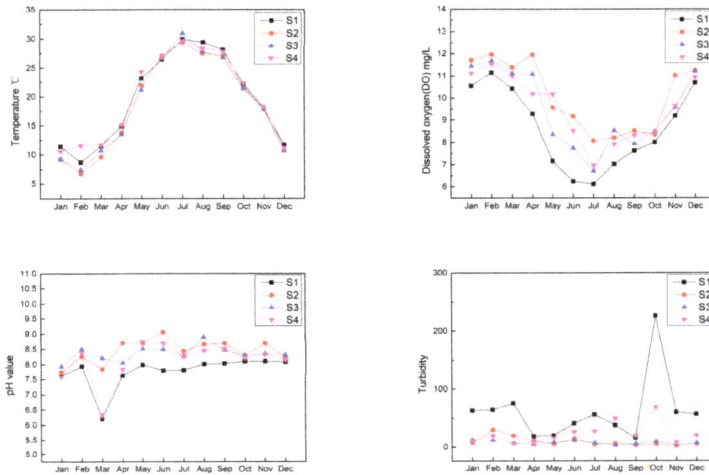

Figure A1. The water quality parameters in sampling sites in this study.

Table A1. The physicochemical properties of the target antibiotic compounds.

Antibiotic	Usage	Molecular Formula	CAS Number	Log Kow	Log Koa	Vapor Pressure (Pascals)
Sulfadiazine	human	$C_{10}H_{10}N_4O_2S$	68-35-9	−0.09	8.1	2.29×10^{-4}
Sulfamonomethoxine	veterinary	$C_{11}H_{12}N_4O_3S$	1220-83-3	0.2	12.706	5.96×10^{-5}
Sulfaquinoxaline	veterinary	$C_{14}H_{12}N_4O_2S$	59-40-5	1.68	14.43	3.91×10^{-6}
Norfloxacin	human/veterinary	$C_{16}H_{18}FN_3O_3$	70458-96-7	−1.03	15.419	1.63×10^{-7}
Ciprofloxacin	human/veterinary	$C_{17}H_{18}FN_3O_3$	85721-33-1	0.28	16.962	7.27×10^{-8}
Ofloxacin	human/veterinary	$C_{18}H_{20}FN_3O_4$	82419-36-1	−0.39	17.301	4.11×10^{-8}
Cefalexin	human/veterinary	$C_{16}H_{19}N_3O_5S$	23325-78-2	−0.08	18.849	7.27×10^{-12}
Penicillin G	human/veterinary	$C_{16}H_{17}N_2O_4SK$	113-98-4	−3.01	6.05	4.96×10^{-13}
Penicillin V	human/veterinary	$C_{16}H_{17}N_2O_5S$	132-98-9	2.09	14.833	3.81×10^{-8}
Tylosin	veterinary	$C_{46}H_{77}NO_{17}$	1401-69-0	1.63	37.257	1.36×10^{-28}
Erythromycin-H_2O	human/veterinary	$C_{37}H_{65}NO_{12}$	23893-13-2	-	-	-
Oxytetracycline	human/veterinary	$C_{22}H_{25}ClN_2O_9$	2058-46-0	−3.6	24.561	2.09×10^{-22}
Tetracycline	human/veterinary	$C_{22}H_{25}ClN_2O_8$	64-75-5	−3.7	25.588	5.33×10^{-23}
Doxycycline	human/veterinary	$C_{22}H_{25}ClN_2O_8$	24390-14-5	-	-	-
Polymix-B	human	$C_{56}H_{100}N_{16}O_{17}S$	1405-20-5	-	-	-
Vancomycin	human/veterinary	$C_{66}H_{75}CL_2N_9O_{24}$	1404-93-9	−0.84	−0.995	-
Lincomycin	human/veterinary	$C_{18}H_{34}N_2O_6S$	154-21-2	0.2	21.111	7.09×10^{-13}

Table A2. Tandem mass spectrometric parameters for the target antibiotics in this study.

Group	Antibiotic	CAS Number	Acronym	Relative Molecule Mass	Precursor Ion M/Z	Product Ion M/Z	Collision Energy E/V
Sulfonamides (SAs)	Sulfadiazine	68-35-9	SDZ	250.05	251.1	156.0, 92.1	14, 26
	Sulfamonomethoxine	1220-83-3	SMM	280.06	281	108, 156, 92	23, 16, 26
	Sulfaquinoxaline	59-40-5	SQX	300.07	301	156, 92.1, 108	14, 24, 21
Fluoroquinolones (FQs)	Norfloxacin	70458-96-7	NFX	319.13	320.1	302.2, 276.2	22, 16
	Ciprofloxacin	85721-33-1	CFX	331.13	332.2	288.2, 245.1	20, 16, 22
	Ofloxacin	82419-36-1	OFX	361.14	362.2	318.0, 261.1	17, 26
β-lactmas	Cefalexin	23325-78-2	LEX	365.10	348.1	106.1, 158, 174	23, 8, 12
	Penicillin G	113-98-4	PEN G	372.05	335	176.1, 160, 114.1	10, 7, 28
	Penicillin V	132-98-9	PEN V	349.09	351.1	229.1, 106.1	14, 14
Macrolides (MLs)	Tylosin	1401-69-0	TYL	915.52	916.6	174, 772.6	34, 28
	Erythromycin-H_2O	23893-13-2	ETM-H_2O	715.45	716.3	158, 558	28, 17
Tetracyclines (TCs)	Oxytetracycline	2058-46-0	OTC	496.12	461.2	443.0, 426.0	11, 18
	Tetracycline	64-75-5	TC	480.13	444.82	410.2, 427.3	19, 10
	Doxycycline	24390-14-5	DC	480.13	445.2	428.2, 154	18, 28
Others	Polymix-B	1405-20-5	POL	1300.72	402	101.1, 120.1	20, 29
	Vancomycin	1404-93-9	VAN	1447.43	724.9	100.1, 144.1	27, 13
	Lincomycin	154-21-2	LIN	406.21	407.2	126.1, 359.3	35, 15
External standard	Norfloxacin-d5	1015856-57-1	NFX-d5	324.34	325.1	307.2, 281.2	22, 16
Internal standards	Ciprofloxacin-d8	1130050-35-9	CFX-d8	339.18	340.2	322.2, 296.2	21, 17
	Amoxicillin-d4	26787-78-0	AMX-d4	369.4	370.2	114, 212.1	19, 10
	Sulfadiazine-d4	1020719-78-1	SDZ-d4	254.28	255.1	160, 96.1	14, 25
	Doxycycline-d3	564-25-0	dox-d3	447.44	448.2	431.3, 323.2	21, 30

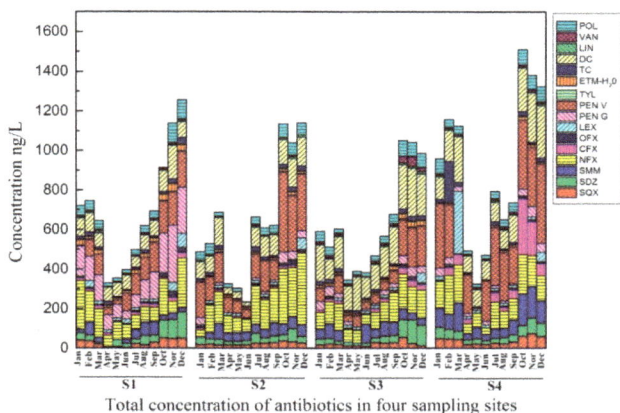

Figure A2. Total concentration of antibiotics in four sampling sites.

Table A3. Toxicity data on algae, invertebrates and fish.

Compounds		Species	Toxicity Data (mg/L)	PNEC (ng/L)	References
Sulfadiazine	Algae	*S. capricornutum*	2.2	2200	[46]
	Invertebrate	N.F	N.F	N.F	N.F
	Fish	N.F	N.F	N.F	N.F
Sulfamonomethoxine	Algae	*S. vacuolatus*	3.82	3820	[47]
	Invertebrate	N.F	2.259	2259	ECOSAR
	Fish	N.F	166.297	166,297	ECOSAR
Sulfaquinoxaline	Algae	N.F	131	131,000	[48]
	Invertebrate	N.F	N.F	N.F	N.F
	Fish	N.F	N.F	N.F	N.F
Norfloxacin	Algae	*M. wesenbergii*	0.038	38	[49]
	Invertebrate	*D. magna*	0.88	880	[50]
	Fish	N.F	20,081.355	20,081,355	ECOSAR
Ciprofloxacin	Algae	*P. subcapitata*	0.002	2	[51]
	Invertebrate	N.F	N.F	N.F	N.F
	Fish	N.F	N.F	N.F	N.F
Ofloxacin	Algae	*M. aeruginosa*	0.021	21	[51]
	Invertebrate	*C. dubia*	3.13	3130	[51]
	Fish	*D. rerio*	>1000	1,000,000	[51]
Cefalexin	Algae	N.F	2.5	2500	[52]
	Invertebrate	N.F	N.F	N.F	N.F
	Fish	N.F	N.F	N.F	N.F
Penicillin G	Algae	N.F	39.032	39,032	ECOSAR
	Invertebrate	N.F	193.241	193,241	ECOSAR
	Fish	N.F	375.923	375,923	ECOSAR
Penicillin V	Algae	N.F	0.006	6	[53]
	Invertebrate	N.F	N.F	N.F	N.F
	Fish	N.F	N.F	N.F	N.F
Tylosin	Algae	*P. subcapitata*	0.95	950	[54]
	Invertebrate	N.F	N.F	N.F	N.F
	Fish	N.F	N.F	N.F	N.F
Erythromycin-H_2O	Algae	*P. subcapitata*	0.02	20	[51]
	Invertebrate	*C. dubia*	0.22	220	[51]
	Fish	*D. rerio*	>1000	1,000,000	[51]
Oxytetracycline	Algae	*M. aeruginosa*	0.23	230	ECOSAR
	Invertebrate	N.F	3.08	30,800	ECOSAR
	Fish	*Oryzias latipes*	50	500,000	ECOSAR
Tetracycline	Algae	*M. aeruginosa*	0.09	90	ECOSAR
	Invertebrate	*B. calyciflorus*	5.6	5600	ECOSAR
	Fish	*Paracheirodon axelrodi*	2.5	25,000	ECOSAR

Table A3. *Cont.*

Compounds		Species	Toxicity Data (mg/L)	PNEC (ng/L)	References
Doxycycline	Algae	*M. aeruginosa*	0.062	62	ECOSAR
	Invertebrate	*C. dubia*	0.5	500	ECOSAR
	Fish	*D. rerio*	2.658	2658	ECOSAR
Polymix-B	Algae	*A. aeruginosa*	2	2000	ECOSAR
	Invertebrate	N.F	N.F	N.F	N.F
	Fish	N.F	N.F	N.F	N.F
Vancomycin	Algae	N.F	0.6	600	[50]
	Invertebrate	N.F	N.F	N.F	N.F
	Fish	N.F	N.F	N.F	N.F
Lincomycin	Algae	*P. subcapitata*	0.07	70	ECOSAR
	Invertebrate	*Thamnocephalus platyurus*	33	33,000	ECOSAR
	Fish	*Danio rerio*	1000	10,000,000	ECOSAR

References

1. Kümmerer, K. Antibiotics in the aquatic environment—A review—Part I. *Chemosphere* **2009**, *75*, 417–434. [CrossRef] [PubMed]
2. Zhang, Q.Q.; Ying, G.G.; Pan, C.G.; Liu, Y.S.; Zhao, J.L. Comprehensive evaluation of antibiotics emission and fate in the river basins of China: Source analysis, multimedia modeling, and linkage to bacterial resistance. *Environ. Sci. Technol.* **2015**, *49*, 6772–6782. [CrossRef] [PubMed]
3. Carlet, J.; Collignon, P.; Goldmann, D.; Goossens, H.; Gyssens, I.C.; Harbarth, S.; Jarlier, V.; Levy, S.B.; N'Doye, B.; Pittet, D.; et al. Society's failure to protect a precious resource: Antibiotics. *Lancet* **2011**, *378*, 369–371. [CrossRef]
4. Mazel, D.; Davies, J. Antibiotic resistance in microbes. *Cell. Mol. Life Sci. CMLS* **1999**, *56*, 742–754. [CrossRef] [PubMed]
5. Batt, A.L.; Kim, S.; Aga, D.S. Comparison of the occurrence of antibiotics in four full-scale wastewater treatment plants with varying designs and operations. *Chemosphere* **2007**, *68*, 428–435. [CrossRef] [PubMed]
6. Xu, W.; Zhang, G.; Li, X.; Zou, S.; Li, P.; Hu, Z.; Li, J. Occurrence and elimination of antibiotics at four sewage treatment plants in the Pearl River Delta (PRD), South China. *Water Res.* **2007**, *41*, 4526–4534. [CrossRef] [PubMed]
7. Li, W.; Shi, Y.; Gao, L.; Liu, J.; Cai, Y. Occurrence of antibiotics in water, sediments, aquatic plants, and animals from Baiyangdian Lake in North China. *Chemosphere* **2012**, *89*, 1307–1315. [CrossRef] [PubMed]
8. Klosterhaus, S.L.; Grace, R.; Hamilton, M.C.; Yee, D. Method validation and reconnaissance of pharmaceuticals, personal care products, and alkylphenols in surface waters, sediments, and mussels in an urban estuary. *Environ. Int.* **2013**, *54*, 92–99. [CrossRef] [PubMed]
9. Brown, K.D.; Kulis, J.; Thomson, B.; Chapman, T.H.; Mawhinney, D.B. Occurrence of antibiotics in hospital, residential, and dairy effluent, municipal wastewater, and the Rio Grande in New Mexico. *Sci. Total Environ.* **2006**, *366*, 772–783. [CrossRef] [PubMed]
10. Michael, I.; Rizzo, L.; McArdell, C.S.; Manaia, C.M.; Merlin, C.; Schwartz, T.; Dagot, C.; Fatta-Kassinos, D. Urban wastewater treatment plants as hotspots for the release of antibiotics in the environment: A review. *Water Res.* **2013**, *47*, 957–995. [CrossRef] [PubMed]
11. Manzetti, S.; Ghisi, R. The environmental release and fate of antibiotics. *Mar. Pollut. Bull.* **2014**, *79*, 7–15. [CrossRef] [PubMed]
12. Hirsch, R.; Ternes, T.; Haberer, K.; Kratz, K.-L. Occurrence of antibiotics in the aquatic environment. *Sci. Total Environ.* **1999**, *225*, 109–118. [CrossRef]
13. Watkinson, A.J.; Murby, E.J.; Costanzo, S.D. Removal of antibiotics in conventional and advanced wastewater treatment: Implications for environmental discharge and wastewater recycling. *Water Res.* **2007**, *41*, 4164–4176. [CrossRef] [PubMed]
14. Musson, S.E.; Townsend, T.G. Pharmaceutical compound content of municipal solid waste. *J. Hazard. Mater.* **2009**, *162*, 730–735. [CrossRef] [PubMed]
15. Focazio, M.J.; Kolpin, D.W.; Barnes, K.K.; Furlong, E.T.; Meyer, M.T.; Zauqq, S.D.; Barber, L.B.; Thurman, M.E. A national reconnaissance for pharmaceuticals and other organic wastewater contaminants in the United States—II) Untreated drinking water sources. *Sci. Total Environ.* **2008**, *402*, 201–216. [CrossRef] [PubMed]

16. Chen, W.; Jiang, L.; Lu, N.; Ma, L. Development of a method for trace level determination of antibiotics in drinking water sources by high performance liquid chromatography-tandem mass spectrometry. *Anal. Methods* **2015**, *7*, 1777–1787. [CrossRef]

17. Adams, C.; Meyer, M.T. Removal of Antibiotics from Surface and Distilled Water in Conventional Water Treatment Processes. *J. Environ. Eng.* **2002**, *128*, 253–260. [CrossRef]

18. Ye, Z.; Weinberg, H.S. Trace analysis of trimethoprim and sulfonamide, macrolide, quinolone, and tetracycline antibiotics in chlorinated drinking water using liquid chromatography electrospray tandem mass spectrometry. *Anal. Chem.* **2007**, *79*, 1135–1144. [CrossRef] [PubMed]

19. Wang, H.; Wang, N.; Wang, B.; Zhao, Q.; Fang, H.; Fu, C.; Tang, C.; Jiang, F.; Zhou, Y.; Chen, Y.; et al. Antibiotics in Drinking Water in Shanghai and Their Contribution to Antibiotic Exposure of School Children. *Environ. Sci. Technol.* **2016**, *50*, 2692–2699. [CrossRef] [PubMed]

20. Jiang, L.; Hu, X.; Yin, D.; Zhang, H.; Yu, Z. Occurrence, distribution and seasonal variation of antibiotics in the Huangpu River, Shanghai, China. *Chemosphere* **2011**, *82*, 822–828. [CrossRef] [PubMed]

21. Chen, K.; Zhou, J.L. Occurrence and behavior of antibiotics in water and sediments from the Huangpu River, Shanghai, China. *Chemosphere* **2014**, *95*, 604–612. [CrossRef] [PubMed]

22. Yan, C.; Yang, Y.; Zhou, J.; Liu, M.; Nie, M.; Shi, H.; Gu, L. Antibiotics in the surface water of the Yangtze Estuary: Occurrence, distribution and risk assessment. *Environ. Pollut.* **2013**, *175*, 22–29. [CrossRef] [PubMed]

23. Shi, H.; Yang, Y.; Liu, M.; Yan, C.; Yue, H.; Zhou, J. Occurrence and distribution of antibiotics in the surface sediments of the Yangtze Estuary and nearby coastal areas. *Mar. Pollut. Bull.* **2014**, *83*, 317–323. [CrossRef] [PubMed]

24. Jin, X.; He, Y.; Kirumba, G.; Hassan, Y.; Li, J. Phosphorus fractions and phosphate sorption-release characteristics of the sediment in the Yangtze River estuary reservoir. *Ecol. Eng.* **2013**, *55*, 62–66. [CrossRef]

25. Chen, L.; Mao, F.; Kirumba, G.C.; Jiang, C.; Manefield, M.; He, Y. Changes in metabolites, antioxidant system, and gene expression in Microcystis aeruginosa under sodium chloride stress. *Ecotoxicol. Environ. Saf.* **2015**, *122*, 126–135. [CrossRef] [PubMed]

26. Xu, C.; Zhang, J.; Bi, X.; Xu, Z.; He, Y.; Gin, K.Y. Developing an integrated 3D-hydrodynamic and emerging contaminant model for assessing water quality in a Yangtze Estuary Reservoir. *Chemosphere* **2017**, *188*, 218–230. [CrossRef] [PubMed]

27. Kexiang, L.; Minghao, S. Simultaneous Determination of 19 Antibiotics in Environmental Water SamplesUsing Solid Phase Extraction Ultra Pressure Liquid Chromatography Coupled with Tandem Mass Spectrometry. *J. Instrum. Anal.* **2010**, *29*, 1209–1214.

28. Chen, H.; Liu, S.; Xu, X.-R. Antibiotics in the coastal environment of the Hailing Bay region, South China Sea: Spatial distribution, source analysis and ecological risks. *Mar. Pollut. Bull.* **2015**, *95*, 365–373. [CrossRef] [PubMed]

29. Watkinson, A.J.; Murby, E.J.; Kolpin, D.W.; Costanzo, S.D. The occurrence of antibiotics in an urban watershed: From wastewater to drinking water. *Sci. Total Environ.* **2009**, *407*, 2711–2723. [CrossRef] [PubMed]

30. Du, J.; Zhao, H.; Liu, S.; Xie, H.; Wang, Y.; Chen, J. Antibiotics in the coastal water of the South Yellow Sea in China: Occurrence, distribution and ecological risks. *Sci. Total Environ.* **2017**, *595*, 521–527. [CrossRef] [PubMed]

31. Boleda, M.R.; Galceran, M.T.; Ventura, F. Validation and uncertainty estimation of a multiresidue method for pharmaceuticals in surface and treated waters by liquid chromatography-tandem mass spectrometry. *J. Chromatogr. A* **2013**, *1286*, 146–158. [CrossRef] [PubMed]

32. Jiang, L.; Hu, X.; Xu, T.; Zhang, H.; Sheng, D.; Yin, D. Prevalence of antibiotic resistance genes and their relationship with antibiotics in the Huangpu River and the drinking water sources, Shanghai, China. *Sci. Total Environ.* **2013**, *458–460*, 267–272. [CrossRef] [PubMed]

33. Xu, W.; Yan, W.; Li, X.; Zou, Y.; Chen, X.; Huang, W.; Miao, L.; Zhang, R.; Zhang, G.; Zou, S. Antibiotics in riverine runoff of the Pearl River Delta and Pearl River Estuary, China: Concentrations, mass loading and ecological risks. *Environ. Pollut.* **2013**, *182*, 402–407. [CrossRef] [PubMed]

34. Tang, J.; Shi, T.; Wu, X.; Cao, H.; Li, X.; Hua, R.; Tang, F.; Yue, Y. The occurrence and distribution of antibiotics in Lake Chaohu, China: Seasonal variation, potential source and risk assessment. *Chemosphere* **2015**, *122*, 154–161. [CrossRef] [PubMed]

35. Pena, A.; Pina, J.; Silva, L.J.; Meisel, L.; Lino, C.M. Fluoroquinolone antibiotics determination in piggeries environmental waters. *J. Environ. Monit. JEM* **2010**, *12*, 642–646. [CrossRef] [PubMed]

36. Kim, S.C.; Carlson, K. Temporal and spatial trends in the occurrence of human and veterinary antibiotics in aqueous and river sediment matrices. *Environ. Sci. Technol.* **2007**, *41*, 50–57. [CrossRef] [PubMed]

37. Karthikeyan, K.G.; Meyer, M.T. Occurrence of antibiotics in wastewater treatment facilities in Wisconsin, USA. *Sci. Total Environ.* **2006**, *361*, 196–207. [CrossRef] [PubMed]

38. Doretto, K.M.; Peruchi, L.M.; Rath, S. Sorption and desorption of sulfadimethoxine, sulfaquinoxaline and sulfamethazine antimicrobials in Brazilian soils. *Sci. Total Environ.* **2014**, *476–477*, 406–414. [CrossRef] [PubMed]

39. Yuan, J.; Li, L. Study on Characteristics of Flow State and Deposit Distribution of Qingcaosha Reservoir Area. *Urban Roads Bridges Flood Control* **2012**, *9*, 181–183.

40. De Souza, S.M.L.; de Vasconcelos, E.C.; Dziedzic, M.; de Oliveira, C.M.R. Environmental risk assessment of antibiotics: An intensive care unit analysis. *Chemosphere* **2009**, *77*, 962–967. [CrossRef] [PubMed]

41. Hernando, M.D.; Mezcua, M.; Fernández-Alba, A.R.; Barceló, D. Environmental risk assessment of pharmaceutical residues in wastewater effluents, surface waters and sediments. *Talanta* **2006**, *69*, 334–342. [CrossRef] [PubMed]

42. Storteboom, H.; Arabi, M.; Davis, J.G.; Crimi, B.; Pruden, A. Tracking antibiotic resistance genes in the South Platte River Basin using molecular signatures of urban, agricultural, and pristine sources. *Environ. Sci. Technol.* **2010**, *44*, 7397–7404. [CrossRef] [PubMed]

43. Liu, S.; Zhao, G.; Zhao, H.; Zhai, G.; Chen, J.; Zhao, H. Antibiotics in a general population: Relations with gender, body mass index (BMI) and age and their human health risks. *Sci. Total Environ.* **2017**, *599*, 298–304. [CrossRef] [PubMed]

44. Zheng, S.; Qiu, X.; Chen, B.; Yu, X.; Liu, Z.; Zhong, G.; Li, H.; Chen, M.; Sun, G.; Hunag, H.; et al. Antibiotics pollution in Jiulong River estuary: Source, distribution and bacterial resistance. *Chemosphere* **2011**, *84*, 1677–1685. [CrossRef] [PubMed]

45. Guo, X.; Li, J.; Yang, F.; Yang, J.; Yin, D. Prevalence of sulfonamide and tetracycline resistance genes in drinking water treatment plants in the Yangtze River Delta, China. *Sci. Total Environ.* **2014**, *493*, 626–631. [CrossRef] [PubMed]

46. Eguchi, K.; Nagase, H.; Ozawa, M.; Endoh, Y.S.; Goto, K.; Hirata, K.; Miyamoto, K.; Yoshimura, H. Evaluation of antimicrobial agents for veterinary use in the ecotoxicity test using microalgae. *Chemosphere* **2004**, *57*, 1733–1738. [CrossRef] [PubMed]

47. Białk-Bielińska, A.; Stolte, S.; Arning, J.; Uebers, U.; Böschen, A.; Stepnowski, P.; Matzke, M. Ecotoxicity evaluation of selected sulfonamides. *Chemosphere* **2011**, *85*, 928–933. [CrossRef] [PubMed]

48. De, L.M.; Fioretto, B.; Poltronieri, C.; Gallina, G. The toxicity of sulfamethazine to Daphnia magna and its additivity to other veterinary sulfonamides and trimethoprim. *Chemosphere* **2009**, *75*, 1519–1524.

49. Ando, T.; Nagase, H.; Eguchi, K.; Hirooka, T.; Nakamura, T.; Miyamoto, K.; Hirata, K. A Novel Method Using Cyanobacteria for Ecotoxicity Test of Veterinary Antimicrobial Agents. *Environ. Toxicol. Chem.* **2010**, *26*, 601–606. [CrossRef]

50. De Souza, S.M.L.; De Vasconcelos, E.C.; Dziedzic, M.; De Oliveira, C.M.R. Environmental risk assessment of antibiotics: An intensive care unit analysis. *Chemosphere* **2009**, *77*, 962–967. [CrossRef] [PubMed]

51. Isidori, M.; Lavorgna, M.; Nardelli, A.; Pascarella, L.; Parrella, A. Toxic and genotoxic evaluation of six antibiotics on non-target organisms. *Sci. Total Environ.* **2005**, *346*, 87–98. [CrossRef] [PubMed]

52. Lin, A.Y.C.; Yu, T.H.; Lin, C.-F. Pharmaceutical contamination in residential, industrial, and agricultural waste streams: Risk to aqueous environments in Taiwan. *Chemosphere* **2008**, *74*, 131–141. [CrossRef] [PubMed]

53. Halling-Sørensen, B. Algal toxicity of antibacterial agents used in intensive farming. *Chemosphere* **2000**, *40*, 731–739. [CrossRef]

54. De Liguoro, M.; Cibin, V.; Capolongo, F.; Halling-Sørensen, B.; Montesissa, C. Use of oxytetracycline and tylosin in intensive calf farming: Evaluation of transfer to manure and soil. *Chemosphere* **2003**, *52*, 203–212. [CrossRef]

water

MDPI

Article

Occurrence, Distribution, and Risk Assessment of Antibiotics in a Subtropical River-Reservoir System

Yihan Chen [1], Hongjie Chen [2,3], Li Zhang [1], Yue Jiang [1], Karina Yew-Hoong Gin [2,3] and Yiliang He [1,4,*]

[1]   School of Environmental Science & Engineering, Shanghai Jiao Tong University, 800 Dongchuan Road, Shanghai 200240, China; chenyhok1987@sjtu.edu.cn (Y.C.); zhangli8157@sina.com (L.Z.); jiangyue0524@sjtu.edu.cn (Y.J.)
[2]   Department of Civil and Environmental Engineering, National University of Singapore, 1 Engineering Drive 2, E1A 07-03, Singapore 117576, Singapore; e0008539@u.nus.edu (H.C.); ceeginyh@nus.edu.sg (K.Y.-H.G.)
[3]   Environmental Research Institute, National University of Singapore, 5A Engineering Drive 1, #02-01, Singapore 117411, Singapore
[4]   China-UK Low Carbon College, Shanghai Jiao Tong University, 800 Dongchuan Road, Shanghai 200240, China
*   Correspondence: ylhe@sjtu.edu.cn; Tel.: +86-21-5474-4008

Received: 25 November 2017; Accepted: 23 January 2018; Published: 26 January 2018

Abstract: Antibiotic pollutions in the aquatic environment have attracted widespread attention due to their ubiquitous distribution and antibacterial properties. The occurrence, distribution, and ecological risk assessment of 17 common antibiotics in this study were preformed in a vital drinking water source represented as a river-reservoir system in South China. In general, 15 antibiotics were detected at least once in the watershed, with the total concentrations of antibiotics in the water samples ranging from 193.6 to 863.3 ng/L and 115.1 to 278.2 μg/kg in the sediment samples. For the water samples, higher rain runoff may contribute to the levels of total concentration in the river system, while perennial anthropic activity associated with the usage pattern of antibiotics may be an important factor determining similar sources and release mechanisms of antibiotics in the riparian environment. Meanwhile, the reservoir system could act as a stable reactor to influence the level and composition of antibiotics exported from the river system. For the sediment samples, hydrological factor in the reservoir may influence the antibiotic distributions along with seasonal variation. Ecological risk assessment revealed that tetracycline and ciprofloxacin could pose high risks in the aquatic environment. Taken together, further investigations should be performed to elaborate the environmental behaviors of antibiotics in the river-reservoir system, especially in drinking water sources.

Keywords: antibiotics; river-reservoir system; water; sediment; risk assessment

1. Introduction

Antibiotics have been extensively and effectively used for several decades not only to relieve symptoms and treat human and animal diseases, but also to promote growths in the livestock, aquaculture and plant agriculture [1,2]. Estimated annual antibiotic consumption in the world ranged from 100,000 to 200,000 tons, and antibiotic consumption is on the rise [3,4]. Based on the market survey, China is considered as the largest producer and user of antibiotics in the world, with approximately 162,000 tons consumed in China (in 2013) [5]. In addition, it has been reported that antibiotics could only be partially metabolized by humans or animals [3]. Then, active substances associated with the antibiotics pass through artificial environments, and may end up in different environmental compartments including water, sediment, and soil, attributed to the fact that most of these chemicals are water soluble and not susceptible to degradation and transformation [6–8]. Consequently, it is inevitable that environmental organisms are more or less exposed to these compounds, which are still active with ecotoxic effects even in a very low concentration. Moreover, antibiotics also can contribute to the dissemination of antibiotic resistance genes (ARGs) and subsequently pose potential risk to human health [9,10].

Aquatic environments play a vital role in maintaining the biogeochemical processes in the world. In recent years, especially natural surface water sources have been universally threatened by the continuous pressure of anthropogenic contamination and are in need for better protection for the drinking water quality. One of the most particular concerns in drinking water sources is the widely detected antibiotics due to their persistence and negative effects [11]. Unfortunately, even advanced drinking water treatment cannot thoroughly remove all antibiotics [12]. Although considerable research has developed regarding the presence of antibiotics in the environment [13–15], there has been comparatively few investigations on their characteristics in drinking water sources. In addition, a large number of reservoirs were built in the upstream rivers for various purposes, including drinking water supply, flood control, irrigation, and the generation of hydropower. The number of reservoirs have increased dramatically over the past several decades, reaching about 16.7 million dams and over 50,000 large dams in the current world [16]. Undoubtedly, these dams have the potential to disrupt the original geochemical processes and ecological connectivity of rivers [16,17]. Based on the geographical distribution from the river to reservoir, river-reservoir systems are commonly used to describe its hybrid environmental properties. Previous studies mainly focused on the dissolved organic matter, pesticides, phthalate and heavy metals [18–21]. However, very few published papers have reported the pollution characteristics of antibiotics in the river-reservoir system and the impact of reservoirs on the spatiotemporal distribution of riverine antibiotics. Therefore, characterizing the differences in the occurrence, distribution and ecological risks of antibiotics from river system to reservoir system will be of great significance for a better understanding of their biogeochemical behaviors along with the environmental gradients.

To achieve these mentioned points, the present work, was experimentally performed in a subtropical river-reservoir system, which was located on the Headwater Region of the Dongjiang River (HRDR). It is also a vital drinking water source for about 40 million people living in several metropolitan cities (e.g., Guangzhou, Shenzhen, and Hong Kong) [10]. Thus, the objectives of this study were to: (1) comprehensively investigate the occurrence, spatiotemporal distribution, and ecological risks of 17 commonly used antibiotics in bulk waters and surface sediments of the HRDR and; (2) perform a comparative study on the dynamic characteristics of antibiotics in the river-reservoir system, as well as to provide new insights into the impact of the reservoir on the biogeochemical behaviors of riverine antibiotics.

2. Materials and Methods

2.1. Field Sites and Sampling

The HRDR is located in southern Jiangxi Province and northern Guangdong Province in China (Figure 1), covering an area of about 5161 km². In general, this catchment consists of two primary rivers located in the anthropic zone and a reservoir (Fengshuba Reservoir) located in the natural environment. Moreover, the rivers of Beiling River (BLR) and Xunwu River (XWR) are characterized by typical nonpoint source pollution from 467,000 rural residents with a large number of livestock breeding [22]. Interestingly, the reservoir could be regarded as a large pool with the average depth of 70 m to impound the upstream waters from the two rivers. As a result, apart from the antibiotic inputs from the rivers, we hypothesized that the reservoir as a natural environment was less directly polluted with antibiotics due to the forestry projects and reservoir resettlement. Detailed description of the basic hydrological information of the catchment can be found in the Supplementary Materials Text S1.

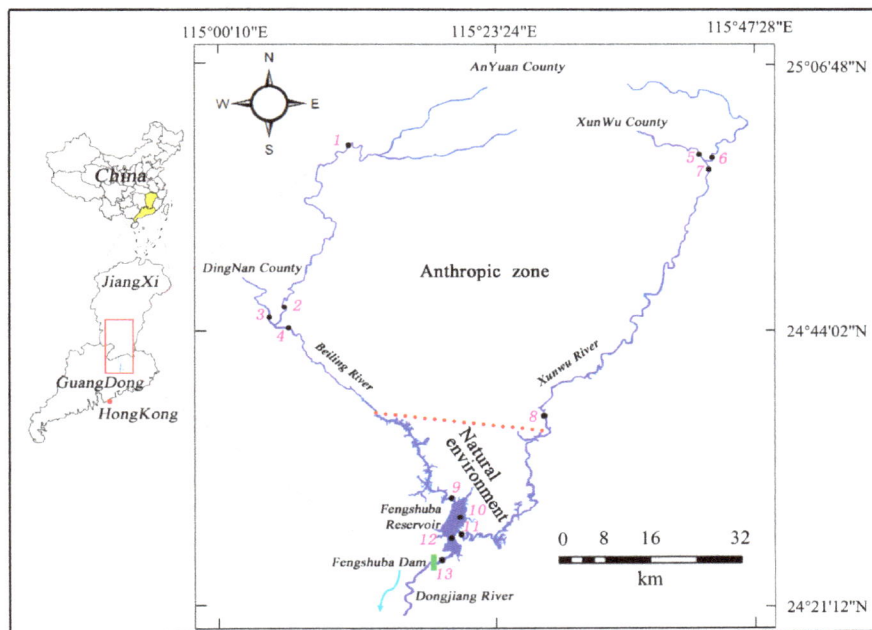

Figure 1. Map showing the sampling sites in the river system (Beiling River (S1–S4), Xunwu River (S5–S8)) and reservoir system (Fengshuba Reservoir (S9–S13)).

According to three typical hydrological periods (Figure S1), three campaigns took place on 18–20 July 2015 (wet-to-dry transition season), 25–27 November 2015 (dry season) and 14–16 March 2016 (wet season), respectively. In detail, eight surface water samples (0.5 m below surface) in the river system (S1–S8) were obtained, but five equal-mixed water samples (S9–S13) were obtained form the surface water (0.5 m below surface), middle water (half of the depth) and bottom water (about 2–4 m above bottom) of the Fengshuba Reservoir (FSBR). In particular, no representative sediment samples were obtained in the rivers due to the fact that rapid water flow was not conducive to the formation of sediment. Accordingly, surface sediment samples (about 15 cm) were only collected by a sediment sampler at the five stations (S9–S13) of FSBR. All the above samples were collected in 5 L acid cleaned glass bottles and were immediately transported to the laboratory in an ultra-low temperature storage tank. Before being analyzed, water samples and sediment samples were kept at 4 °C and −80 °C in the dark until extraction, respectively.

2.2. Chemicals and Standards

Antibiotic standards of three sulfonamides (including sulfadiazine (SDZ), sulfamonomethoxine (SMM) and sulfaquinoxaline (SQX)), three fluoroquinolones (including norfloxacin (NOR), ciprofloxacin (CIP) and ofloxacin (OFC)), four β-lactamases (including amoxicillin (AMX), cefalexin (CLX), penicillin G (PENG) and penicillin V (PENV)), three tetracyclines (including oxytetracycline (OTC), tetracycline (TC) and doxycycline (DC)) and four others (including tylosin (TYL), erythromycin-H_2O (ETM-H_2O), lincomycin (LIN) and vancomycin (VAN)) were purchased from the Dr. Ehrenstorfer (Augsburg, Germany). Furthermore, isotopic standard of ciprofloxacin-D8 is a surrogate standard for all the selected antibiotics, and norfloxacin-D5 was chosen as an internal standard for the quantification of all the samples. These two isotopic standards were purchased from the TRC (Toronto, ON, Canada). The basic physicochemical characteristics and usages of these antibiotics are shown in Supplementary Materials Tables S2 and S3.

Methanol, acetonitrile, and formic acid (HCOOH) were all of high-performance liquid chromatography (HPLC) grade and purchased from the Thermo Fisher Scientific Inc. (Loughborough, UK). Phosphate-citric acid and disodium ethylenediamine tetraacetate (Na$_2$EDTA) were all of analytical grade and obtained from the Shanghai Sangon Biotechnology Co. Ltd. (Shanghai, China). Ultra pure water (10–18 MΩ·cm, 25 °C) was obtained from a Milli-Q system (Millipore, Bedford, MA, USA).

2.3. Sample Preparation

About 1 L triplicate water samples were filtered through 0.45 μm glass fiber filter (ANPEL Corp., Shanghai, China) and subsequently subjected to solid phase extraction (SPE) with Poly-Sery hydrophile-lipophile balance (HLB) cartridge (6 cc/200 mg, ANPEL Corp., Shanghai, China) following our previous method [23]. In particular, prior to extraction, all the filtered waters were acidified to pH = 3.0 with 6 mol/L HCl, and then spiked with 200 mg of Na$_2$EDTA and 50 ng of ciprofloxacin-D8. After extraction, the loaded cartridges were eluted with 12 mL methanol, subsequently the eluates were then concentrated to near dryness under a gentle stream of nitrogen. Then, 500 μL mobile phase (methanol and 50 ng of internal standard) were added to redissolve these target compounds, and stored in the refrigerator at −40 °C.

The sediments were freeze-dried and then sieved through a 100-mesh sieve (<150 μm) for further processing. Each sediment sample (2 g dry weight) was weighted into a 30 mL glass tube, followed by addition of 20 ng of ciprofloxacin-D8. Then, the samples were fully mixed and stored in a refrigerator at 4 °C overnight. Ten milliliters phosphate-citric acid (pH = 3.0 and 10 mL) and acetonitrile were added together into each glass tube, followed by mixing using a vortex oscillator for 20 min, then, in an ultrasonic extractor for 20 min. All the glass tubes were centrifuged at 12,000 rpm for 10 min to get the supernatants. In particular, the extraction processes were then repeated twice and the supernatants from the three extractions were combined together. All these supernatants were piped into a 200 mL round bottom flask and evaporated at 40 °C to remove organic solvents on a rotary evaporator. The concentrated residues in the flask were diluted to 200 mL with ultra-pure water to make sure that the organic solvent in the solution was less than 5% [24]. Then, the following method was equivalent to the above processes for the water SPE.

2.4. Chemical Analysis

For all water samples, the measurement of water properties including basic indictors (e.g., temperature, pH, dissolve oxygen (DO), and oxidation reduction potential (ORP)) and wet-to-dry transition parameters (e.g., dissolved organic carbon (DOC), dissolved total nitrogen (DTN), and dissolved total phosphorus (DTP)) could be found in Supplementary Materials Text Section 2. In addition, the contents of total carbon (TOC), total nitrogen (TN) and total phosphorus (TP) in the sediment samples were in accordance with standard methods [25]. Heavy metals (e.g., V, Ag, As, Cd, Co, Cr, Cu, Mo, Ni, Pb, Sn and Zn) for the sediment samples (<150 μm) were quantified by inductive coupled plasma-mass spectroscopy (ICP-MS) according to previous methods [18].

The antibiotics extracted from these samples were analyzed using ultra-high-performance liquid chromatography (UHPLC, Agilent 1290 Infinity, Santa Clara, CA, USA) coupled with tandem mass spectrometry (MS/MS, Agilent 6490 Triple Quadrupole, Santa Clara, CA, USA) [23]. An Agilent C18 column (50 mm × 2.1 mm, 1.8 μm) was used to separate and quantify the target analytes. All the target analytes were separated using a gradient method and is described in detail as follows: 0.1% formic acid in Milli-Q water (mobile phase A) and acetonitrile (mobile phase B) were prepared and the gradient elution started with 5% mobile phase B and kept isocratic for 1.3 min, and then linearly rose to 40% mobile phase B at 8 min and held until 10 min, subsequently dropped to 5% mobile phase B at 12 min. Flow rate of the mobile phase was 0.3 mL/min, the injection volume was 10 mL and the column temperature was maintained at room temperature (25 ± 2 °C). Mass spectrometric analysis was operated in the positive electrospray ionization multiple reaction monitoring (MRM) mode. Tandem mass spectrometric parameters for each antibiotic are summarized in Supplementary Materials Table S4. In addition, detailed information on the quality assurance and control are shown in Supplementary Materials Text Section 3, Table S5.

2.5. Ecological Risk Assessment

The ecological risk quotient (RQ) was calculated for each antibiotic using the ratio between the measured environmental concentration (MEC) and predicted no effect concentration (PNEC). The PNEC of each antibiotic was calculated by acute and chronic aquatic toxicity, dividing the lowest short-term L(E)C$_{50}$ or long-term non-observable effect concentration (NOEC) respectively through the division of an appropriate factor (AF). The appropriate assessment factors were used based on the methods and principles as described in detail in previous studies [26]. However, when NOEC values were unavailable, LC50 or EC50 values were used instead. In addition, typical taxons (algae, aquatic invertebrates and fish) from three trophic levels were chosen to represent the food chain in aquatic ecosystems. For most of the antibiotics, L(E)C$_{50}$ and NOEC values were obtained from the ECOSAR V1.10. In general, RQs were classified into three levels of risk: when RQ $\geq$ 1.0, high risk; $0.1 \leq$ RQ < 1.0, medium risk; RQ < 0.1, low risk [27].

2.6. Data Analysis

Statistical analyses and graphic visualizations including means $\pm$ standard deviation (SD), one-way ANOVA, and correlation analysis were performed using IBM SPSS 21.0 (IBM, Chicago, IL, USA) and OriginPro 2016 software (OriginLab, Northampton, MA, USA). All the test results with *p*-level $\leq$ 0.05 were considered to be statistically significant for pair comparisons. Spatial mapping of sampling sites, distributions and calculations of the mass balance of the antibiotics in the reservoir were performed using ArcGIS 10.2 (ESRI, Redlands, LA, USA) based on the Kriging analysis. Ecological risk assessment of the antibiotics were performed using HemI 1.0 [28].

3. Results and Discussion

3.1. Antibiotics in the Water Phase of the HRDR

Among the 17 target antibiotics, 15 antibiotics from five categories were detected in the water samples of the HRDR (Table 1). Five antibiotics (SDZ, SMM, SQX, DC, and Lin) were the most frequently detected compounds in 100% of the samples. NOR, CIP, OFC, CLX, PENG, PENV, OTC, and TC showed the second highest detection frequencies greater than 80%. These results suggest that the drinking water source has been severely disrupted by anthropic activities. The total concentrations of antibiotics in the water samples ranged from 193.6 ng/L located in the reservoir system to 863.3 ng/L located in the river system (Supplementary Materials Figure S2). The concentrations of Σ sulfonamides, Σ fluoroquinolones, Σ beta-lactamases, Σ tetracyclines and Σ others (defined as the sum of the corresponding antibiotics) ranged from 11.6 to 108.6 ng/L, from 26.8 to 597.2 ng/L, from 7.6 to 193.2 ng/L, from 9.1 to 300.7 ng/L, and from 0.3 to 15.4 ng/L, respectively. Additionally, fluoroquinolones were the dominant compounds in the water samples of the river system, which is consistent with previous studies in other water environments (Supplementary Materials Figure S3) [29,30]. In comparison, the information found in the water samples of the reservoir system was inconsistent with those in the river system. Taken together, these differences in the detection frequencies and levels of various antibiotics in the river-reservoir system may be attributed to the usage patterns of antibiotics in the surrounding catchments, as well as to the biogeochemical processes of antibiotics along with different hydrologic gradients, such as photo-degradation, adsorption, and biodegradation [3,27,31].

Table 1. Concentrations of antibiotics in the water samples of the HRDR (ng/L).

Classification	Compound	Range	Mean	Median	Detection Rate (%)
Sulfonamides	sulfadiazine	1.7–83.8	15.9	10.3	100 (39/39)
	sulfamonomethoxine	2.2–66.0	18.6	14.7	100 (39/39)
	sulfaquinoxaline	0.4–6.0	2.68	2.7	100 (39/39)
Fluoroquinolones	norfloxacin	<LOQ–156.3	62.3	58.9	97.4 (38/39)
	ciprofloxacin	<LOQ–442.1	169.2	156.7	94.9 (37/39)
	ofloxacin	<LOQ–17.2	7.1	6.7	97.4 (38/39)
Beta-lactamases	cefalexin	<LOQ–25.7	7.4	5.6	97.4 (38/39)
	penicillin G	<LOQ–97.0	19.2	10.8	92.3 (36/39)
	penicillin V	<LOQ–115.8	42.7	35.4	92.3 (36/49)
Tetracyclines	oxytetracycline	<LOQ–135.5	49.7	41.0	94.9 (37/39)
	tetracycline	<LOQ–111.5	44.9	43.3	92.3 (36/39)
	doxycycline	0.8~256.4	20.6	9.1	100 (39/39)
Others	tylosin	<LOQ–1.6	0.6	0.6	74.4 (29/39)
	erythromycin-H_2O	<LOQ–6.9	0.7	0.3	79.5 (31/38)
	lincomycin	0.13–10.4	1.4	0.7	100 (39/39)

Note: <LOD = values were below the level of detection (LOD).

For fluoroquinolones, the maximum concentration of CIP was found at site S3 (dry season) with the value of 442.1 ng/L, followed by site S5 (wet season) with the value of 417.6 ng/L (Supplementary Materials Figure S2). Unfortunately, the two sites of S3 and S5 are located in the rivers receiving directly integrated wastewater from the cities of Dingnan (urban population of 39,700) and Xunwu (urban population of 99,498) without effective domestic wastewater treatment facilities (Figure 1). Fluoroquinolones are used extensively in both human and veterinary medicines, and nearly 5340 tons of CIP were consumed in China in 2013 [5]. However, untreated wastewater related to high population density and the livestock units along the urban rivers may contribute to the CIP levels in the receiving rivers [32]. In general, it could be concluded that non-routine episodes associated with untreated effluents, landfills and medical wastewaters from urban inhabitants would exacerbate the pollution of antibiotics [33,34]. Furthermore, compared with previous studies, the mean CIP concentration for the water samples (169.2 ± 107.9 ng/L) was of similar levels to some other aquatic environments, such as the Wangyang River in China (205.5 ng/L) [35], Lebanese rivers (108 ng/L) [36], and the rivers in Northern Pakistan (110 ng/L) [37]. However, it was far greater than that in the water samples of Seine River (France) [38], Yangtze Estuary (China) [39], and Pearl River (China) [40], as well as other catchments in China (Poyang Lake, Chao Lake, and Liao River) [27,41,42]. Taken together, our results further support the concept that the CIP makes a major contribution to the burden of antibiotic pollutions and CIP contamination was at a moderate level in the HRDR, hence meriting preferential control.

For the water samples in the river system, the total concentration of antibiotics in the dry season (368.8 ± 66.9 ng/L) was significantly lower than that in the wet-to-dry transition season (567.0 ± 110.3 ng/L) and wet season (636.0 ± 138.8 ng/L) ($p < 0.05$) (Figure 2), suggesting that frequent rain runoff may contribute to the levels of total concentration especially in the rivers characterized as non-point source pollution watershed (Supplementary Materials Figure S2) [22,43–45]. However, this seasonal variation trend is inconsistent with previous investigations [31,39]. Furthermore, it was notable that the antibiotic compositions in the rivers did not exhibit a significant difference with the seasonal variation (analysis of similarities (ANOSIM), $p > 0.05$) (Supplementary Materials Figure S3), indicating that perennial anthropic activity, but not the seasons or hydrologic conditions, may be an important factor influencing the usage pattern of these antibiotics. That is to say, this result also suggests that these antibiotics probably have similar sources and release mechanisms from anthropogenic activities in the riparian environment [46]. In addition, for the river-reservoir system, the total concentrations of antibiotics in the river system was markedly higher than those in the reservoir system along with the seasons ($p < 0.05$) (Figure 2), implying that river input was a potential important source of antibiotics in the reservoir. Also, this decreasing trend form the river system to reservoir system further suggests that the reservoir could be considered as a stable reactor to reduce the levels of total antibiotics from the river system. In addition, the antibiotic compositions changed significantly from the river system to the reservoir system in the dry and wet seasons (ANOSIM, $p < 0.05$). One plausible explanation for this is that long hydraulic retention time (HRT), increased transparency, and large surface area in the reservoir system, compared to those in the river system, would contribute to the attenuation in the total antibiotic concentrations of river system, probably via various geochemical processes (e.g., biotransformation, photolysis, sorption, and dispersion) [11,47]. In contrast, both the levels and compositions of antibiotics in the reservoir did not exhibit significant differences along with the seasons ($p > 0.05$). On the whole, these results suggest that hydrologic conditions or river input could not be the predominant factor influencing the environmental behaviors of antibiotics in the reservoir system. Meanwhile, it also implies that the reservoir can act as an ideal ecological barrier, which probably has a great capacity of water environment to balance and regulate the level and composition of antibiotics exported from the catchment.

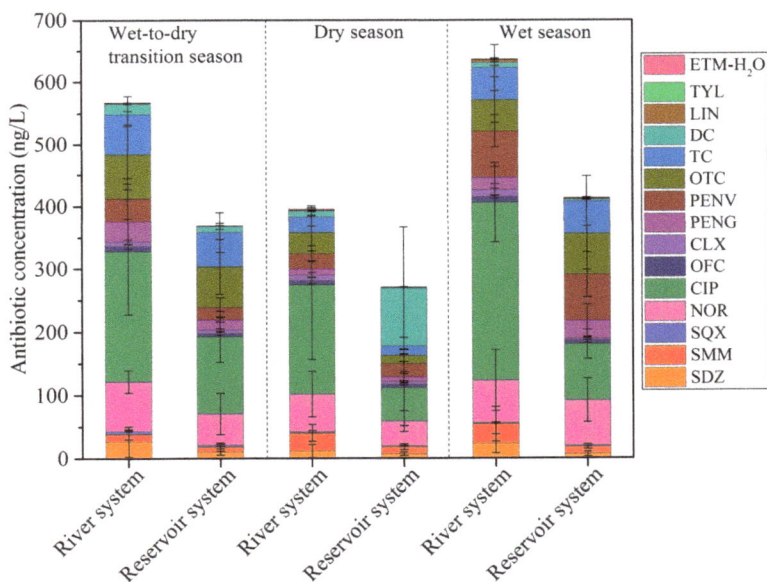

Figure 2. Accumulated concentrations of antibiotics in the river system (S1–S8) and reservoir system (S9–S13) over the three seasons. Error bars represent standard error of the mean.

3.2. Antibiotics in the Sediment Phase of the FSBR

Among the target antibiotics, 16 antibiotics were detected in the sediment samples of the FSBR (Table 2), with total concentrations ranging from 115.1 to 278.2 µg/kg (Figure 3, Supplementary Materials Figure S4). The 11 antibiotic compounds (e.g., SDZ, NOR, CLX, OTC, and LIN) were detected with a high detection frequency of 100% (Table 2). The average concentrations of five categories decreased in the order: tetracyclines (88.8 ± 34.2 µg/kg) > fluoroquinolones (53.7 ± 33.6 µg/kg) > β-lactamases (33.7 ± 23.4 µg/kg) > sulfonamides (18.1 ± 12.8 µg/kg) > others (11.5 ± 9.8 µg/kg). The tetracyclines in the sediments showed the highest concentration levels, which is consistent with previous results in other catchments in China, such as the Huangpu River [48], Liao River [49],

Jiulong River and Nanliu River [50]. In addition, OTC, NOR, PENV, SMM and LIN were the dominant compounds in the corresponding categories, respectively (Table 2). It also followed the order: OTC > NOR > PENV > LIN > SMM. However, this finding in the sediments was markedly different from those in the waters. Probably, the reason for this is that total organic matter and cation exchange capacity in the sediment exert a greater affinity with the OTC, but the SMM may have relatively higher water solubility and weaker affinity to the sediments [51,52]. Furthermore, the mean concentration of OTC (44.6 ± 27.4 µg/kg) in the sediments was obviously higher than that in the Liao River (29.1 µg/kg) [49], Huangpu River (6.9 µg/kg) [48], and Yellow River Delta (4.9 µg/kg) [53], but comparable to that in the Taihu Lake [54]. However, it was dramatically lower than that in the Wangyang River (36,148.3 µg/kg) [35] and southern Baltic Sea [55]. In general, these results support the view that the OTC pollution is at a moderate level in the sediment phase of FSBR.

Table 2. Concentrations of antibiotics in the sediment samples of the FSBR (µg/kg).

Classification	Compound	Range	Mean	Median	Detection Rate (%)
Sulfonamides	sulfadiazine	0.7–19.5	7.6	3.8	100 (15/15)
	sulfamonomethoxine	1.5–20.1	7.8	5.5	100 (15/15)
	sulfaquinoxaline	0.8–6.6	2.8	1.9	100 (15/15)
Fluoroquinolones	norfloxacin	9.7–132.3	30.3	22.0	100 (15/15)
	ciprofloxacin	6.1–27.4	16.4	18.4	100 (15/15)
	ofloxacin	<LOQ–13.5	6.9	5.0	93.3 (14/15)
Beta-lactamases	cefalexin	3.7–21.9	12.6	10.8	100 (15/15)
	penicillin G	<LOQ–28.9	4.8	<LOQ	33.3 (5/15)
	penicillin V	<LOQ–73.2	16.3	11.3	80 (12/15)
Tetracyclines	oxytetracycline	12.2–102.4	44.6	44.2	100 (15/15)
	tetracycline	6.0–95.7	40.0	30.4	100 (15/15)
	doxycycline	0.8–20.9	5.2	3.1	100 (15/15)
Others	tylosin	<LOQ–4.8	0.6	<LOQ	46.7 (7/15)
	erythromycin-H_2O	0.4–9.1	3.0	2.0	100 (15/15)
	lincomycin	0.2–24.7	6.4	6.0	100 (15/15)
	vancomycin	<LOQ–7.0	1.4	1.1	60 (9/15)

Note: <LOD = values were below the level of detection (LOD).

In addition, it can be easily found that spatial distribution of the total concentration of the antibiotics exhibit a similar trend between the wet-to-dry transition season and wet season, which is contrary to the change trend in the dry season (Figure 3, Supplementary Materials Figure S4). Coincidentally, paired-sample t-tests indicated that antibiotic compositions in the dry season was significantly different from those in the wet-to-dry transition season (ANOSIM, $p < 0.05$) or wet season (ANOSIM, $p < 0.01$). However, no significant differences could be found between the wet-to-dry transition season and wet season ($p > 0.05$) (Supplementary Materials Figure S4). Based on these, it was notable that the change trend in the antibiotic levels and compositions of the sediments showed a consistent result along with the seasonal variations. Meanwhile, the hydraulic retention time in the dry season (November, 2015; HRT = 246 days) is much longer than that in the wet-to-dry transition season (July, 2015; HRT = 122 days) or wet season (March, 2016; HRT = 49 days) (Supplementary Materials Table S1). It can be concluded that hydrological factors in the reservoir may influence the antibiotic distributions in the sediments [52]. Taken together, generally, these results further suggest that longer HRT may contribute to the accumulation of antibiotics in the internal region of FSBR (Figure 3b), however shorter HRT contributes to the accumulation of antibiotics in the inlet or outlet region of FSBR (Figure 3a,c). Unfortunately, it was difficult to determine the complex mechanisms associated with the hydrological and chemical factors in the fate and transport of antibiotics in the reservoir studied, hence meriting further investigations.

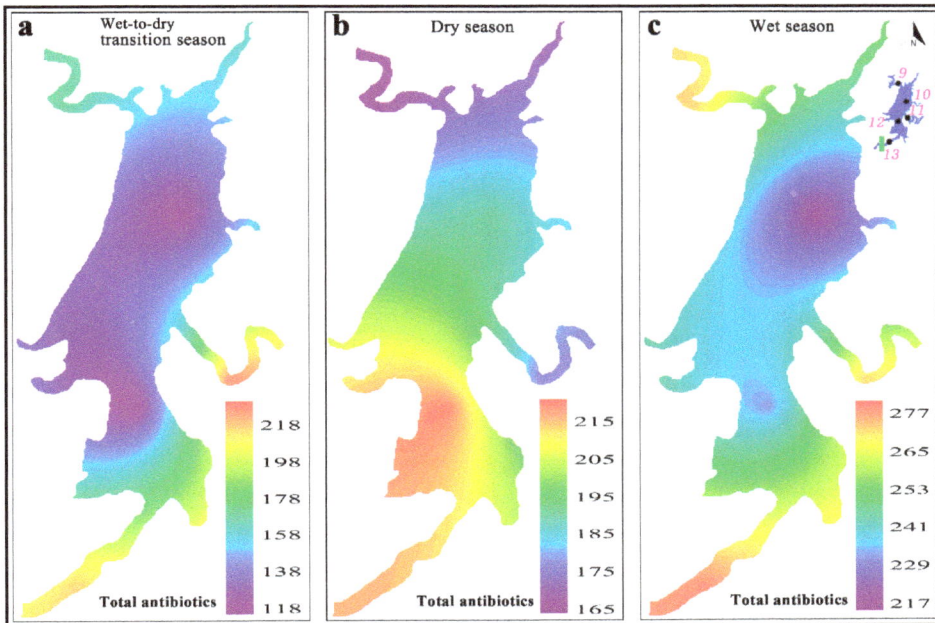

Figure 3. Contour maps showing the spatiotemporal distributions of total antibiotics in the sediment phase of FSBR (µg/kg). (**a**) Wet-to-dry transition season. (**b**) Dry season. (**c**) Wet season.

3.3. Relationships between Basic Parameters and Antibiotic Concentrations

Owing to the above-mentioned differences between the river system and reservoir system, the water and sediment samples were separately performed to explore the correlations between the basic parameters and antibiotic concentrations (Figure 4). For the water samples, the levels of NOR showed significantly positive relationships with the DTN ($r^2 = 0.55$, $p < 0.01$) and DTP ($r^2 = 0.45$, $p < 0.05$), indicating that these compounds may have an identical pollutant source from anthropogenic emissions in the river watersheds. In China, it has recently been reported that the total usage of NOR was 5440 tons, with only 18.62% consumed by humans and the rest consumed by animals. Furthermore, nitrogen and phosphorus pollutions are closely related to nonpoint source pollution (e.g., disperse domestic sewage, animal husbandry, and manure-fertilizer) in the rural catchment [22,52]. Based on these, it can be further inferred that nonpoint source pollution, especially untreated livestock wastewater and agricultural pollution, may contribute to most of the level of NOR in the river system. In addition, a positive relationship was found between the levels of SMM and TYL ($r^2 = 0.47$, $p < 0.05$) in the river system. Considering the fact that the TYL is only a veterinary antibiotic, it can be concluded that the SMM may be closely linked to the livestock and poultry farming in the catchment, which is consistent with the survey in China [5]. In contrast, all these relationships were not found in the reservoir system. Meanwhile, it was obvious that the relationship profiles in the river system was markedly different from that in the reservoir system. These results further suggest that the reservoir may have a potency to regulate the migration and transformation of antibiotics. On the other hand, compared to the pristine reservoir, it also implies that the concentration and distribution of antibiotics in the aquatic environment may be more related to human activities [56].

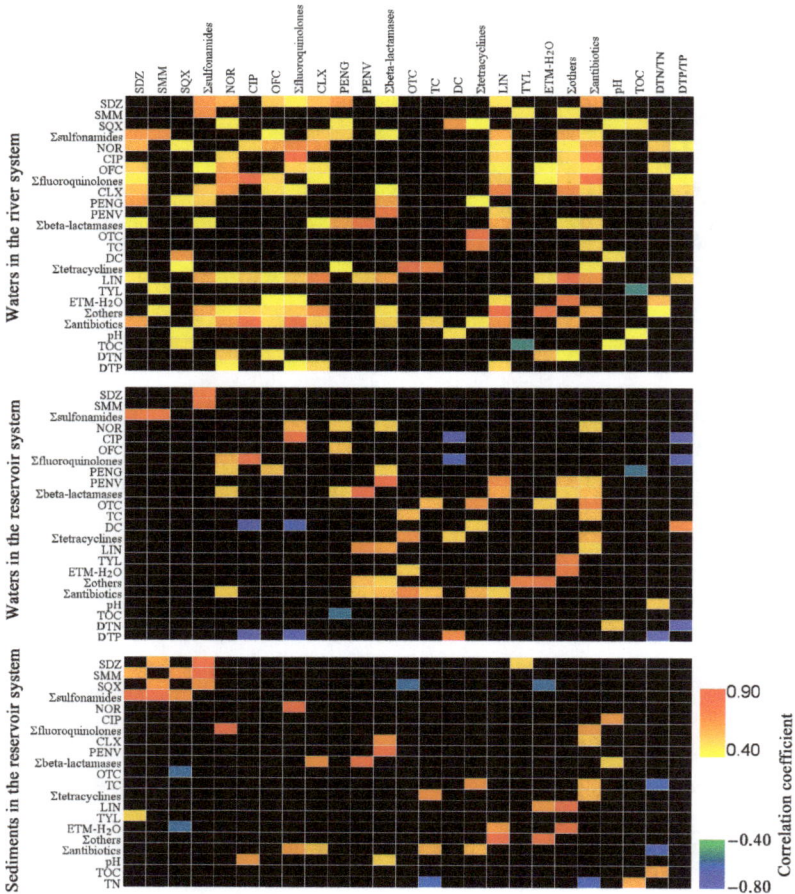

Figure 4. Pearson's correlation coefficients for antibiotic concentrations and basic parameters in the water and sediment samples in the river-reservoir system ($p < 0.05$). Bottom right shows the color band indicator of the correlation coefficient. Black boxes and no box shown in each line in the figure delineate no significant correlation for corresponding parameters ($p > 0.05$).

For the sediment samples, the levels of SMM showed significant relationships with the SDX ($r^2 = 0.63$, $p < 0.05$) and SQX ($r^2 = 0.67$, $p < 0.01$) (Figure 4). However, no significant relationships could be found between the sulfonamides and sediment property (e.g., pH, TOC, TN, and TP). These results suggest that similar chemical structures, but not the sediment property, may play an important role in the fate and residue of the selected sulfonamides in the sediments [57]. In addition, a positive relationship could be found between the level of CIP and pH ($r^2 = 0.66$, $p < 0.01$), probably suggesting that higher pH could accelerate the accumulation of CIP in the sediment particulates [58]. The levels of NOR showed significantly positive relationships with the heavy metals (e.g., As, Cd, Co, Cu, Mo, and Pb) ($p < 0.05$), respectively. This phenomenon associated with the concurrent presences of NOR and heavy metals may be attributed to the common mechanism of electrostatic adsorption in the sediments [59,60]. In comparison, no significant relationships were found between the heavy metals and other antibiotics. However, the heavy metal pollution could also enhance the level of bacterial antibiotic resistance in the environment [9]. Hence, the co-selection potential of heavy metals and antibiotics in the dissemination of ARGs should not be neglected in any case.

In general, no significant relationships were found between the levels of total antibiotics and DOC/TOC in the three environmental compartments (Figure 4), suggesting that organic matter may not be the pivotal factor controlling the levels and distributions of antibiotics in the HRDR. A similar phenomenon was also found in the Baiyangdian Lake [61]. However, this finding is inconsistent with previous studies in the Yangtze Estuary [39], South Yellow Sea [29], Honghu Lake and East Dongting Lake [62].

3.4. Pseudo-Partitioning Coefficients of Antibiotics in the FSBR

The pseudo-partitioning coefficients (P-PC) is of great significance to get a better understanding of the dynamics of antibiotics between the sediment and water phases. In the study, P-PCs were calculated from the antibiotic concentrations in the sediments divided by their corresponding concentrations in the bulk waters, and the organic carbon normalized pseudo-partitioning coefficients (K_{oc}) were also calculated from the P-PCs divided by their corresponding fraction organic carbon content in sediments [48]. In general, the P-PCs and K_{oc} values were highly variable in the reservoir. For instance, the P-PC ranged from 145.46 to 5993.60 L/kg for SDZ, from 195.32 to 1470.89 L/kg for NOR, and from 15.61 to 8742.34 L/kg for DC. This suggests that NOR accumulates more easily in the sediments compared to SDZ and NOR, which is consistent with their corresponding chemical property associated with the Log K_{ow} (Supplementary Materials Table S2). Furthermore, the K_{oc} values ranged from 15,376.54 to 218,229.40 L/kg-oc for SDZ, from 18,172.64 to 140,804.50 L/kg-oc for NOR, and from 1973.08 to 817,805.80 L/kg-oc for DC. Taken together, such a variability suggests that although organic carbon content in sediments is an important parameter in sediment-water interactions of antibiotics, these different degrees of sediment-water interactions in the reservoir are likely driven by multiple factors (e.g., chemical characteristic, hydrological factor and mineral surface), rather than any single factor [48,52,53]. In other words, when the environmental conditions change, antibiotics adsorbed by the sediment phase would be released into the aqueous phase again.

3.5. Ecological Risk Assessment

Considering the fact that toxicity data of these antibiotics in sediment phase is very scarce, ecological risk assessment of the antibiotics were only evaluated in the water phase. The PNEC values of antibiotics were calculated by different assessment factors (Supplementary Materials Table S6), and risk quotients (RQs) corresponding to the spatiotemporal distribution of each antibiotic concentration detected in this study (shown in Figure 5a). In general, the RQs of ten antibiotics including SMM, SQX, OFC, CLX, PENG, PENV, DC, LIN, TYL, and ETM-H$_2$O were found to be below 0.1, indicating that these antibiotics may pose low ecological risks in the HRDR. The RQs of NOR and CIP ranged from 0.1 to 1.0, potentially suggesting that they pose medium ecological risks to the aquatic ecosystem. However, the RQs of TC were markedly higher than 1.0 due to the fact that some fishes are more sensitive to the TC (Supplementary Materials Table S6). This result suggests that the TC may pose a high ecological risk to the aquatic ecosystem and should be given priority controls.

Most importantly, selection pressure from antibiotics in the aquatic environment may accelerate the evolution and dissemination of ARGs [9]. Nevertheless, few minimal selective concentrations were currently determined from experimental research or natural ecosystems. Here, the PNEC (resistance selection) values of available antibiotics were obtained from an estimated research (Supplementary Materials Table S7) [63]. Likewise, risk quotients (RQs) of the antibiotics on potential selection for resistant bacteria corresponding to the spatiotemporal distribution of available antibiotics were shown in Figure 5b. The RQs of seven antibiotics including OFC, CLX, TC, DC, LIN, TYL, and ETM-H$_2$O were found to be below 0.1, implying that these antibiotics may not be selective for resistant bacteria. In addition, it is notable that the RQs of CIP are generally higher than 1. This suggests that the water concentrations of CIP may have a high probably to exert selection pressure for resistant bacterial [63]. Hence, CIP should also be given priority controls considering this severe situation. Strikingly, antibiotics in aquatic ecosystems could have great impacts not only in aquatic organisms, but also in the dynamics of bacterial population and ARGs [64,65]. Unfortunately, the ecological risk assessments associated with the ARGs and bacterial population dynamics are still scarce. Taken together, more attention should be given in order to combat these challenges, especially in the drinking water sources that are closely linked to human health.

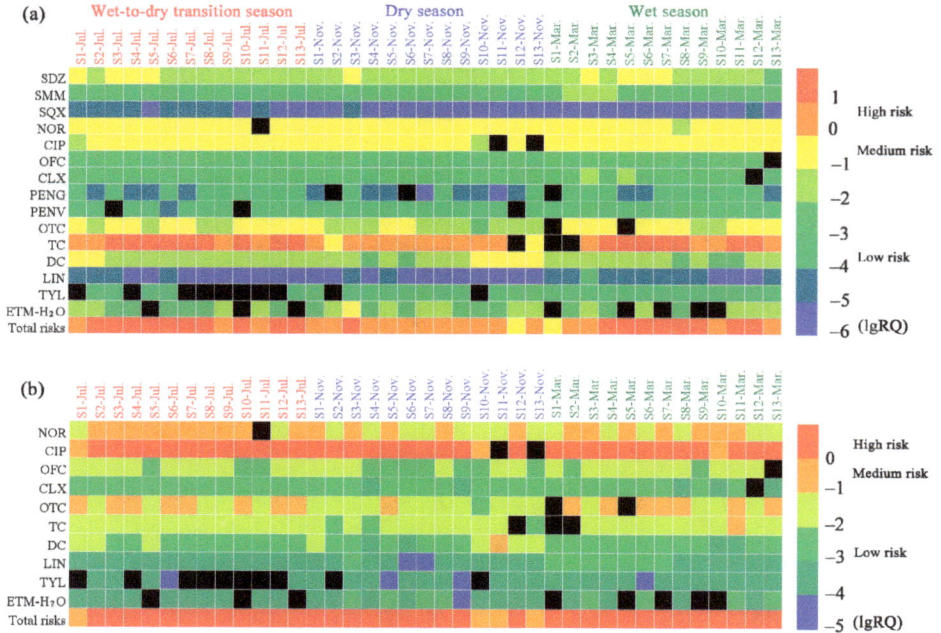

Figure 5. Calculated risk quotients (RQs) for the detected antibiotics in the water samples with the three seasons. (**a**) Ecological risk of the antibiotics on the typical taxons. (**b**) Ecological risk of the antibiotics on the selection for resistant bacterial. Bottom right shows the color band indicator of the level of ecological risk based on a log scale. Black boxes delineate no ecological risks due to the corresponding antibiotics undetected.

4. Conclusions

In this study, the occurrence and distribution of 17 antibiotics were investigated in a river-reservoir system, a key drinking water source in South China. Generally, 15 antibiotics were detected at least once in the watershed, with the total concentrations of antibiotics in the water samples ranging from 193.59 to 863.27 ng/L and 115.14 to 278.16 µg/kg in the sediment samples. For the waters, higher rain runoff may contribute to the levels of total antibiotics, while perennial anthropic activity may be an important factor influencing the usage pattern of antibiotics in the river system. However, the reservoir system can act as a stable reactor to balance and regulate the level and composition of antibiotics exported from the catchment. For the sediments, hydrological factors in the reservoir may influence the antibiotic distributions along with seasonal variation. Meanwhile, pH and heavy metals may control the accumulation of CIP and NOR in the sediment phase, respectively. Overall, tetracycline and ciprofloxacin could pose high risks in the aquatic environment, and should be preferentially controlled.

Supplementary Materials: The following are available online at http://www.mdpi.com/2073-4441/10/2/104/s1, Figure S1: Monthly variations of rainfall and inflowing runoff for the HRDR from April 2015 to March 2016;

Figure S2: Antibiotic concentrations in the 39 water samples in the HRDR during the three seasons (July-2015, November-2015, and March-2016); Figure S3: Spatiotemporal variations of antibiotic compositions in the 39 water samples of the HRDR during the three seasons (July-2015, November-2015, and March-2016); Figure S4: Antibiotic concentrations and compositions in the 15 sediment samples of the FSBR with the three seasons. Table S1: Description of the basic hydrological information of Fengshuba Reservoir. Hydraulic retention times were calculated as volume of reservoir divided reservoir outflow; Table S2: The basic physicochemical characteristics of the target antibiotics; Table S3: Usage of selected antibiotics in China in 2013; Table S4: HPLC-MS/MS parameters for analysis of the analytes by MRM; Table S5: Method quantification limit (MQL), and recoveries of the antibiotics; Table S6: Ecotoxicity endpoints for fish, aquatic invertebrates, and algae and related PNEC values (µg/L) for these antibiotics; Table S7: Minimal selective concentrations (MSCs) for the available antibiotic (in µg/L).

Acknowledgments: The authors are grateful for the financial support from the National Science and Technology Major Projects of Water Pollution Control and Management of China (2014ZX07206001), and Singapore under its Campus for Research Excellence and Technological Enterprise (CREATE) programme (E2S2-CREATE project CS-B: Challenge of Emerging Contaminants on Environmental Sustainability in Megacities).

Author Contributions: Yihan Chen designed the experiments, collected and analyzed the field data, and wrote the manuscript. Hongjie Chen and Yue Jiang mainly performed the antibiotic detections. Li Zhang helped with the detections of heavy metals in the sediments. Karina Yew-Hoong Gin and Yiliang He provided advice on the experiments and helped with preparation of the manuscript.

Conflicts of Interest: The authors declare no conflicts of interest.

References

1. Zuccato, E.; Castiglioni, S.; Bagnati, R.; Melis, M.; Fanelli, R. Source, occurrence and fate of antibiotics in the Italian aquatic environment. *J. Hazard. Mater.* **2010**, *179*, 1042–1048. [CrossRef] [PubMed]
2. Le Page, G.; Gunnarsson, L.; Snape, J.; Tyler, C.R. Integrating human and environmental health in antibiotic risk assessment: A critical analysis of protection goals, species sensitivity and antimicrobial resistance. *Environ. Int.* **2017**, *109*, 155–169. [CrossRef] [PubMed]
3. Kummerer, K. Antibiotics in the aquatic environment—A review—Part I. *Chemosphere* **2009**, *75*, 417–434. [CrossRef] [PubMed]
4. Laxminarayan, R.; Duse, A.; Wattal, C.; Zaidi, A.K.M.; Wertheim, H.F.L.; Sumpradit, N.; Vlieghe, E.; Hara, G.L.; Gould, I.M.; Goossens, H.; et al. Antibiotic resistance—The need for global solutions. *Lancet Infect. Dis.* **2013**, *13*, 1057–1098. [CrossRef]
5. Zhang, Q.Q.; Ying, G.G.; Pan, C.G.; Liu, Y.S.; Zhao, J.L. Comprehensive evaluation of antibiotics emission and fate in the river basins of China: Source analysis, multimedia modeling, and linkage to bacterial resistance. *Environ. Sci. Technol.* **2015**, *49*, 6772–6782. [CrossRef] [PubMed]
6. Gbylik-Sikorska, M.; Posyniak, A.; Mitrowska, K.; Gajda, A.; Błądek, T.; Śniegocki, T.; Żmudzki, J. Occurrence of veterinary antibiotics and chemotherapeutics in fresh water, sediment, and fish of the rivers and lakes in Poland. *Bull. Vet. Inst. Pulawy* **2014**, *58*, 399–404. [CrossRef]
7. Zhou, Y.; Niu, L.; Zhu, S.; Lu, H.; Liu, W. Occurrence, abundance, and distribution of sulfonamide and tetracycline resistance genes in agricultural soils across China. *Sci. Total Environ.* **2017**, *599–600*, 1977–1983. [CrossRef] [PubMed]
8. Li, J.; Cheng, W.; Xu, L.; Strong, P.J.; Chen, H. Antibiotic-resistant genes and antibiotic-resistant bacteria in the effluent of urban residential areas, hospitals, and a municipal wastewater treatment plant system. *Environ. Sci. Pollut. Res. Int.* **2015**, *22*, 4587–4596. [CrossRef] [PubMed]
9. Grenni, P.; Ancona, V.; Barra Caracciolo, A. Ecological effects of antibiotics on natural ecosystems: A review. *Microchem. J.* **2018**, *136*, 25–39. [CrossRef]
10. Su, H.C.; Pan, C.G.; Ying, G.G.; Zhao, J.L.; Zhou, L.J.; Liu, Y.S.; Tao, R.; Zhang, R.Q.; He, L.Y. Contamination profiles of antibiotic resistance genes in the sediments at a catchment scale. *Sci. Total Environ.* **2014**, *490*, 708–714. [CrossRef] [PubMed]
11. Li, W.C. Occurrence, sources, and fate of pharmaceuticals in aquatic environment and soil. *Environ. Pollut.* **2014**, *187*, 193–201. [CrossRef] [PubMed]
12. Jones, O.A.; Lester, J.N.; Voulvoulis, N. Pharmaceuticals: A threat to drinking water? *Trends Biotechnol.* **2005**, *23*, 163–167. [CrossRef] [PubMed]

13. Zhang, M.; Shi, Y.; Lu, Y.; Johnson, A.C.; Sarvajayakesavalu, S.; Liu, Z.; Su, C.; Zhang, Y.; Juergens, M.D.; Jin, X. The relative risk and its distribution of endocrine disrupting chemicals, pharmaceuticals and personal care products to freshwater organisms in the Bohai Rim, China. *Sci. Total Environ.* **2017**, *590–591*, 633–642. [CrossRef] [PubMed]

14. Kim, H.Y.; Lee, I.S.; Oh, J.E. Human and veterinary pharmaceuticals in the marine environment including fish farms in Korea. *Sci. Total Environ.* **2017**, *579*, 940–949. [CrossRef] [PubMed]

15. Dinh, Q.T.; Moreau-Guigon, E.; Labadie, P.; Alliot, F.; Teil, M.J.; Blanchard, M.; Chevreuil, M. Occurrence of antibiotics in rural catchments. *Chemosphere* **2017**, *168*, 483–490. [CrossRef] [PubMed]

16. Lehner, B.; Liermann, C.R.; Revenga, C.; Vörösmarty, C.; Fekete, B.; Crouzet, P.; Döll, P.; Endejan, M.; Frenken, K.; Magome, J.; et al. High-resolution mapping of the world's reservoirs and dams for sustainable river-flow management. *Front. Ecol. Environ.* **2011**, *9*, 494–502. [CrossRef]

17. Vicente-Serrano, S.M.; Zabalza-Martínez, J.; Borràs, G.; López-Moreno, J.I.; Pla, E.; Pascual, D.; Savé, R.; Biel, C.; Funes, I.; Martín-Hernández, N.; et al. Effect of reservoirs on streamflow and river regimes in a heavily regulated river basin of Northeast Spain. *Catena* **2017**, *149*, 727–741. [CrossRef]

18. Mwanamoki, P.M.; Devarajan, N.; Niane, B.; Ngelinkoto, P.; Thevenon, F.; Nlandu, J.W.; Mpiana, P.T.; Prabakar, K.; Mubedi, J.I.; Kabele, C.G.; et al. Trace metal distributions in the sediments from river-reservoir systems: Case of the Congo River and Lake Ma Vallee, Kinshasa (Democratic Republic of Congo). *Environ. Sci. Pollut. Res. Int.* **2015**, *22*, 586–597. [CrossRef] [PubMed]

19. Sun, J.; Huang, J.; Zhang, A.; Liu, W.; Cheng, W. Occurrence of phthalate esters in sediments in Qiantang River, China and inference with urbanization and river flow regime. *J. Hazard. Mater.* **2013**, *248–249*, 142–149. [CrossRef] [PubMed]

20. Sun, Q.; Jiang, J.; Zheng, Y.; Wang, F.; Wu, C.; Xie, R.R. Effect of a dam on the optical properties of different-sized fractions of dissolved organic matter in a mid-subtropical drinking water source reservoir. *Sci. Total Environ.* **2017**, *598*, 704–712. [CrossRef] [PubMed]

21. Temoka, C.; Wang, J.X.; Bi, Y.H.; Deyerling, D.; Pfister, G.; Henkelmann, B.; Schramm, K.W. Concentrations and mass fluxes estimation of organochlorine pesticides in Three Gorges Reservoir with virtual organisms using in situ prc-based sampling rate. *Chemosphere* **2016**, *144*, 1521–1529. [CrossRef] [PubMed]

22. Wu, Y.; Chen, J. Investigating the effects of point source and nonpoint source pollution on the water quality of the East River (Dongjiang) in South China. *Ecol. Indic.* **2013**, *32*, 294–304. [CrossRef]

23. Tran, N.H.; Chen, H.; Do, T.V.; Reinhard, M.; Ngo, H.H.; He, Y.; Gin, K.Y. Simultaneous analysis of multiple classes of antimicrobials in environmental water samples using spe coupleSd with uhplc-esi-ms/ms and isotope dilution. *Talanta* **2016**, *159*, 163–173. [CrossRef] [PubMed]

24. Zhou, L.J.; Ying, G.G.; Liu, S.; Zhao, J.L.; Chen, F.; Zhang, R.Q.; Peng, F.Q.; Zhang, Q.Q. Simultaneous determination of human and veterinary antibiotics in various environmental matrices by rapid resolution liquid chromatography-electrospray ionization tandem mass spectrometry. *J. Chromatogr. A* **2012**, *1244*, 123–138. [CrossRef] [PubMed]

25. Administration, S.E.P. *Water and Wastewater Monitoring Analysis Method of Editorial Committee: Water and Wastewater Monitoring Analysis Method*, 4th ed.; China Environmental Science Press: Beijing, China, 2002.

26. Papadakis, E.N.; Tsaboula, A.; Kotopoulou, A.; Kintzikoglou, K.; Vryzas, Z.; Papadopoulou-Mourkidou, E. Pesticides in the surface waters of lake vistonis basin, greece: Occurrence and environmental risk assessment. *Sci. Total Environ.* **2015**, *536*, 793–802. [CrossRef] [PubMed]

27. Ding, H.; Wu, Y.; Zhang, W.; Zhong, J.; Lou, Q.; Yang, P.; Fang, Y. Occurrence, distribution, and risk assessment of antibiotics in the surface water of poyang lake, the largest freshwater lake in china. *Chemosphere* **2017**, *184*, 137–147. [CrossRef] [PubMed]

28. Deng, W.; Wang, Y.; Liu, Z.; Cheng, H.; Xue, Y. Hemi: A toolkit for illustrating heatmaps. *PLoS ONE* **2014**, *9*, e111988. [CrossRef] [PubMed]

29. Du, J.; Zhao, H.; Liu, S.; Xie, H.; Wang, Y.; Chen, J. Antibiotics in the coastal water of the south yellow sea in china: Occurrence, distribution and ecological risks. *Sci. Total Environ.* **2017**, *595*, 521–527. [CrossRef] [PubMed]

30. Yao, L.; Wang, Y.; Tong, L.; Li, Y.; Deng, Y.; Guo, W.; Gan, Y. Seasonal variation of antibiotics concentration in the aquatic environment: A case study at jianghan plain, central china. *Sci. Total Environ.* **2015**, *527–528*, 56–64. [CrossRef] [PubMed]

31. Ma, R.; Wang, B.; Yin, L.; Zhang, Y.; Deng, S.; Huang, J.; Wang, Y.; Yu, G. Characterization of pharmaceutically active compounds in beijing, china: Occurrence pattern, spatiotemporal distribution and its environmental implication. *J. Hazard. Mater.* **2017**, *323*, 147–155. [CrossRef] [PubMed]

32. Zhang, N.S.; Liu, Y.S.; Van den Brink, P.J.; Price, O.R.; Ying, G.G. Ecological risks of home and personal care products in the riverine environment of a rural region in south china without domestic wastewater treatment facilities. *Ecotoxicol. Environ. Saf.* **2015**, *122*, 417–425. [CrossRef] [PubMed]

33. Wu, D.; Huang, Z.; Yang, K.; Graham, D.; Xie, B. Relationships between antibiotics and antibiotic resistance gene levels in municipal solid waste leachates in Shanghai, China. *Environ. Sci. Technol.* **2015**, *49*, 4122–4128. [CrossRef] [PubMed]

34. Gothwal, R.; Shashidhar, T. Antibiotic pollution in the environment: A review. *CLEAN Soil Air Water* **2015**, *43*, 479–489. [CrossRef]

35. Jiang, Y.; Li, M.; Guo, C.; An, D.; Xu, J.; Zhang, Y.; Xi, B. Distribution and ecological risk of antibiotics in a typical effluent-receiving river (Wangyang River) in North China. *Chemosphere* **2014**, *112*, 267–274. [CrossRef] [PubMed]

36. Mokh, S.; El Khatib, M.; Koubar, M.; Daher, Z.; Al Iskandarani, M. Innovative SPE-LC-MS/MS technique for the assessment of 63 pharmaceuticals and the detection of antibiotic-resistant-bacteria: A case study natural water sources in Lebanon. *Sci. Total Environ.* **2017**, *609*, 830–841. [CrossRef] [PubMed]

37. Khan, G.A.; Berglund, B.; Khan, K.M.; Lindgren, P.E.; Fick, J. Occurrence and abundance of antibiotics and resistance genes in rivers, canal and near drug formulation facilities—A study in Pakistan. *PLoS ONE* **2013**, *8*, e62712. [CrossRef] [PubMed]

38. Tamtam, F.; Mercier, F.; Le Bot, B.; Eurin, J.; Tuc Dinh, Q.; Clement, M.; Chevreuil, M. Occurrence and fate of antibiotics in the Seine river in various hydrological conditions. *Sci. Total Environ.* **2008**, *393*, 84–95. [CrossRef] [PubMed]

39. Yan, C.; Yang, Y.; Zhou, J.; Liu, M.; Nie, M.; Shi, H.; Gu, L. Antibiotics in the surface water of the Yangtze Estuary: Occurrence, distribution and risk assessment. *Environ. Pollut.* **2013**, *175*, 22–29. [CrossRef] [PubMed]

40. Xu, W.; Yan, W.; Li, X.; Zou, Y.; Chen, X.; Huang, W.; Miao, L.; Zhang, R.; Zhang, G.; Zou, S. Antibiotics in riverine runoff of the Pearl River delta and Pearl River Estuary, China: Concentrations, mass loading and ecological risks. *Environ. Pollut.* **2013**, *182*, 402–407. [CrossRef] [PubMed]

41. Dong, D.; Zhang, L.; Liu, S.; Guo, Z.; Hua, X. Antibiotics in water and sediments from Liao river in Jilin province, China: Occurrence, distribution, and risk assessment. *Environ. Earth Sci.* **2016**, *75*, 1202. [CrossRef]

42. Tang, J.; Shi, T.; Wu, X.; Cao, H.; Li, X.; Hua, R.; Tang, F.; Yue, Y. The occurrence and distribution of antibiotics in lake Chaohu, China: Seasonal variation, potential source and risk assessment. *Chemosphere* **2015**, *122*, 154–161. [CrossRef] [PubMed]

43. Zhang, S.; Pang, S.; Wang, P.; Wang, C.; Han, N.; Liu, B.; Han, B.; Li, Y.; Anim-Larbi, K. Antibiotic concentration and antibiotic-resistant bacteria in two shallow urban lakes after stormwater event. *Environ. Sci. Pollut. Res. Int.* **2016**, *23*, 9984–9992. [CrossRef] [PubMed]

44. Stoob, K.; Singer, H.P.; Mueller, S.R.; Schwarzenbach, R.P.; Stamm, C.H. Dissipation and transport of veterinary sulfonamide antibiotics after manure application to grassland in a small catchment. *Environ. Sci. Technol.* **2007**, *41*, 7349–7355. [CrossRef] [PubMed]

45. Gray, J.L.; Borch, T.; Furlong, E.T.; Davis, J.G.; Yager, T.J.; Yang, Y.Y.; Kolpin, D.W. Rainfall-runoff of anthropogenic waste indicators from agricultural fields applied with municipal biosolids. *Sci. Total Environ.* **2017**, *580*, 83–89. [CrossRef] [PubMed]

46. Chen, B.; Liang, X.; Huang, X.; Zhang, T.; Li, X. Differentiating anthropogenic impacts on args in the Pearl River Estuary by using suitable gene indicators. *Water Res.* **2013**, *47*, 2811–2820. [CrossRef] [PubMed]

47. Paiga, P.; Santos, L.H.; Ramos, S.; Jorge, S.; Silva, J.G.; Delerue-Matos, C. Presence of pharmaceuticals in the Lis River (Portugal): Sources, fate and seasonal variation. *Sci. Total Environ.* **2016**, *573*, 164–177. [CrossRef] [PubMed]

48. Chen, K.; Zhou, J.L. Occurrence and behavior of antibiotics in water and sediments from the Huangpu River, Shanghai, China. *Chemosphere* **2014**, *95*, 604–612. [CrossRef] [PubMed]

49. Bai, Y.; Meng, W.; Xu, J.; Zhang, Y.; Guo, C. Occurrence, distribution and bioaccumulation of antibiotics in the Liao River basin in China. *Environ. Sci. Process. Impacts* **2014**, *16*, 586–593. [CrossRef] [PubMed]

50. Zhu, Y.G.; Zhao, Y.; Li, B.; Huang, C.L.; Zhang, S.Y.; Yu, S.; Chen, Y.S.; Zhang, T.; Gillings, M.R.; Su, J.Q. Continental-scale pollution of estuaries with antibiotic resistance genes. *Nat. Microbiol.* **2017**, *2*, 16270. [CrossRef] [PubMed]
51. Guo, X.; Feng, C.; Zhang, J.; Tian, C.; Liu, J. Role of dams in the phase transfer of antibiotics in an urban river receiving wastewater treatment plant effluent. *Sci. Total Environ.* **2017**, *607–608*, 1173–1179. [CrossRef] [PubMed]
52. Luo, Y.; Xu, L.; Rysz, M.; Wang, Y.; Zhang, H.; Alvarez, P.J. Occurrence and transport of tetracycline, sulfonamide, quinolone, and macrolide antibiotics in the Haihe River basin, China. *Environ. Sci. Technol* **2011**, *45*, 1827–1833. [CrossRef] [PubMed]
53. Zhao, S.; Liu, X.; Cheng, D.; Liu, G.; Liang, B.; Cui, B.; Bai, J. Temporal-spatial variation and partitioning prediction of antibiotics in surface water and sediments from the intertidal zones of the Yellow River delta, China. *Sci. Total Environ.* **2016**, *569–570*, 1350–1358. [CrossRef] [PubMed]
54. Xu, J.; Zhang, Y.; Zhou, C.; Guo, C.; Wang, D.; Du, P.; Luo, Y.; Wan, J.; Meng, W. Distribution, sources and composition of antibiotics in sediment, overlying water and pore water from Taihu Lake, China. *Sci. Total Environ.* **2014**, *497–498*, 267–273. [CrossRef] [PubMed]
55. Siedlewicz, G.; Bialk-Bielinska, A.; Borecka, M.; Winogradow, A.; Stepnowski, P.; Pazdro, K. Presence, concentrations and risk assessment of selected antibiotic residues in sediments and near-bottom waters collected from the polish coastal zone in the Southern Baltic Sea—Summary of 3 years of studies. *Mar. Pollut. Bull.* **2017**. [CrossRef] [PubMed]
56. Osorio, V.; Larranaga, A.; Acena, J.; Perez, S.; Barcelo, D. Concentration and risk of pharmaceuticals in freshwater systems are related to the population density and the livestock units in Iberian Rivers. *Sci. Total Environ.* **2016**, *540*, 267–277. [CrossRef] [PubMed]
57. Li, Y.; Li, Q.; Zhou, K.; Sun, X.L.; Zhao, L.R.; Zhang, Y.B. Occurrence and distribution of the environmental pollutant antibiotics in Gaoqiao mangrove area, China. *Chemosphere* **2016**, *147*, 25–35. [CrossRef] [PubMed]
58. Vasudevan, D.; Bruland, G.L.; Torrance, B.S.; Upchurch, V.G.; MacKay, A.A. Ph-dependent ciprofloxacin sorption to soils: Interaction mechanisms and soil factors influencing sorption. *Geoderma* **2009**, *151*, 68–76. [CrossRef]
59. Tamtam, F.; van Oort, F.; Le Bot, B.; Dinh, T.; Mompelat, S.; Chevreuil, M.; Lamy, I.; Thiry, M. Assessing the fate of antibiotic contaminants in metal contaminated soils four years after cessation of long-term waste water irrigation. *Sci. Total Environ.* **2011**, *409*, 540–547. [CrossRef] [PubMed]
60. Zhang, J.; Li, Z.; Ge, G.; Sun, W.; Liang, Y.; Wu, L. Impacts of soil organic matter, ph and exogenous copper on sorption behavior of norfloxacin in three soils. *J. Environ. Sci. China* **2009**, *21*, 632–640. (In Chinese) [CrossRef]
61. Cheng, D.; Liu, X.; Wang, L.; Gong, W.; Liu, G.; Fu, W.; Cheng, M. Seasonal variation and sediment-water exchange of antibiotics in a shallower large lake in north China. *Sci. Total Environ.* **2014**, *476–477*, 266–275. [CrossRef] [PubMed]
62. Yang, Y.Y.; Cao, X.H.; Lin, H.; Wang, J. Antibiotics and antibiotic resistance genes in sediment of Honghu Lake and east Dongting Lake, China. *Microb. Ecol.* **2016**, *72*, 791–801. [CrossRef] [PubMed]
63. Bengtsson-Palme, J.; Larsson, D.G. Concentrations of antibiotics predicted to select for resistant bacteria: Proposed limits for environmental regulation. *Environ. Int.* **2016**, *86*, 140–149. [CrossRef] [PubMed]
64. Martínez, J.L. Antibiotics and antibiotic resistance genes in natural environments. *Science* **2008**, *321*, 365–367. [CrossRef] [PubMed]
65. Larsson, D.G. Antibiotics in the environment. *Upsala J. Med. Sci.* **2014**, *119*, 108–112. [CrossRef] [PubMed]

water

MDPI

Article

The Effect of Primary, Secondary, and Tertiary Wastewater Treatment Processes on Antibiotic Resistance Gene (ARG) Concentrations in Solid and Dissolved Wastewater Fractions

Jennipher Quach-Cu [1], Bellanira Herrera-Lynch [1], Christine Marciniak [1,2], Scott Adams [1], April Simmerman [1] and Ryan A. Reinke [1,*]

[1] Water Quality Laboratory, Los Angeles County Sanitation Districts, Whittier, CA 90601, USA; jquach-cu@lacsd.org (J.Q.-C.); BLynch@lacsd.org (B.H.-L.); webchristine@gmail.com (C.M.); sadams@lacsd.org (S.A.); asimmerman@lacsd.org (A.S.)
[2] Vantari Genetics, Irvine, CA 92618, USA
*   Correspondence: rreinke@lacsd.org; Tel.: +1-562-908-4288 (ext. 3059)

Received: 20 November 2017; Accepted: 27 December 2017; Published: 4 January 2018

Abstract: Wastewater treatment plants (WWTPs) have been identified as potential sources of antibiotic resistance genes (ARGs) but the effects of tertiary wastewater treatment processes on ARGs have not been well characterized. Therefore, the objective of this study was to determine the fate of ARGs throughout a tertiary-stage WWTP. Two ARGs, *sul1* and *bla*, were quantified via quantitative polymerase chain reaction (qPCR) in solids and dissolved fractions of raw sewage, activated sludge, secondary effluent and tertiary effluent from a full-scale WWTP. Tertiary media filtration and chlorine disinfection were studied further with the use of a pilot-scale media filter. Results showed that both genes were reduced at each successive stage of treatment in the dissolved fraction. The solids-associated ARGs increased during activated sludge stage and were reduced in each subsequent stage. Overall reductions were approximately four $\log_{10}$ with the tertiary media filtration and disinfection providing the largest decrease. The majority of ARGs were solids-associated except for in the tertiary effluent. There was no evidence for positive selection of ARGs during treatment. The removal of ARGs by chlorine was improved by filtration compared to unfiltered, chlorinated secondary effluent. This study demonstrates that tertiary-stage WWTPs with disinfection can provide superior removal of ARGs compared to secondary treatment alone.

Keywords: antibiotic resistance genes; wastewater treatment; tertiary media filtration

---

1. Introduction

Antibiotic resistant bacteria (ARB) are having profound effects on the treatment of human diseases. In the United States, the number of ARB-related hospitalizations continues to increase [1] with an estimated cost of up to $30 billion annually [2]. In recent years the importance of ARB has come to the forefront of many scientific disciplines including environmental microbiology. Antibiotic resistant genes in the environment have the potential to spread into the human population presenting a possible public health problem. Although the full impact of environmental transfer of antibiotic resistance genes (ARGs) on public health is not currently known, several reports have highlighted the presence of ARB and ARGs in wastewater treatment plants [3–8], agricultural feedlots [9–11] soils [12], rivers and lakes [9,13–18] raising concerns about potential public health impacts from these sources. As the number and cost of ARB-related illnesses continues to grow, multifaceted efforts are needed to control ARB in the clinic along with further investigation into potential impacts arising from environmental sources.

The presence of antibiotics and ARB in wastewater treatment plants (WWTPs) and water reclamation plant (WRP) effluents are of specific interest. The primary reasons for concern include the presence of clinically relevant ARB and ARGs in raw sewage entering these facilities, the potential for ineffective removal and/or selection of ARG/ARB by WWTPs and, the possibility of human contact with treated waters [13,17,19]. Additionally, some studies have revealed the presence of ARB in treated wastewater effluents and have shown that ARGs and ARB are more prevalent downstream from WWTPs [3,15,20–25] which suggest WWTPs have the potential to influence the ARB population in receiving waters. Almost all of these reports, however, focused on the impact of facilities using traditional primary and secondary stage treatments with or without disinfection. Currently, the effects of tertiary wastewater treatment processes, such as media filtration, on ARB and ARGs have not been well characterized.

Conventional activated sludge wastewater treatment is divided into distinct stages. The first or primary stage consists of physical removal of oils and greases along with sedimentation of large particles. The secondary phase of treatment utilizes microbial organisms to reduce the amount of organic matter in the wastewater. Typically, processes in addition to traditional secondary treatment which are employed to further improve water quality are referred to as tertiary-stage treatment [26]. Filtration, a commonly used type of tertiary treatment, has been shown to be effective in reducing the concentrations of viable indicator bacteria and viruses such as bacteriophage, substantially decreasing the number of hazardous microbes in final effluent waters [27]. Furthermore, tertiary filtration yields additional reductions in suspended solids and biological oxygen demand producing a higher quality effluent. Consequently, disinfection of tertiary effluents can be more efficient due to lower chlorine demand and higher transmittance of UV light compared to secondary effluent [28].

The public health and environmental hazards associated with ARGs are related to their capacity to be transferred between bacteria coupled with positive selective pressure occurring from the pervasive use of antibiotics both clinically and agriculturally. Specifically, the transfer of ARGs between bacteria horizontally is one of the main factors that have led to the rapid spread of antibiotic resistance across the globe [29,30]. The three main mechanisms by which ARGs are acquired by bacteria include: the direct exchange of genetic material between two viable bacteria (conjugation), through phage infections (transduction) and by uptake of extracellular, free DNA (transformation). Data has been presented indicating that horizontal gene transfer (HGT) of resistant genes occurs at a higher frequency among more closely related bacteria [31].

When considering the potential environmental impact of ARGs related to anthropogenic activities, it is important to recognize that ARGs occur naturally and can readily be found in environmental matrices unaffected by human activities [32–35]. Antibiotic resistance genes and ARB have been found in 30,000 year old permafrost as well as in remote caves isolated from humans for over four million years [36,37] signifying that these genes are commonly found in the environment and have evolved over thousands to millions of years prior to the therapeutic use of antibiotics by humans. Moreover, research has shown that most of the functional ARGs found in WWTPs are specific to those matrices and are not widely disseminated into human populations or the environment [38,39]. Therefore, it is likely that, at any given time, only a small proportion of the total wastewater treatment plant (WWTP) resistome may be associated with potential adverse public health effects however; the possibility exists for the sudden mobilization and horizontal transfer of new ARGs from within the wastewater resistome which could result in additional ARGs entering the human population. It stands to reason, that assessing wastewater matrices not only for the presence of ARGs but their ability to be transferred would result in data that is more relevant to public health. Subsequently, WWTPs that reduce the fraction of clinically relevant ARGs that can be horizontally transferred would lower the probability of treated effluent waters contributing to this public health crisis. Unfortunately, obtaining a direct measurement of HGT events in wastewater has been difficult to achieve making the evaluation of treatment options more challenging.

In water-based matrices, HGT mechanisms can be associated with either the cellular or extra-cellular fractions. The cellular fraction consists of ARGs existing within intact and/or viable bacterial cells or attached to sediments and particulates. Conversely, extracellular ARGs include bacterial DNA that has been excreted or released upon death of the organism and genes contained within viral particles. In aqueous solutions, the extracellular ARGs exist dissolved in solution whereas cell-associated ARGs can settle with other solids. The separation of cellular and extra-cellular ARGs in water can easily be achieved using centrifugation that partitions the components necessary for different HGT pathways into either the pellet or the supernatant. Specifically, HGT by conjugation would be primarily associated with viable bacteria partitioned to the solids fraction (pellet) whereas the dissolved phase (supernatant) would contain the components necessary for transformation (extra-cellular DNA) and transduction (phage particles). Quantifying the amount of ARGs in different fractions would identify where the majority of ARGs reside and consequently, which HGT pathways are likely to be involved in the transfer of ARGs at a particular treatment stage. Furthermore, such information could also be used to identify processes that are better at removing ARGs from each of the particular fractions.

This study was designed to determine the fate of dissolved and solids-associated ARGs through different stages of a tertiary WRP and to identify treatment processes that result in the best removal of ARGs. A qPCR approach was utilized to quantify two ARGs throughout wastewater treatment: the first, targets the sulfonamide resistance gene *sul1*, and a second targets SHV and TEM-type β-lactamase ($bla_{SHV/TEM}$) resistance genes. Furthermore, a pilot-scale solid media filter was used to provide a more detailed characterization of tertiary filtration on ARGs.

2. Materials and Methods

2.1. Sample Collection and Processing

Samples were obtained from a tertiary WRP designed for an average flow of 235 million L/day of raw sewage (daily dry weather ranges from 94 to 340 million L/day). The treatment scheme included primary sedimentation, activated sludge with nitrification and denitrification (NDN), secondary clarification (flocculation/settling), tertiary media filtration (anthracite coal/sand/gravel) and, sequential chlorine disinfection (free chlorine with a minimum required residual of 1.0 mg/L followed by chloramine with typical residual concentrations from 1.0 to 3.0 mg/L).

Unconcentrated samples were collected in sterile one-liter Nalgene bottles. Sodium thiosulfate (one milliliter of a 10% weight/volume solution) was added to dechlorinate the water. Samples were collected at the WRP from incoming raw sewage, activated sludge, clarified secondary effluent, and disinfected final effluent. Samples were collected for each water type on at least three separate occasions.

2.2. Collection of Hollow Fiber Filtration (HFF) Concentrated Wastewater Samples

Antibiotic resistance genes from secondary and final effluent waters were concentrated using a one-step HFF procedure. Briefly, approximately 10^9 copies of AdvIPC:pSMART plasmid were added to 100 L of each matrix (secondary or final effluent) then pumped through an Asahi Kasei REXEED^{TM}-25S filter (Asahi Kasei, Oita, Japan) in a recirculation configuration with pressure within the filter maintained between four and eight psi. Bacteria and nucleic acids were eluted by recirculation of 0.05 M Glycine containing 0.01% Antifoam A (pH 7.0) for 10 to 30 min. New tubing was used with each sample to prevent interferences associated with reused tubing (biofilm, DNA, etc.). The original samples were collected in 50 L plastic carboys that had been pre-washed with hypochlorite (to remove DNA), dechlorinated, and then autoclaved. Secondary effluent was collected directly from the WRP secondary settling tank and tertiary-treated disinfected effluent (final effluent) was collected from a channel immediately after dechlorination but prior to entering the distribution system. Sodium thiosulfate was added to each carboy after collection and total chlorine concentrations were measured using the Hach DPD (*N*, *N*-diethyl-p-phenylenediamine) colorimetric procedure (Hach Co., Loveland, CO, USA) according to method SM4500-Cl [40].

2.3. Separation of Solids and Dissolved Fractions

Each sample collected was separated into a solids-associated and dissolved fraction via centrifugation. Specifically, one milliliter of each sample was placed in a sterile, DNA-free tube and subjected to centrifugal force (9600× *g*) for five minutes. Larger volume samples (50 mL) collected during the pilot-scale filtration experiments were separated by centrifugal force (6000× *g*) for ten minutes. The supernatants (dissolved fraction) were decanted into a sterile DNA-free tube with the pellets (solids-associated fraction) remaining in the original tube and stored at −80 °C until the DNA was extracted.

2.4. DNA Extraction

The DNA extraction of the fractionated wastewater samples was performed with the QIAPrep® spin miniprep (solids fraction) and the QIAquick® gel extraction kit (dissolved fraction) according to the manufacturer's instructions (Qiagen Inc., Hilden, Germany). Elution volumes for the full-scale plant samples were 40 µL and 30 µL for the pilot-scale filtration experiments. Samples collected from the full-scale WRP were extracted on the same day that the samples were collected. Samples collected from the pilot-scale dual-media filter experiments were extracted within one week of collection. The DNA was divided into individual single-use aliquots and stored at −80 °C.

2.5. AdvIPC:pGEM-T and AdvIPC:pSMART Vectors

Two plasmids were constructed to contain a DNA sequence not found in nature [41]. The synthetic DNA sequence (termed Adv IPC; synthesized by Integrated DNA Technologies Inc., Coralville, IA, USA) was inserted into the pGEM-Teasy vector (Promega Co., Madison, WI, USA) and pSMART® GC LK vector systems (Lucigen Co., Middleton, WI, USA) using standard molecular biological techniques including restriction digestion and ligation. The pGEM-Teasy vector was used with the pilot-scale filtration experiments whereas the pSMART® vector, which did not contain a β-lactamase gene, was used as an internal control for the HFF concentration of secondary and final effluents. See supplementary materials for additional sequence information and plasmid map.

2.6. Quantitative PCR Primers/Probes and Plasmids

Quantitative PCR primers and probes targeting the β-lactamase and sulfonamide resistance genes were used for the analysis of wastewater matrices. Additionally, qPCR assays were utilized to assess total bacterial biomass (*16S* rRNA gene) and to determine recovery/loss of DNA (AdvIPC) during sample concentration by HFF. A sample processing control (SPC) was used to monitor DNA extraction and qPCR inhibition in all samples analyzed. The SPC consisted of approximately 120 ng of salmon sperm DNA (Ambion, Waltham, MA, USA) added into each sample prior to DNA extraction which was subsequently quantified by qPCR. If inhibition was identified, the sample was diluted and re-analyzed. Primer and probe sequences for all qPCRs are listed in Table 1. Each *sul1* qPCR consisted of 1× SsoAdvanced™ Universal SYBR® Green Supermix (BioRad, Hercules, CA, USA), 120 nM primer and 1 µL template in a final volume of 25 µL. The *bla*$_{SHV/TEM}$ and *16S* qPCRs had a final concentration of 1× SsoAdvanced™ Universal Probes Supermix (BioRad, Hercules, CA, USA), 320 nM each primer and probe, 3 µL template in a total volume of 25 µL. The AdvIPC qPCR contained the same reaction constituents as the *bla*$_{SHV/TEM}$ but the final probe concentration was 200 nM. The SPC qPCR consisted of 1× SsoAdvanced™ Universal Probes Supermix, 1 µM each primer, 800 nM probe, 2 µL template in a final volume of 25 µL. The qPCR reactions were performed and analyzed using either a RotorGene (Qiagen Inc., Hilden, Germany) or LightCycler® 480 (Roche, Basel, Switzerland) real-time thermocycler. The qPCR efficiency, limit of detection (LOD), and regression statistics (mean square error for the LightCycler® 480 software and r^2 for the RotorGene software) were determined for each reaction (see Supplementary Materials). The specificity of the *bla*$_{SHV/TEM}$ PCR was confirmed by DNA sequence analysis of the amplified products. All primers and probes were procured from Integrated DNA Technologies (Coralville, IA, USA). Data analyses were performed using the software associated with each respective thermocycler (LightCycler® 480 software version 1.5.1.62, Roche Diagnostics Corporation, Indianapolis, IN, USA, and RotorGene Q software version 2.1.0, Qiagen Inc., Hilden, Germany). Full-scale WRP samples were extracted in triplicate and each extract analyzed singularly (RotorGene) or in triplicate (LightCycler® 480 analyses) by qPCR. Samples were only deemed positive if at least two of the three extractions resulted in a positive signal. Pilot-scale dual-media filtration samples were extracted and analyzed singularly (RotorGene). Turbidity readings were also taken for the full-scale WRP samples and are provided in the Supplementary Materials. The qPCR cycling conditions along with the average efficiencies and linearity data are provided in the Supplementary Materials along with additional information on the development of the qPCRs.

Table 1. Primer and probe sequences.

Target	Primer/Probe	Sequence [1]	Reference
blaSHV/TEM	β-Lac-qPCR-F β-Lac-qPCR-R β-Lac-qPCR Probe	5'-GCCATAACCATGAGYGATAAC-3' 5'-TTATCRGCAATAAACCAGCC-3' 5'FAM-TCATTCAGCTCCGKTTCCCA-BHQ-1-3'	This study
sul1	sul I-FW sul I-RV	5'-CGCACCGGAAACATCGCTGCAC-3' 5'-TGAAGTTCCGCCGCAAGGCTCG-3'	[42]
Bacterial 16S rDNA	Bac 1055f Bac 1392r 16S Taq1115	5'-ATGGCTGTCGTCAGCT-3' 5'-ACGGGCGGTGTGTAC-3' 5'-FAM-CAACGAGCGCAACCC-TAMRA-3'	[43]
Salmon Sperm DNA	Samn qPCR-F Samn qPCR-R Samn qPCR-Probe	5'-GGTTCCGCAGCTGGG-3' 5'-CCGAGCCGTCCTGGTCTA-3' 5'FAM-AGTCGCAGGCGGCCACCGT -BHQ-1-3'	[44]
AdvIPC	Adv Hex RT-Forward Adv Hex RT-Reverse Adv IPC Probe	5'-GGAYGCCTCGGAGTACCTGAG-3' 5'-ACiGTGGGGTTTCTRAACTTGTT-3' 5'Cy5-CACCGACGGCGAGACCGACTTT-BHQ2-3'	[45] [41]

Note: [1] The base "i" designates inosine; "K" denotes G/T; "R" indicates A/G; "Y" indicates C/T.

2.7. Pilot-Scale Dual-Media Filtration

A pilot-scale dual-media filter was constructed using a PVC tube (15.24 cm diameter) and the same depth of sand (31 cm) and anthracite coal (61 cm) as is used in the full-scale WRP filter beds. The flow was controlled via an electric pump and monitored with a flow meter. The filter was operated at hydraulic loading rate of 3.0 to 3.8 liters per minute (L/min), (166 L/(min m^2) to 209 L/(min m^2)) similar to full-scale filters used at the WRP (122–204 L/(min m^2)).

The pilot-scale filtration experiments were conducted using a laboratory modified plasmid (AdvIPC:pGEM-T) added to secondary effluent. Briefly, 379 L of clarified secondary effluent from a WRP was added to a 2271 L polyethylene tank and dosed with the AdvIPC:pGEM-T plasmid to achieve a final concentration of 1.95×10^8 copies/mL in the secondary effluent. Samples were collected at a total of six locations/time points: (1) prior to the addition of plasmid, (2) after addition but before filtration, (3–5) filtrate at 20, 90 and 100 min. from the start of the filtration and (6) after backwashing the filter. Dissolved and solids-associated fractions were evaluated for exogenous plasmid concentrations in each of the samples as described above. Additionally, the filtration experiments were performed under both "dirty" and "clean" conditions: (1) using a filter that had been in service for over four weeks to simulate a bio-fouled filter, and (2) using a filter treated with high-dose chlorine to simulate a clean filter condition.

The concentration of the exogenous plasmid was determined by qPCR in each of the samples upon collection and again after chlorine treatment (10 mg/L for 45 min followed by dechlorination). Liquid hypochlorite was prepared using Clorox as a stock solution (5.84% available chlorine) and diluted with ultrapure water (double reverse osmosis, carbon filtered, UV disinfected). The Clorox was stored in a light impenetrable container at room temperature. Chlorine was diluted and used within two days. Total chlorine concentrations were measured prior to treatment via the Hach DPD colorimetric method. Treatment doses were based on the total chlorine concentrations measured using the DPD method.

2.8. Statistical Analyses

Statistical analyses were performed using either Microsoft Excel 2010 (version 14.0.7190.5000, Microsoft Corporation, Redmond, WA, USA) or SigmaPlot version 11.0 software (Systat Software, Inc., San Jose, CA, USA).

3. Results

3.1. Quantification of Dissolved and Solids-Associated ARGs and Bacterial 16S DNA Through Primary, Secondary and Tertiary Wastewater Treatment Processes

The quantity of SHV/TEM type β-lactamase genes ($bla_{SHV/TEM}$) and *sul1* sulfonamide resistance genes was determined by qPCR analysis for both the solids-associated and dissolved fractions from each sample. Additionally, a qPCR targeting the *16S* rRNA gene DNA (*16S* rDNA) was used to estimate total bacterial biomass in each water type analyzed. The average qPCR efficiencies were 95%, 86% and 99% with LODs of 70, 6, and 57 copies/μL for the $bla_{SHV/TEM}$, *sul1*, and *16S* rDNA reactions, respectively (Supplementary Materials).

3.1.1. Antibiotic Resistance Gene Concentrations throughout a Full-Scale WRP

All three gene targets were readily detected in the raw sewage however; concentrations of $bla_{SHV/TEM}$ in the secondary effluent, and *sul1* and $bla_{SHV/TEM}$ in the final effluent were found to be below the detection limit of the method in all unconcentrated samples analyzed ($n = 6$). In response, a HFF method was used to concentrate ARGs from secondary and final effluents (by approximately 200-fold). A laboratory modified plasmid (AdvIPC:pGEM-T) added to each final effluent sample prior to concentration was used to evaluate recovery of dissolved DNA. The results showed an average recovery after concentration of 57% which demonstrated that the method could be used as an effective means to concentrate ARGs in these waters. Consequently, *sul1* was detected within the quantifiable range in HFF concentrated secondary and final effluents and $bla_{SHV/TEM}$ was detected in the concentrated secondary effluent but remained below detection in all final effluent samples. This demonstrated that the HFF procedure provided increased sensitivity for ARGs and permitted the quantitative assessment of ARGs in secondary and final effluents.

The quantities of both ARGs and the *16S* DNA decreased through each stage of treatment in the dissolved fractions (Figure 1 and Table 2). In particular, *sul1* genes were reduced by 1.46 $\log_{10}$ (raw to activated sludge), 1.28 $\log_{10}$ (activated sludge to secondary effluent) and 2.05 $\log_{10}$ (secondary effluent to final effluent). Reductions were also observed for the $bla_{SHV/TEM}$ genes: 0.83 $\log_{10}$ (raw to activated sludge), 2.01 $\log_{10}$ (activated sludge to secondary effluent) and $\geq$0.24 $\log_{10}$ decrease (secondary effluent to final effluent; $bla_{SHV/TEM}$ gene concentrations in all HFF concentrated final effluent samples were below the detection limit, resulting in the "greater than" value). This data indicates that a large proportion of the dissolved ARG DNA was degraded or removed during the various treatment processes.

Table 2. Log reductions in dissolved ARGs and *16S* total bacterial rDNA during wastewater treatment.

Gene Target	Raw to AS [1]	Raw to SE [1]	Raw to FE [1]	AS to SE [1]	SE to FE [1]
$bla_{SHV/TEM}$	0.83 [3]	2.84	$\geq$3.08 [2]	2.01	$\geq$0.24 [2]
sul1	1.46	2.74	4.79	1.28	2.05 [3]
16S	0.63	1.24	3.99	0.61	2.76

Notes: [1] AS: activated sludge; SE: HFF concentrated secondary effluent; FE: HFF concentrated final effluent; [2] All samples tested were below the detectable limit of the qPCR and were assigned the value of the assay's detection limit for log removal calculations; [3] Three of nine $bla_{SHV/TEM}$ AS replicates and one of six *sul1* FE replicates were negative by qPCR analysis. Negative samples were assigned the LOD in log removal calculations.

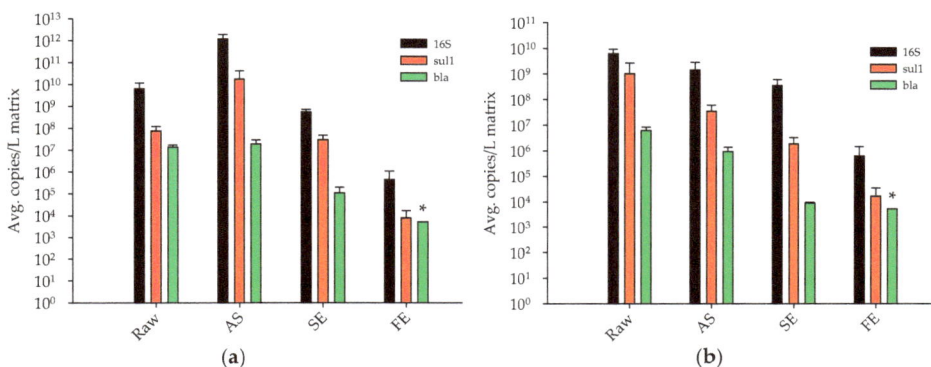

Figure 1. Concentrations of $bla_{SHV/TEM}$, $sul1$, and bacterial $16S$ rDNA in different wastewater treatment matrices. (**a**) The solids-associated qPCR concentrations; (**b**) The dissolved fraction qPCR concentrations. Error bars represent one standard deviation from the mean. Bar colors indicate the following samples, black: bacterial $16S$, red: $sul1$, green: $bla_{SHV/TEM}$. The * symbol denotes matrices where all samples analyzed were below the LOD with the corresponding bar representing the detection limit for that particular gene. Raw: raw sewage, AS: activated sludge, SE: secondary effluent, FE: disinfected final effluent.

With reference to the solids-associated fraction, the $16S$, $bla_{SHV/TEM}$, and $sul1$ genes all increased between the raw sewage and activated sludge stages (Figure 1 and Table 3) but not to the same extent. The $sul1$ and $16S$ DNA exhibited a similar rise (2.38 vs. 2.26 $\log_{10}$) whereas the $bla_{SHV/TEM}$ was substantially smaller (0.14 $\log_{10}$). Specifically, there was approximately an order of magnitude difference between the increase in the $bla_{SHV/TEM}$ and $sul1$ (0.14 log vs. 2.38 log, Table 3) or the $16S$ DNA (0.14 vs. 2.26, Table 3). These differences between the increase in $bla_{SHV/TEM}$ and either the $sul1$ or the $16S$ DNA were statistically significant (ANOVA $p \leq 0.05$; Supplementary Materials). While the $sul1$ did show a slightly larger increase compared to the $16S$ DNA, it was not statistically significant (ANOVA $p \geq 0.05$; Supplementary Materials). The amount of increase was specific to the individual ARG, suggesting that the ARG distributions among bacteria in the activated sludge differ.

Table 3. Log reductions in solids-associated ARGs and 16S bacterial rDNA during wastewater treatment.

Gene Target	Raw to AS [1]	Raw to SE [1]	Raw to FE [1]	AS to SE [1]	SE to FE [1]
$bla_{SHV/TEM}$	−0.14	2.11	≥3.42 [2]	2.25	≥1.31 [2]
$sul1$	−2.38	0.40	3.98	2.78	3.58 [3]
$16S$	−2.26	1.06	4.16	3.32	3.10 [3]

Notes: [1] AS: activated sludge; SE: HFF concentrated secondary effluent; FE: HFF concentrated final effluent; [2] All samples tested were below the detectable limit of the qPCR and were assigned the value of the assay's detection limit for log removal calculations; [3] Three of six $sul1$ FE replicates and two of six $16S$ AS replicates were negative by qPCR analysis. Negative samples were assigned the LOD in log removal calculations.

The complete WRP treatment process (from raw to final effluent) resulted in an overall decrease in $sul1$ genes by 4.79 $\log_{10}$ in the dissolved fraction and 3.98 $\log_{10}$ in the solids fractions with similar reductions observed in the total bacterial biomass (Tables 2 and 3). The $bla_{SHV/TEM}$ concentrations were reduced by greater than three $\log_{10}$ throughout the treatment process although, it was not possible to determine exactly how much more than three $\log_{10}$ because all final effluent samples were below the LOD and thus the method's detection limit was used to calculate the reductions (Tables 2 and 3). Tertiary filtration/disinfection produced the largest stage to stage decrease in $sul1$ (both fractions) and $16S$ rDNA (dissolved fraction). Moreover, $sul1$ genes are some of the most prevalent ARGs found in wastewater, yet despite being concentrated over 200-fold by HFF, half of the solids fraction (three of six) and one of the dissolved final effluent samples remained below the LOD. This indicated that the tertiary filtration and disinfection stage was instrumental in achieving the large ARG reductions observed over the complete treatment process (Figure 1 and Tables 2 and 3).

The qPCR analyses from the full-scale WRP samples demonstrated that the *sul1* genes were consistently found in higher concentrations compared to the $bla_{SHV/TEM}$ for all treated and untreated wastewater samples in both the solids and dissolved fractions (Figure 1). The different sample types had different proportions of ARGs associated with each fraction. The raw sewage, activated sludge and secondary effluent solids fractions on average contained higher counts of both ARGs compared to the dissolved fraction (Figure 2) although, one of the raw sewage replicates did show a larger *sul1* concentration in the dissolved fraction. The highest proportion of solids-associated ARGs occurred in the activated sludge with over 100-fold more ARG observed in the solids fraction compared to the dissolved ARGs (Figure 2). In comparison, the tertiary-treated effluent contained nearly equal amounts of dissolved and solids-associated ARGs (Figure 2). The solids-associated to dissolved ARG ratio decreased successively from the activated sludge through to the final effluent (Figure 2). The larger proportion of dissolved ARGs in the final effluent may be the result of DNA released during chlorine disinfection.

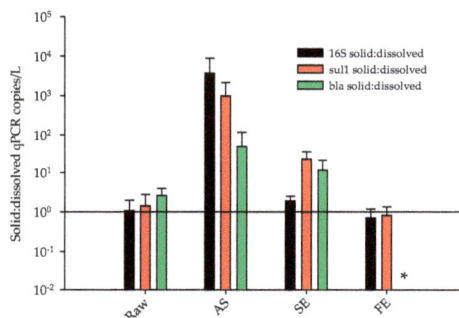

Figure 2. The proportion of ARGs observed in the solids and dissolved fractions during different wastewater treatment stages. The plot shows the ratio of solids to dissolved qPCR copies for total bacteria (black bars), *sul1* (red bars) and $bla_{SHV/TEM}$ (green bars) in raw sewage (Raw), activated sludge (AS), secondary effluent (SE), and disinfected final effluent (FE). The bars represent the average of at least three separate samples with error bars indicating one standard deviation from the mean. The horizontal line is provided for reference and denotes equal ratios of solids and dissolved genes. The * symbol indicates that $bla_{SHV/TEM}$ concentrations in the final effluent were all below the LOD.

3.1.2. Determination of Total Bacteria Biomass Concentrations in Different Wastewater Treatment Processes by qPCR

The total bacterial biomass concentrations were estimated using qPCR targeting a region of the *16S* rRNA gene that is well conserved among bacterial species [43]. Native solids and dissolved fractions (not concentrated by HFF) were all positive for the *16S* rDNA qPCR target (Figure 1). The proportion of solids and dissolved *16S* rDNA were similar in the raw sewage, secondary and final effluents with a large increase occurring in the solids-associated ARGs in the activated sludge samples (Figure 2). The dissolved *16S* rDNA decreased in concentration at each consecutive stage from raw sewage to the final effluent (Figure 1). In contrast, an increase in solids-associated *16S* rDNA was observed between the raw and activated sludge with reductions occurring in the secondary and final effluents (Figure 1; Tables 2 and 3). Viable bacterial concentrations are known to increase during the activated sludge stage of treatment and the increase in the solids-associated *16S* qPCR copies coupled with the concurrent decrease in dissolved *16S* DNA suggested that the viable fraction of bacterial cells did indeed partition to the solids-associated fraction. The total reduction in *16S* DNA was approximately four-$\log_{10}$ from the raw sewage to final effluent concentrations for both fractions similar to what was observed with the *sul1* (Tables 2 and 3).

3.2. Tertiary Wastewater Treatment Processes did not Result in Positive Selection for bla$_{SHV/TEM}$ or sul1 ARGs

Examination of the qPCR data from different WRP treatment stages provided no evidence that the ARGs analyzed in this project were being selected for during the treatment process. Normalizing the ARG concentrations to the total bacterial biomass (*16S* bacterial rDNA qPCR) provided an assessment of how the concentration of each gene changed in relation to the total bacterial population during wastewater treatment. The solids fractions were used to represent intact bacteria in each water type. The ratio of ARG: *16S* rDNA was calculated for each matrix and used to determine if the ARGs increased or decreased during wastewater treatment.

The *sul1* and *16S* gene concentrations trended together in the solids fractions of each matrix tested, resulting in *sul1:16S* ratios that were within one order of magnitude for all water types (Table 4). The differences were not statistically significant (see supplementary material) with the exception of a decrease in the ratio from secondary effluent to final effluent (Table 4). Conversely, the solids fraction *bla$_{SHV/TEM}$:16S* ratios decreased by over two orders of magnitude between the raw sewage (range: 1.7×10^{-3} to 1.4×10^{-2}) and activated sludge (range: 4.2×10^{-6} to 4.3×10^{-5}; Table 4). This indicated that the total bacterial biomass in the activated sludge contained a significantly smaller proportion of the *bla$_{SHV/TEM}$* gene compared to the raw sewage (Table 4); the difference was statistically relevant (*t*-test $p \leq 0.05$; Supplementary Materials) suggesting its presence was not being selected for during the biological phase of treatment. In the secondary effluent, the *bla$_{SHV/TEM}$* ratio was slightly higher than the activated sludge (difference not statistically significant, *t*-test $p \geq 0.05$; Supplementary Materials) however; the secondary effluent ratio (range: 1.1×10^{-4} to 2.9×10^{-4}) was still significantly lower (by 1.52 log units) compared to the raw sewage (Table 4 and Supplementary Materials). The qPCR data from final effluent samples showed that all *bla$_{SHV/TEM}$* were below the LOD and therefore ratios could not be accurately calculated. Overall, the qPCR data demonstrated that, not only did the quantity of ARGs decrease during WRP treatment but, so did the proportion of bacteria carrying the *bla$_{SHV/TEM}$* genes. The lack of a significant increase in either the *bla$_{SHV/TEM}$:16S* or *sul1:16S* ratio suggests that these ARGs are not being selected for throughout the treatment process. The dissolved fractions, although unlikely to contain significant amounts of intact bacteria, also did not show any increase in ARG:*16S* ratio between the raw and secondary or final effluents (Table 4).

Table 4. Ratio of ARGs to bacterial *16S* qPCR concentrations in wastewater matrices (average ± standard deviation).

ARG:16S [1]	Raw Sewage (n = 3)	Activated Sludge (n = 9)	Secondary Effluent (n = 3) [2]	Final Effluent (n = 6) [3]
bla:16S solids	$5.8 \times 10^{-3} \pm 7.1 \times 10^{-3}$	$2.0 \times 10^{-5} \pm 1.4 \times 10^{-5}$	$1.7 \times 10^{-4} \pm 9.8 \times 10^{-5}$	BD
sul1:16S solids	$2.0 \times 10^{-2} \pm 1.7 \times 10^{-2}$	$1.3 \times 10^{-2} \pm 8.3 \times 10^{-3}$	$6.3 \times 10^{-2} \pm 5.3 \times 10^{-2}$	$2.9 \times 10^{-2} \pm 2.9 \times 10^{-2}$
bla:16S dissolved	$1.1 \times 10^{-3} \pm 5.8 \times 10^{-4}$	$3.2 \times 10^{-3} \pm 4.2 \times 10^{-3}$	$3.3 \times 10^{-5} \pm 1.7 \times 10^{-5}$	BD
sul1:16S dissolved	$2.2 \times 10^{-1} \pm 3.8 \times 10^{-1}$	$4.8 \times 10^{-2} \pm 6.0 \times 10^{-2}$	$8.2 \times 10^{-3} \pm 6.7 \times 10^{-3}$	$2.6 \times 10^{-2} \pm 2.9 \times 10^{-2}$

Notes: [1] Designates the ARG ratio. "Solids" term indicates the concentrations in the pelleted fraction for each matrix. "Dissolved" term indicates the concentrations in the supernatant fractions for each matrix; [2] Samples were concentrated via HFF prior to DNA extraction; [3] Samples were collected after dual media filtration of clarified secondary effluent and chlorine disinfection. All samples were concentrated by HFF prior to DNA extraction. BD: All concentrations for *bla$_{SHV/TEM}$* in the final effluent were below the detection limit of the assay.

3.3. Evaluation of Tertiary Filtration and Disinfection Processes for the Removal of ARGs

Analysis of full-scale WRP samples demonstrated that β-lactamase gene concentration was below the LOD in all HFF concentrated final effluent samples (from both fractions) and *sul1* was not detected in the solids fraction in half of the final effluent samples. The lack of detection of ARGs in the final effluent even after HFF concentration suggested that large amounts of dissolved ARGs were being removed during the tertiary filtration and/or chlorination processes. A closer inspection of the tertiary filtration and chlorine disinfection was performed using a pilot-scale dual-media filter which provided a controlled filtration process that mimicked the configuration and operation of the full-scale WRP filters. The pilot-scale filters allowed precise control over operational parameters as well as the ability to characterize the removal of ARG containing DNA. Specifically, a known quantity of a laboratory modified plasmid was added to a defined volume of secondary effluent (379 L) which was then pumped into the pilot-scale filter unit at a flow rate similar to the full-scale filters used at the WRP. The plasmid concentration was measured in samples collected before and after filtration using qPCR. Given that the concentration of modified plasmid added to the secondary effluent was known, and that the volume applied to the filter was defined, the amount of plasmid removed per unit volume could be determined. In addition, bench-scale chlorination experiments were conducted on the pre- and post- filtration samples to determine the effect of chlorination on the exogenous plasmid.

The presence of naturally occurring ARGs in secondary effluent precluded the use of an ARG-specific qPCR for these experiments. Therefore, a laboratory-modified plasmid was constructed which contained a $bla_{SHV/TEM}$ ARG as well as an exogenous DNA sequence that does not occur in wastewater. A qPCR assay targeting the exogenous sequence (not the ARG) was used to specifically quantify the laboratory-modified plasmid through the filtration and disinfection experiments thus, avoiding any interference by naturally occurring genes.

Shortly after addition of the modified plasmid to the secondary effluent, a majority of the plasmid was detected in the dissolved fraction with a smaller portion present in the solids-associated fraction. In particular, samples collected approximately ten minutes after addition (but prior to filtration) showed the plasmid concentration in the dissolved fraction averaged 1.23×10^8 copies/mL ($n = 7$) compared to 8.79×10^6 copies/mL in the solids-associated fraction (Supplementary Materials). The difference in plasmid concentrations between the two fractions was statistically significant (t-test p-value 0.007). This demonstrates that a small portion of the free DNA (less than 10%) had a propensity to associate with solid particles in secondary effluent.

The pilot-scale filtration experiments showed that tertiary filtration coupled with chlorine disinfection removed plasmid ARGs better than chlorination alone. Reduction in plasmid concentration by the filtration and chlorination processes was determined by subtracting the log transformed pre-filtration data from the post-filtration sample data with and without chlorine treatment. The decrease in plasmid concentration due solely to the filtration process was determined by analyzing samples before and after tertiary filtration without chlorine treatment. In the solids-associated fraction, this resulted in a $0.7 \pm 0.5 \log_{10}$ reduction ($n = 7$, Table 5) after approximately 341 of the 379 L had been filtered. Similar reductions of $0.9 \pm 0.3 \log_{10}$ were observed in the dissolved partition (Table 6). In contrast, chlorination caused in a more dramatic reduction in both the dissolved and solids-associated plasmid concentrations. The chlorine-treated secondary effluent (prior to filtration) reduced the dissolved plasmid concentration by $3.4 \pm 1.6 \log_{10}$ and the solids-associated by $2.6 \pm 1.4 \log_{10}$ (Tables 5 and 6, Figure 3). Even larger plasmid reductions were observed after chlorine treatment of the filtered samples. In particular, the plasmid concentrations were reduced to below detectable limits in all of the tertiary filtered effluents treated with chlorine (samples collected after 90 min. of filtration, $n = 7$), resulting in average log removal values of greater than 5.2 and 4.4-$\log_{10}$ for the dissolved and solids-associated fractions, respectively (Tables 5 and 6, Figure 3). The LOD was used to calculate the log removal values for samples below detection thus, the data represent the minimum removal that was achieved; the actual reductions may be substantially larger. Comparing the effects of chlorine on the pre- and post-filtered samples, after accounting for the amount that was removed by the filter alone, showed that the filtered chlorinated samples removed approximately an order of magnitude or greater of plasmid ARGs compared to the pre-filtered chlorinated samples (Tables 5 and 6). These data indicate that combining filtration and chlorination processes produced a synergistic effect resulting in additional removal of extracellular ARGs compared to chlorine treated pre-filtered samples.

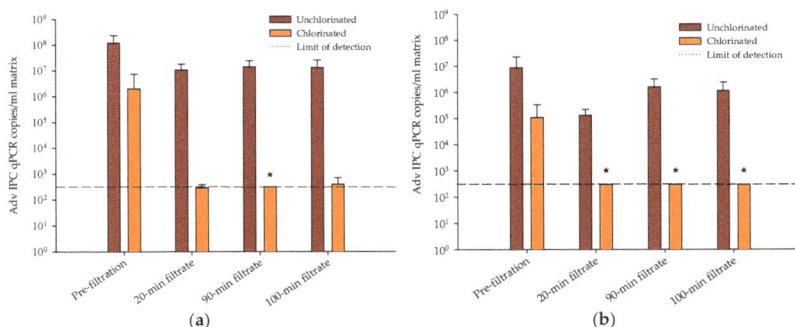

Figure 3. The removal of an antibiotic resistance plasmid by filtration and chlorination processes. An exogenous laboratory-modified plasmid was added into secondary effluent from a WRP and filtered through a dual-media pilot-scale filter with or without chlorine treatment. (**a**) dissolved fraction qPCR data (**b**) solids-associated fraction qPCR data. Pre-filtration refers to samples taken after addition of the plasmid but prior to filtration. The 20, 90 and 100 min. filtrate designations refer to effluent coming through the filter 20, 90 and 100 min. after the start of filtration. The data represent the average of at least seven independent filtration experiments. The maroon bars represent the unchlorinated samples and the orange bars represent the chlorine treated samples, the dashed line represents the limit of detection for the assay. Error bars represent one standard deviation from the mean. The * symbol indicates that all replicate samples were below the LOD of the assay with the corresponding bar representing the detection limit value.

Table 5. Comparison of exogenous plasmid concentrations in the solids fraction before and after filtration and/or disinfection.

Sample [1]	Average Log Difference ($n = 7$) [2]	Rank Sum Test p-Value [3]
SE vs. chlorinated filtrate	>4.4 ± 0.6	<0.001
SE vs. chlorinated SE	2.6 ± 1.4	<0.001
SE vs. filtrate (no chlorine)	0.7 ± 0.5	0.097
Chlorinated SE vs. chlorinated filtrate	>1.8 ± 1.6	0.004
Filtrate vs. chlorinated filtrate	>3.7 ± 0.6	<0.001

Notes: [1] Statistical comparisons were performed on the concentrations of plasmids between two different matrices. SE: secondary effluent prior to filtration but after plasmid addition; filtrate: secondary effluent collected after it had passed through the filter (90-min time point). "Chlorinated" designates the samples were treated with 10 mg/L chlorine for 45 min; [2] The qPCR data (copies/mL) was log transformed for each sample and the water type listed second was subtracted from the first type listed to yield the log difference. The average of the seven individual log differences is shown with the standard deviation; [3] Statistical relevance between the two matrices using the Mann-Whitney rank sum test was indicated by $p < 0.05$.

Table 6. Comparison of exogenous plasmid concentrations in the dissolved fraction before and after filtration and/or disinfection.

Sample [1]	Average Log Difference ($n = 8$) [2]	Rank Sum Test p-Value [3]
SE vs. chlorinated filtrate	>5.2 ± 0.9	<0.001
SE vs. chlorinated SE	3.4 ± 1.6	<0.001
SE vs. filtrate (no chlorine)	0.9 ± 0.3	0.007
Chlorinated SE vs. Chlorinated filtrate	>1.8 ± 1.8	0.038
Filtrate vs. chlorinated filtrate	>4.3 ± 0.8	<0.001

Notes: [1] Statistical comparisons were performed between two different matrices. SE: secondary effluent prior to filtration but after plasmid addition; filtrate: secondary effluent collected after it had passed through the filter. "Chlorinated" designates the samples were treated with 10 mg/L chlorine for 45 min; [2] The qPCR data (copies/µL) was log transformed for each sample and subtracted between two treatments to yield the log difference. The averaged result of the eight individual log differences is shown with the standard deviation; [3] Statistical relevance between the two matrices using the Mann-Whitney rank sum test was indicated by $p < 0.05$.

Gross inhibition of target amplification, as determined using the salmon sperm DNA, was not detected in any of the reactions. In addition, the control samples collected before the addition of the plasmid all tested negative for the AdvIPC target amplicon indicating that the unique sequence was indeed not present in this matrix. Finally, no difference was found with regards to plasmid removal, between the experiments performed after the filters had been in service for greater than two weeks (bio-fouled conditions) or after the filters had been treated with a high dose of chlorine to remove fouling matter (data not shown).

4. Discussion

Antibiotic resistance genes have been identified as contaminants of emerging concern in environmental matrices and could be subject to regulatory monitoring requirements in the future. However, ARGs have been found in every environment on earth and their presence alone does not necessarily signify a major public health concern. The primary factors which have led to the current public health crisis with respect to antibiotic resistance are the horizontal transfer of resistance genes to antibiotic susceptible bacteria coupled with the prodigious use of antibiotics which provided a positive selective pressure resulting in the rapid dissemination of AR across the globe. Therefore, it becomes important to not only identify clinically relevant ARGs in the environment but to also characterize their ability to be transferred between bacteria.

The ARGs chosen for this study represent both clinically relevant and transmissible genes suggesting they are appropriate indicators of ARGs in wastewater. In particular, β-lactams are some of the most prescribed antibiotics worldwide [46]; moreover, *bla* ARGs are prevalent in human pathogens as well as in wastewater [47,48] and have shown the ability to undergo HGT [49]. The *sul1* gene is known to be associated with class I integrons on conjugative plasmids [50,51] and is one of the most abundant ARGs in wastewater [52].

Previous data combined from several reports show that concentrations of ARGs can vary throughout wastewater treatment depending on the ARG studied, the extent of treatment, geographical location, and operational parameters (reviewed in [52]). The results presented herein show that the concentrations of $bla_{SHV/TEM}$ and *sul1* were similar to what has been reported by others for raw wastewater, activated sludge, secondary and tertiary effluent although the *sul1* concentrations were slightly higher in the activated sludge than what has been reported previously [52]. The tertiary WRP examined in this study produced ARG reductions of approximately four-$\log_{10}$ in the final disinfected effluent compared to the raw sewage concentrations which are some of the largest overall reductions reported for activated sludge–type WWTPs [53–57]. The full extent of the β-lactamase reductions in the disinfected final effluent from the full-scale WRP, could not be precisely quantified because the values for all samples analyzed were below the detectable limit of the qPCR assay.

Although *sul1* and $bla_{SHV/TEM}$ concentrations were greatly reduced after activated sludge treatment, the ARGs were still present in both fractions in the secondary effluent. In contrast, the $bla_{SHV/TEM}$ genes could not be detected in tertiary-treated final effluent (after chlorination) even upon concentrating the water samples over 200-fold. The *sul1* gene was detected in higher numbers throughout each of the treatment stages compared to the $bla_{SHV/TEM}$ yet, was only detected in three of six concentrated final effluent samples for the solids fraction and five of six dissolved fraction samples. This indicated that substantial removal occurred during the tertiary filtration and disinfection processes (solids: >2-$\log_{10}$ and dissolved: >3-$\log_{10}$ reduction). Pilot-scale filtration experiments were incorporated to further study the media filtration and disinfection processes. The results indicated that tertiary filtration coupled with chlorine disinfection provided a synergistic benefit with respect to removal of ARGs compared to chlorination alone. More specifically, chlorine mediated reduction of ARGs was more effective on filtered effluents compared to non-filtered effluents. The increased effectiveness of the chlorine in the filtered water may reflect the removal of substances that increase chlorine demand during filtration. As a consequence, more chlorine would be available in the filtered water compared to the unfiltered secondary effluent resulting in greater degradation of the

ARGs. A previous report found that disinfection of tertiary-treated effluent waters provided less than a one-log$_{10}$ reduction in ARG concentration [55]. Several factors may explain the discrepancy between the results presented here and Munir et al. (2011): the WRP studied here incorporated biological ammonia removal through a nitrification and denitrification process lowering ammonia levels in the effluent resulting in less chloramine formation. Therefore, the removal rates documented here may differ for WWTPs that do not incorporate ammonia removal processes. Specifically, other investigators have reported that chloramines are less effective in reducing ARG concentrations in wastewater compared to free chlorine [58]. It should be noted that the effects of ammonia on ARG removal were not specifically evaluated as part of this project. Additionally, the geographic locations of the WWTPs differed in the two studies and the operational parameters were not reported; both factors could have contributed to differences in the presence and removal of ARGs. Taken together, these data indicate that WWTPs that include the use of tertiary filtration with disinfection can provide additional ARG reductions and thereby further minimize any potential public health and environmental impacts compared to those without filtration.

The pilot-scale tertiary filtration and chlorine disinfection experiments resulted in a removal of at least five log$_{10}$ of dissolved ARG plasmid. Very little ARG removal was attributed to the filtration process itself (less than 0.9 log$_{10}$) whereas the chlorinated, filtered effluent decreased the ARG concentration to undetectable limits, signifying a substantial role for chlorine in the overall removal. Treatment of secondary effluents with chlorine successfully reduced ARG concentrations however; tertiary filtration was shown to enhance the chlorine mediated removal of ARGs and provided at least an additional order of magnitude more reduction compared to chlorinated secondary effluent alone. The full-scale WRP data showed an average removal of greater than two log$_{10}$ between secondary and tertiary filtered, disinfected effluents indicating that further optimization of the filtration and disinfection processes could produce additional ARG reductions. The use of pilot-scale processes offered some experimental advantages not available at the full-scale level. In particular, the pilot-scale filtration permitted the study of ARG removal using a single homogenous plasmid added into the secondary effluent immediately prior to filtration whereas, the full-scale plant experiments detected ARGs from multiple sources of DNA (plasmids, genomic, phage, etc.) that could interact with elements present throughout the treatment process. Furthermore, chlorination of the pilot-scale samples was performed under controlled conditions in the laboratory where the precise concentration and contact time were monitored along with the use of fresh liquid chlorine thus, maximizing its oxidative potential. Future studies focused on the optimization of ARG removal by full-scale filtration and chlorination processes would provide additional information on how to further reduce these constituents.

Currently, it is difficult to directly assess the total HGT activity in wastewater ecosystems. While there are culture-based methods available that can approximate transfer in particular model organisms or by certain mechanisms, they have drawbacks [59]. The use of non-culture-based methods such as PCR offer the ability to identify ARGs and transmissible elements from essentially all sources in a sample and because HGT mechanisms are associated with cellular or extracellular fractions, PCR analysis of ARGs in each fraction can provide information concerning which HGT pathways have a greater likelihood of leading to ARG transfer. This approach assumes that higher concentrations of ARGs in a particular fraction correlate with an increased chance of HGT occurring by the pathways associated with each fraction. In support of this, transformation rates in water have been shown to increase with increasing gene concentrations [60] and the same logic would be expected to apply with transduction-mediated HGT. With respect to the solids-associated fraction, conjugation rates in water have been shown to be dependent on the number of cells containing the transferrable gene [61].

While not a direct assessment of HGT, analyzing the cellular and extracellular fractions for ARGs provided some insight into where these genes reside during different treatment stages as well as which HGT mechanisms may have a higher probability of occurring in each water type. For example, the ratio of solids to dissolved ARGs was at least 100-fold higher in the activated sludge compared to the tertiary-treated, disinfected effluent, which would suggest that HGT via conjugation would be more likely to occur in the activated sludge compared to the final effluent. Note that there are multiple factors that can affect the horizontal transfer of genes in the environment and while the fractionation experiments give some insight into the probability of one transfer pathway occurring over another it is not a direct analysis of HGT. Given the importance of HGT with regard to spread of AR, a more thorough study targeting HGT in these waters would be beneficial.

The concentrations of dissolved and solids-associated ARGs differed between the treatment processes. The dissolved concentrations of both ARGs were highest in the raw sewage and showed reductions at each successive stage of treatment (raw > activated sludge > secondary effluent > tertiary effluent). In contrast, the concentrations of both ARGs and *16S* DNA in the solids fraction increased in the activated sludge compared to the raw sewage. The *sul1* and *16S* genes increased in a similar fashion while the *bla*$_{SHV/TEM}$ demonstrated a smaller increase than the other two genes (statistically significant) suggesting the genes may be present in separate microbial populations. This trend was not observed in subsequent stages as both ARGs were reduced substantially in the solids fraction of the secondary and final effluents (raw < activated sludge > secondary effluent > final effluent).

These results would be consistent with a model in which ARG concentrations increase in conjunction with the expansion of microbial populations during the biological phase of treatment (activated sludge) followed by a decrease in the amount of solids associated ARGs in the secondary effluent resulting from floc settling and solids removal. Finally, an increase in the percentage of dissolved ARGs observed after tertiary filtration and disinfection could result from the chlorine-mediated destruction of bacterial cells releasing additional DNA in the aqueous fraction. Reductions in viable indicator bacterial concentrations after tertiary filtration and disinfection have been well documented and, coupled with data that demonstrate the presence of extracellular *Bacteroides* DNA in the absence of the viable bacteria after wastewater treatment [62], lend support to this theory. Additionally, a recent disinfection study using enterococci containing the *vanA* resistance gene demonstrated that *vanA* DNA could be detected after chlorination of secondary effluent when the concentration of viable enterococci was reduced to below detectable limits [63]. What effect, if any, the release of dissolved DNA has on the presence and dissemination of ARGs in waters that receive treated effluents is currently unknown.

Presently there is no consensus as to whether ARGs are selected for or against during wastewater treatment with published reports illustrating both scenarios [17,20,24,54,64–69]. However, the activated sludge treatment stage in particular, has been proposed as a potential source for ARG transfer [56]. This study examined samples from each stage of treatment and evaluated the data for positive and negative selection by comparing the two target ARGs to the number of *16S* rDNA genes (an indicator of the total bacterial biomass). The ratio of *sul1* to *16S* did not change significantly throughout each stage of treatment (for both fractions) suggesting this ARG was neither selected for or against. The *bla*$_{SHV/TEM}$:*16S* ratio in the solids fraction decreased between the raw sewage and activated sludge matrices in a statistically relevant manner indicating that the number of *bla*$_{SHV/TEM}$ genes in relation to the total bacterial biomass went down compared to the raw sewage. The solids fraction ratio increased slightly in the secondary effluent compared to the activated sludge (not statistically significant) but was still below the ratio observed in the raw sewage. A comparison of the dissolved fractions showed little change between the raw and activated sludge *bla*$_{SHV/TEM}$:*16S* but decreased significantly in the secondary effluent (statistically significant). The fact that the ARGs did not increase at a significantly larger rate than the total bacterial population suggests that they were not selected for during the treatment processes with respect to the overall bacterial populations present at each stage. In total, the data showed no evidence that the two ARGs analyzed in this project were being selected for during wastewater treatment. It should be noted that not all ARGs were evaluated in this study. While the two resistant genes selected for this project represent highly prevalent and clinically relevant genes, the possibility cannot be ruled out that other genes might be selected for during wastewater treatment. In addition, operational parameters have been shown to affect ARG concentrations [19,70] thus ARG reductions may differ between facilities operating under different conditions.

5. Conclusions

1. The full-scale tertiary stage WRP reduced concentrations of *sul1* by approximately four-log$_{10}$ from the raw sewage. In addition, the *bla$_{SHV/TEM}$* ARG was reduced to below detectable limits in the final effluent (removal of greater than three log$_{10}$).

2. The percentage of ARGs that partitioned with the solids and dissolved phases differed between treatment processes.

3. Positive selection for *sul1* or *bla$_{SHV/TEM}$* ARGs, in reference to the total bacterial biomass, was not observed throughout the treatment process.

4. Tertiary media filtration and chlorine disinfection were the most effective treatment processes with respect to ARG reductions.

5. Pilot-scale dual-media filter experiments demonstrated that tertiary filtration enhanced chlorine mediated reduction of an ARG containing plasmid compared to chlorine treatment of secondary effluent.

6. This data demonstrated that tertiary filtration and disinfection can result in additional removal of ARGs compared to non-filtered disinfected secondary effluent.

Supplementary Materials: The following are available online at www.mdpi.com/2073-4441/10/1/37/s1, Figures S1, S2; Tables S1–S12. References [71,72] are cited in the supplementary materials.

Author Contributions: R.A.R. and A.S. conceived and designed the experiments; J.Q.-C., B.H.-L., S.A., C.M. and R.A.R. performed the experiments; R.A.R. analyzed the data and wrote the paper.

Conflicts of Interest: The authors declare no conflict of interest.

References

1. Mainous, A.G., 3rd; Diaz, V.A.; Matheson, E.M.; Gregorie, S.H.; Hueston, W.J. Trends in hospitalizations with antibiotic-resistant infections: U.S., 1997–2006. *Public Health Rep.* **2011**, *126*, 354–360. [CrossRef] [PubMed]
2. Phelps, C.E. Bug/drug resistance: Sometimes less is more. *Med. Care* **1989**, *27*, 194–203. [CrossRef] [PubMed]
3. Auerbach, E.A.; Seyfried, E.E.; McMahon, K.D. Tetracycline resistance genes in activated sludge wastewater treatment plants. *Water Res.* **2007**, *41*, 1143–1151. [CrossRef] [PubMed]
4. Grabow, W.O.; Prozesky, O.W. Drug resistance of coliform bacteria in hospital and city sewage. *Antimicrob. Agents Chemother.* **1973**, *3*, 175–180. [CrossRef] [PubMed]
5. Grabow, W.O.K.; Prozesky, O.W.; Smith, L.S. Drug resistant coliforms call for review of water quality standards. *Water Res.* **1974**, *8*, 1–9. [CrossRef]
6. Linton, K.B.; Richmond, M.H.; Bevan, R.; Gillespie, W.A. Antibiotic resistance and R factors in coliform bacilli isolated from hospital and domestic sewage. *J. Med. Microbiol.* **1974**, *7*, 91–103. [CrossRef] [PubMed]
7. Niemi, M.; Sibakov, M.; Niemela, S. Antibiotic resistance among different species of fecal coliforms isolated from water samples. *Appl. Environ. Microbiol.* **1983**, *45*, 79–83. [PubMed]
8. Reinthaler, F.F.; Posch, J.; Feierl, G.; Wnst, G.; Haas, D.; Ruckenbauer, G.; Mascher, F.; Marth, E. Antibiotic resistance of *E. coli* in sewage and sludge. *Water Res.* **2003**, *37*, 1685–1690. [CrossRef]
9. Chee-Sanford, J.C.; Aminov, R.I.; Krapac, I.J.; Garrigues-Jeanjean, N.; Mackie, R.I. Occurrence and diversity of tetracycline resistance genes in lagoons and groundwater underlying two swine production facilities. *Appl. Environ. Microbiol.* **2001**, *67*, 1494–1502. [CrossRef] [PubMed]
10. Koike, S.; Krapac, I.G.; Oliver, H.D.; Yannarell, A.C.; Chee-Sanford, J.C.; Aminov, R.I.; Mackie, R.I. Monitoring and source tracking of tetracycline resistance genes in lagoons and groundwater adjacent to swine production facilities over a 3-year period. *Appl. Environ. Microbiol.* **2007**, *73*, 4813–4823. [CrossRef] [PubMed]
11. Sayah, R.S.; Kaneene, J.B.; Johnson, Y.; Miller, R. Patterns of antimicrobial resistance observed in *Escherichia coli* isolates obtained from domestic- and wild-animal fecal samples, human septage, and surface water. *Appl. Environ. Microbiol.* **2005**, *71*, 1394–1404. [CrossRef] [PubMed]
12. Schmitt, H.; Stoob, K.; Hamscher, G.; Smit, E.; Seinen, W. Tetracyclines and tetracycline resistance in agricultural soils: Microcosm and field studies. *Microb. Ecol.* **2006**, *51*, 267–276. [CrossRef] [PubMed]

13. Baquero, F.; Martinez, J.L.; Canton, R. Antibiotics and antibiotic resistance in water environments. *Curr. Opin. Biotechnol.* **2008**, *19*, 260–265. [CrossRef] [PubMed]
14. Feary, T.W.; Sturtevant, A.B.J.; Lankford, J. Antibiotic-resistant coliforms in fresh and salt water. *Arch. Environ. Health* **1972**, *25*, 215–220. [PubMed]
15. Goni-Urriza, M.; Capdepuy, M.; Arpin, C.; Raymond, N.; Caumette, P.; Quentin, C. Impact of an urban effluent on antibiotic resistance of riverine *Enterobacteriaceae* and *Aeromonas* spp. *Appl. Environ. Microbiol.* **2000**, *66*, 125–132. [CrossRef] [PubMed]
16. Goñi-Urriza, M.; Pineau, L.; Capdepuy, M.; Roques, C.; Caumette, P.; Quentin, C. Antimicrobial resistance of mesophilic *Aeromonas* spp. Isolated from two european rivers. *J. Antimicrob. Chemother.* **2000**, *46*, 297–301. [CrossRef] [PubMed]
17. Iwane, T.; Urase, T.; Yamamoto, K. Possible impact of treated wastewater discharge on incidence of antibiotic resistant bacteria in river water. *Water Sci. Technol.* **2001**, *43*, 91–99. [PubMed]
18. Smith, H.W. Incidence of river water of *Escherichia coli* containing r factors. *Nature* **1970**, *228*, 1286–1288. [CrossRef] [PubMed]
19. Kim, S.; Aga, D.S.; Jensen, J.N.; Weber, A.S. Effect of sequencing batch reactor operation on presence and concentration of tetracycline-resistant organisms. *Water Environ. Res.* **2007**, *79*, 2287–2297. [CrossRef] [PubMed]
20. Zhang, Y.; Marrs, C.F.; Simon, C.; Xi, C. Wastewater treatment contributes to selective increase of antibiotic resistance among *Acinetobacter* spp. *Sci. Total Environ.* **2009**, *407*, 3702–3706. [CrossRef] [PubMed]
21. Pruden, A.; Arabi, M.; Storteboom, H.N. Correlation between upstream human activities and riverine antibiotic resistance genes. *Environ. Sci. Technol.* **2012**, *46*, 11541–11549. [CrossRef] [PubMed]
22. Martins da Costa, P.; Vaz-Pires, P.; Bernardo, F. Antimicrobial resistance in *Enterococcus* spp. Isolated in inflow, effluent and sludge from municipal sewage water treatment plants. *Water Res.* **2006**, *40*, 1735–1740. [CrossRef] [PubMed]
23. LaPara, T.M.; Burch, T.R.; McNamara, P.J.; Tan, D.T.; Yan, M.; Eichmiller, J.J. Tertiary-treated municipal wastewater is a significant point source of antibiotic resistance genes into Duluth-Superior Harbor. *Environ. Sci. Technol.* **2011**, *45*, 9543–9549. [CrossRef] [PubMed]
24. Czekalski, N.; Berthold, T.; Caucci, S.; Egli, A.; Burgmann, H. Increased levels of multiresistant bacteria and resistance genes after wastewater treatment and their dissemination into lake geneva, Switzerland. *Front. Microbiol.* **2012**, *3*, 106. [CrossRef] [PubMed]
25. Bockelmann, U.; Dorries, H.H.; Ayuso-Gabella, M.N.; Salgot de Marcay, M.; Tandoi, V.; Levantesi, C.; Masciopinto, C.; Van Houtte, E.; Szewzyk, U.; Wintgens, T.; et al. Quantitative PCR monitoring of antibiotic resistance genes and bacterial pathogens in three European artificial groundwater recharge systems. *Appl. Environ. Microbiol.* **2009**, *75*, 154–163. [CrossRef] [PubMed]
26. U.S. Environmental Protection Agency. *Guidelines for Water Reuse*; U.S. Environmental Protection Agency: Washington, DC, USA, 2004; p. 450.
27. Rose, J.B.; Farrah, S.R.; Harwood, V.J.; Levine, A.; Lukasik, J.; Menendez, P.; Scott, T.M. *Reduction of Pathogens, Indicator Bacteria, and Alternative Indicators by Wastewater Treatment and Reclamation Processes*; Water Environment Research Foundation: Alexandria, VA, USA, 2004.
28. Asano, T. *Wastewater Reclamation and Reuse: Water Quality Management Library*; Taylor & Francis: Milton Park, UK, 1998.
29. Mazel, D.; Davies, J. Antibiotic resistance in microbes. *Cell Mol. Life Sci.* **1999**, *56*, 742–754. [CrossRef] [PubMed]
30. Von Wintersdorff, C.J.H.; Penders, J.; van Niekerk, J.M.; Mills, N.D.; Majumder, S.; van Alphen, L.B.; Savelkoul, P.H.M.; Wolffs, P.F.G. Dissemination of antimicrobial resistance in microbial ecosystems through horizontal gene transfer. *Front. Microbiol.* **2016**, *7*, 173. [CrossRef] [PubMed]
31. Hu, Y.; Yang, X.; Li, J.; Lv, N.; Liu, F.; Wu, J.; Lin, I.Y.; Wu, N.; Weimer, B.C.; Gao, G.F.; et al. The bacterial mobile resistome transfer network connecting the animal and human microbiomes. *Appl. Environ. Microbiol.* **2016**, *82*, 6672–6681. [CrossRef] [PubMed]
32. Allen, H.K.; Donato, J.; Wang, H.H.; Cloud-Hansen, K.A.; Davies, J.; Handelsman, J. Call of the wild: Antibiotic resistance genes in natural environments. *Nat. Rev. Microbiol.* **2010**, *8*, 251–259. [CrossRef] [PubMed]
33. Aminov, R.I. The role of antibiotics and antibiotic resistance in nature. *Environ. Microbiol.* **2009**, *11*, 2970–2988. [CrossRef] [PubMed]

34. Martinez, J.L. The role of natural environments in the evolution of resistance traits in pathogenic bacteria. *Proc. Biol. Sci.* **2009**, *276*, 2521–2530. [CrossRef] [PubMed]

35. Martinez, J.L. Environmental pollution by antibiotics and by antibiotic resistance determinants. *Environ. Pollut.* **2009**, *157*, 2893–2902. [CrossRef] [PubMed]

36. Bhullar, K.; Waglechner, N.; Pawlowski, A.; Koteva, K.; Banks, E.D.; Johnston, M.D.; Barton, H.A.; Wright, G.D. Antibiotic resistance is prevalent in an isolated cave microbiome. *PLoS ONE* **2012**, *7*, e34953. [CrossRef] [PubMed]

37. D'Costa, V.M.; King, C.E.; Kalan, L.; Morar, M.; Sung, W.W.; Schwarz, C.; Froese, D.; Zazula, G.; Calmels, F.; Debruyne, R.; et al. Antibiotic resistance is ancient. *Nature* **2011**, *477*, 457–461. [CrossRef] [PubMed]

38. Munck, C.; Albertsen, M.; Telke, A.; Ellabaan, M.; Nielsen, P.H.; Sommer, M.O.A. Limited dissemination of the wastewater treatment plant core resistome. *Nat. Commun.* **2015**, *6*. [CrossRef] [PubMed]

39. Negreanu, Y.; Pasternak, Z.; Jurkevitch, E.; Cytryn, E. Impact of treated wastewater irrigation on antibiotic resistance in agricultural soils. *Environ. Sci. Technol.* **2012**, *46*, 4800–4808. [CrossRef] [PubMed]

40. Pei, R.; Kim, S.C.; Carlson, K.H.; Pruden, A. Effect of river landscape on the sediment concentrations of antibiotics and corresponding antibiotic resistance genes (ARG). *Water Res.* **2006**, *40*, 2427–2435. [CrossRef] [PubMed]

41. Harms, G.; Layton, A.C.; Dionisi, H.M.; Gregory, I.R.; Garrett, V.M.; Hawkins, S.A.; Robinson, K.G.; Sayler, G.S. Real-time PCR quantification of nitrifying bacteria in a municipal wastewater treatment plant. *Environ. Sci. Technol.* **2003**, *37*, 343–351. [CrossRef] [PubMed]

42. U.S. Environmental Protection Agency. *Method A: Enterococci in Water by Taqman Quantitative Polymerase Chain Reaction (qPCR) Assay*; EPA-821-R-10-004; U.S. EPA: Washington, DC, USA, 2010; Volume 2010.

43. Jothikumar, N.; Cromeans, T.L.; Hill, V.R.; Lu, X.; Sobsey, M.D.; Erdman, D.D. Quantitative real-time PCR assays for detection of human adenoviruses and identification of serotypes 40 and 41. *Appl. Environ. Microbiol.* **2005**, *71*, 3131–3136. [CrossRef] [PubMed]

44. Malorny, B.; Paccassoni, E.; Fach, P.; Bunge, C.; Martin, A.; Helmuth, R. Diagnostic real-time PCR for detection of Salmonella in food. *Appl. Environ. Microbiol.* **2004**, *70*, 7046–7052. [CrossRef] [PubMed]

45. Association, A.P.H. *Standard Methods for the Examination of Water and Wastewater*; General Books: Vancouver, BC, Canada, 2013.

46. Hicks, L.A.; Bartoces, M.G.; Roberts, R.M.; Suda, K.J.; Hunkler, R.J.; Taylor, T.H., Jr.; Schrag, S.J. US outpatient antibiotic prescribing variation according to geography, patient population, and provider specialty in 2011. *Clin. Infect. Dis.* **2015**, *60*, 1308–1316. [CrossRef] [PubMed]

47. Livermore, D.M. Beta-lactamases in laboratory and clinical resistance. *Clin. Microbiol. Rev.* **1995**, *8*, 557–584. [PubMed]

48. Yang, Y.; Zhang, T.; Zhang, X.-X.; Liang, D.-W.; Zhang, M.; Gao, D.-W.; Zhu, H.-G.; Huang, Q.-G.; Fang, H.H. Quantification and characterization of β-lactam resistance genes in 15 sewage treatment plants from East Asia and North America. *Appl. Microbiol. Biotechnol.* **2012**, *95*, 1351–1358. [CrossRef] [PubMed]

49. Paterson, D.L.; Bonomo, R.A. Extended-spectrum β-lactamases: A clinical update. *Clin. Microbiol. Rev.* **2005**, *18*, 657–686. [CrossRef] [PubMed]

50. Gillings, M.; Boucher, Y.; Labbate, M.; Holmes, A.; Krishnan, S.; Holley, M.; Stokes, H.W. The evolution of class 1 integrons and the rise of antibiotic resistance. *J. Bacteriol.* **2008**, *190*, 5095–5100. [CrossRef] [PubMed]

51. Gillings, M.R.; Gaze, W.H.; Pruden, A.; Smalla, K.; Tiedje, J.M.; Zhu, Y.-G. Using the class 1 integron-integrase gene as a proxy for anthropogenic pollution. *ISME J.* **2015**, *9*, 1269–1279. [CrossRef] [PubMed]

52. Olivieri, A.W.; Crook, J.; Anderson, M.A.; Bull, R.J.; Drewes, J.E.; Haas, C.N.; Jakubowski, W.; McCarty, P.L.; Nelson, J.B.R.; Sedlak, D.L.; et al. *Expert Panel Final Report: Evaluation of the Feasibility of Developing Uniform Water Recycling Criteria for Direct Potable Reuse*; California State Water Resources Control Board: Sacramento, CA, USA, 2016; pp. 165–194.

53. Chen, H.; Zhang, M. Effects of advanced treatment systems on the removal of antibiotic resistance genes in wastewater treatment plants from Hangzhou, China. *Environ. Sci. Technol.* **2013**, *47*, 8157–8163. [CrossRef] [PubMed]

54. Lachmayr, K.L.; Kerkhof, L.J.; Dirienzo, A.G.; Cavanaugh, C.M.; Ford, T.E. Quantifying nonspecific TEM beta-lactamase (blaTEM) genes in a wastewater stream. *Appl. Environ. Microbiol.* **2009**, *75*, 203–211. [CrossRef] [PubMed]

55. Munir, M.; Wong, K.; Xagoraraki, I. Release of antibiotic resistant bacteria and genes in the effluent and biosolids of five wastewater utilities in Michigan. *Water Res.* **2011**, *45*, 681–693. [CrossRef] [PubMed]

56. Rizzo, L.; Manaia, C.; Merlin, C.; Schwartz, T.; Dagot, C.; Ploy, M.C.; Michael, I.; Fatta-Kassinos, D. Urban wastewater treatment plants as hotspots for antibiotic resistant bacteria and genes spread into the environment: A review. *Sci. Total Environ.* **2013**, *447*, 345–360. [CrossRef] [PubMed]

57. Zhang, T.; Zhang, M.; Zhang, X.; Fang, H.H. Tetracycline resistance genes and tetracycline resistant lactose-fermenting Enterobacteriaceae in activated sludge of sewage treatment plants. *Environ. Sci. Technol.* **2009**, *43*, 3455–3460. [CrossRef] [PubMed]

58. Zhang, Y.; Zhuang, Y.; Geng, J.; Ren, H.; Zhang, Y.; Ding, L.; Xu, K. Inactivation of antibiotic resistance genes in municipal wastewater effluent by chlorination and sequential UV/chlorination disinfection. *Sci. Total Environ.* **2015**, *512–513*, 125–132. [CrossRef] [PubMed]

59. Aminov, R.I. Horizontal gene exchange in environmental microbiota. *Front. Microbiol.* **2011**, *2*, 158. [CrossRef] [PubMed]

60. Baur, B.; Hanselmann, K.; Schlimme, W.; Jenni, B. Genetic transformation in freshwater: *Escherichia coli* is able to develop natural competence. *Appl. Environ. Microbiol.* **1996**, *62*, 3673–3678. [PubMed]

61. Guo, M.T.; Yuan, Q.B.; Yang, J. Distinguishing effects of ultraviolet exposure and chlorination on the horizontal transfer of antibiotic resistance genes in municipal wastewater. *Environ. Sci. Technol.* **2015**, *49*, 5771–5778. [CrossRef] [PubMed]

62. Bae, S.; Wuertz, S. Discrimination of viable and dead fecal Bacteroidales bacteria by quantitative PCR with propidium monoazide. *Appl. Environ. Microbiol.* **2009**, *75*, 2940–2944. [CrossRef] [PubMed]

63. Furukawa, T.; Jikumaru, A.; Ueno, T.; Sei, K. Inactivation effect of antibiotic-resistant gene using chlorine disinfection. *Water* **2017**, *9*, 547. [CrossRef]

64. Borjesson, S.; Matussek, A.; Melin, S.; Lofgren, S.; Lindgren, P.E. Methicillin-resistant Staphylococcus aureus (MRSA) in municipal wastewater: An uncharted threat? *J. Appl. Microbiol.* **2010**, *108*, 1244–1251. [CrossRef] [PubMed]

65. Borjesson, S.; Melin, S.; Matussek, A.; Lindgren, P.E. A seasonal study of the mecA gene and *Staphylococcus aureus* including methicillin-resistant *S. aureus* in a municipal wastewater treatment plant. *Water Res.* **2009**, *43*, 925–932. [CrossRef] [PubMed]

66. Garcia, S.; Wade, B.; Bauer, C.; Craig, C.; Nakaoka, K.; Lorowitz, W. The effect of wastewater treatment on antibiotic resistance in *Escherichia coli* and *Enterococcus* sp. *Water Environ. Res.* **2007**, *79*, 2387–2395. [CrossRef] [PubMed]

67. Nagulapally, S.R.; Ahmad, A.; Henry, A.; Marchin, G.L.; Zurek, L.; Bhandari, A. Occurrence of ciprofloxacin-, trimethoprim-sulfamethoxazole-, and vancomycin-resistant bacteria in a municipal wastewater treatment plant. *Water Environ. Res.* **2009**, *81*, 82–90. [CrossRef] [PubMed]

68. Rijal, G.K.; Zmuda, J.T.; Gore, R.; Abedin, Z.; Granato, T.; Kollias, L.; Lanyon, R. Antibiotic resistant bacteria in wastewater processed by the metropolitan water reclamation district of greater chicago system. *Water Sci. Technol.* **2009**, *59*, 2297–2304. [CrossRef] [PubMed]

69. Luczkiewicz, A.; Fudala-Ksiazek, S.; Jankowska, K.; Quant, B.; Olanczuk-Neyman, K. Diversity of fecal coliforms and their antimicrobial resistance patterns in wastewater treatment model plant. *Water Sci. Technol.* **2010**, *61*, 1383–1392. [CrossRef] [PubMed]

70. Kim, S.; Jensen, J.N.; Aga, D.S.; Weber, A.S. Fate of tetracycline resistant bacteria as a function of activated sludge process organic loading and growth rate. *Water Sci. Technol.* **2007**, *55*, 291–297. [CrossRef] [PubMed]

71. Bustin, S.A.; Benes, V.; Garson, J.A.; Hellemans, J.; Huggett, J.; Kubista, M.; Mueller, R.; Nolan, T.; Pfaffl, M.W.; Shipley, G.L.; et al. The MIQE guidelines: Minimum information for publication of quantitative real-time pcr experiments. *Clin. Chem.* **2009**, *55*, 611–622. [CrossRef] [PubMed]

72. Kralik, P.; Ricchi, M. A basic guide to real time PCR in microbial diagnostics: Definitions, parameters, and everything. *Front Microbiol.* **2017**, *8*, 108. [CrossRef] [PubMed]

water

MDPI

Article

Pseudomonas aeruginosa Psl Exopolysaccharide Interacts with the Antimicrobial Peptide LG21

Joyce Seow Fong Chin [1], Sheetal Sinha [2,3,4], Anjaiah Nalaparaju [2], Joey Kuok Hoong Yam [1], Zhiqiang Qin [5], Luyan Ma [6], Zhao-Xun Liang [2], Lanyuan Lu [2], Surajit Bhattacharjya [2] and Liang Yang [1,2,*]

[1]   Singapore Centre for Environmental Life Sciences Engineering (SCELSE), Nanyang Technological University, Singapore 637551, Singapore; chinsf@ntu.edu.sg (J.S.F.C.); joeyyam@ntu.edu.sg (J.K.H.Y.)
[2]   School of Biological Sciences, Nanyang Technological University, Singapore 639798, Singapore; SHEETAL003@e.ntu.edu.sg (S.S.); anjai@ntu.edu.sg (A.N.); ZXLiang@ntu.edu.sg (Z.-X.L.); LYLU@ntu.edu.sg (L.L.); surajit@ntu.edu.sg (S.B.)
[3]   Advanced Environmental Biotechnology Centre, Nanyang Environment and Water Research Institute, Nanyang Technological University, 1 Cleantech Loop, Singapore 637141, Singapore
[4]   Interdisciplinary Graduate School, Nanyang Technological University, 50 Nanyang Avenue, Singapore 639798, Singapore
[5]   Departments of Genetics, Louisiana State University Health Sciences Center, Louisiana Cancer Research Center, 1700 Tulane Ave., New Orleans, LA 70112, USA; zqin@lsuhsc.edu
[6]   State Key Laboratory of Microbial Resources in Institute of Microbiology, Chinese Academy of Sciences (IMCAS), Beijing 100101, China; luyanma27@im.ac.cn
*   Correspondence: yangliang@ntu.edu.sg

Received: 9 June 2017; Accepted: 4 September 2017; Published: 16 September 2017

Abstract: Biofilm formation by opportunistic pathogens serves as one of the major causes of chronic and persistent infections. Bacterial cells in the biofilms are embedded in their self-generated protective extracellular polymeric substances (EPS), which include exopolysaccharides, large adhesin proteins and extracellular DNA. In this study, we identified an antimicrobial peptide (AMP) LG21 that is able to interact specifically with the Psl exopolysaccharide of *Pseudomonas aeruginosa*, thus it can be used as a diagnostic tool for *P. aeruginosa* biofilms. Molecular dynamics simulation analysis showed that residues numbered from 15 to 21 (WKRKRFG) in LG21 are involved in interacting with Psl. Our study indicates that host immune systems might detect and interact with microbial biofilms through AMPs. Engineering biofilm EPS-targeting AMPs might provide novel strategies for biofilm detection and treatment.

Keywords: *Pseudomonas aeruginosa*; Psl; exopolysaccharide; antimicrobial peptide (AMP); biofilm; EPS

1. Introduction

Antimicrobial peptides (AMPs) serve as an essential component of the innate immune system to defend against invading pathogens [1]. AMPs are amphipathic molecules that can directly interact with bacterial cell wall components such as lipopolysaccharide (LPS) and compromise the cell wall integrity [2]. AMPs are also able to target microbial intracellular components such as DNA and RNA [3]. In addition to directly targeting microbial cells, host-derived AMPs are known to modulate the innate immune response and boost the host's capacity for bacterial clearance [4].

Microbial pathogens have successfully evolved multiple strategies to survive from AMP attack. For example, numerous bacterial species have developed AMP sensing mechanisms, which regulate modifications of the cell surface upon AMP exposure [5–7]. Extracellular proteases secreted by bacterial cells have been shown to degrade AMPs and contribute to AMP resistance [8,9]. In addition, microbial cells are able to form surface-attached biofilm communities, which represent a distinct lifestyle with increased resistance towards antimicrobials including AMPs [10].

Biofilms consist of microbial cells entrapped by their self-generated extracellular polymeric substance (EPS), such as extracellular DNA, proteins, and exopolysaccharides [11]. EPS serves as a physical shield to protect biofilm cells against harmful conditions such as host immune clearance and antimicrobial treatment. Recently, certain EPS components were shown to interact with bacterial signaling molecules [12] and modulate gene expressions [13]. Biofilm EPS components might interact with AMPs and modulate their functions.

Pseudomonas aeruginosa is an opportunistic pathogen that causes a wide range of nosocomial infections [14]. *P. aeruginosa* is well-known to form biofilms during infections, which prolong hospitalization and increase the recurrence risk [15,16]. The EPS of *P. aeruginosa* biofilms formed by different strains might contain three exopolysaccharides, alginate, Pel and Psl, which play important roles in biofilm structure maintenance and functions [17,18]. Among these three exopolysaccharides, Psl appears to be the most rigid material and crosslinks *P. aeruginosa* cells, leading to microcolony formation [19]. *P. aeruginosa* small colony variants that over-synthesize Psl have often been observed in clinical settings after acquiring mutations in the *wspF* gene [20,21]. Psl was shown to protect biofilm cells against antibiotic treatment and phagocytosis [17,22]. Previous genetic and biochemical analysis showed that Psl shares conserved structure components (e.g., D-mannose, D-glucose and L-rhamnose) with the LPS [23] and found antibodies to Psl were cross-reactive with LPS [24]. Thus, we hypothesized that AMPs might be able to interact with Psl in a manner similar to LPS.

In this study, we screened a local AMP peptide library to identify AMPs that are able to interact with the *P. aeruginosa* Psl. We identified an AMP, LG21, that is able to specifically bind Psl. Our study provided evidence that AMPs could be developed as potential biofilm matrix-targeting compounds.

2. Results and Discussion

2.1. LG21 Stains Psl Positive P. Aeruginosa Biofilms

Psl exopolysaccharide is a critical structural component of *P. aeruginosa* biofilms. We screened biofilms formed by *P. aeruginosa* wild-type PAO1 (Pel+Psl+), its Psl deficient Δ*pslBCD* mutant (Pel+Psl-), and Pel deficient Δ*pelA* mutant (Pel-Psl+) against our local fluorescent-tagged AMP library. Through this screening, the rhodamine-tagged LG21 was found to strongly stain the Psl+ biofilms (formed by the PAO1 and Δ*pelA* mutant) but not the Psl- biofilm (formed by the Δ*pslBCD* mutant) (Figure 1). Fluorescent signals of the rhodamine-tagged LG21 colocalized well with another well-known Psl stain, TR-ConA [25] (Figure 1). Interestingly, both LG21 and TR-ConA also stain the Δ*cdrA* mutant [26], which is unable to produce the Psl-affiliated matrix component CdrA (Figure 1). This result suggests that LG21 might be able to interact with Psl directly.

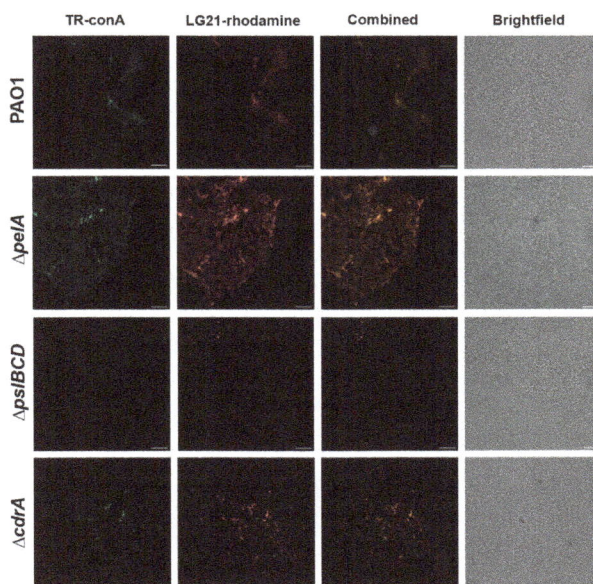

Figure 1. Microscopy images of TR-conA and LG21-rhodamine stained *P. aeruginosa* wild-type and mutant biofilms. Column 1 depicts the confocal images of the red fluorescence representing the presence of TR-conA. Column 2 depicts the confocal images of the green fluorescence representing the presence of LG21-rhodamine. The merged images are shown in column 3. Bright field images of the biofilms are shown in column 4. Labels of the *P. aeruginosa* strains are indicated in the left of each row. The experiments were performed in triplicate, and the representative image of each condition is shown as the result. Scale bar, 10 μm.

2.2. PslG Treatment Abolishes Binding of LG21 to Psl+ Biofilms

To further investigate the binding specificity of LG21 to Psl+ *P. aeruginosa* biofilms, we used rhodamine-tagged LG21 to stain biofilms formed by the WFPA801 strain, a PAO1 derivative strain with an arabinose-inducible *psl* promoter [19]. The WFPA801 strain synthesizes Psl in the presence of arabinose in a dose-dependent manner. Rhodamine-tagged LG21 was found to strongly stain the WFPA801 biofilms cultivated in the presence of 0.5% arabinose and above and the fluorescent signal colocalized well with TR-ConA (Figure 2). Furthermore, we tested whether treatment of the Psl+ biofilms by using PslG, a glycosyl hydrolase that specifically degrades Psl [27], is able to reduce binding of LG21 to *P. aeruginosa* biofilms. As we expected, treatment of the WFPA801 biofilms grown at 2% arabinose by 50 nM PslG for 30 minutes dramatically reduced the binding of Rhodamine-tagged LG21 and TR-ConA to WFPA801 biofilms and a 45-minute treatment by PslG completely abolished the binding (Figure 2).

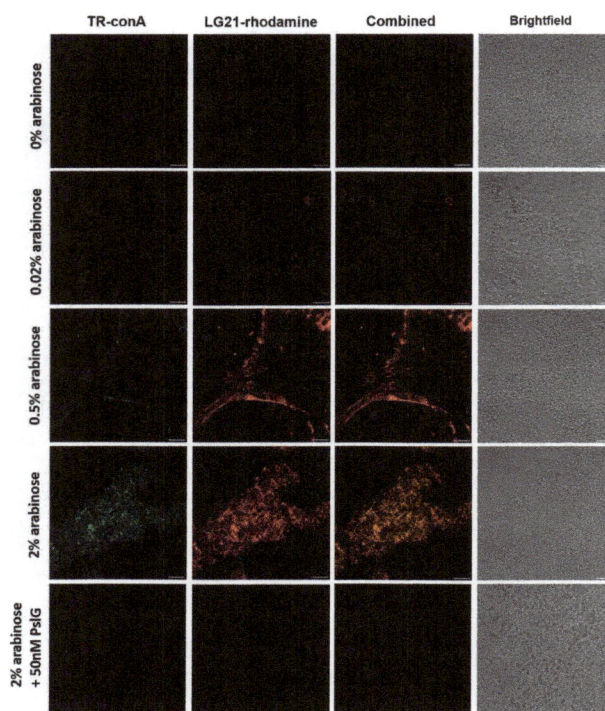

Figure 2. Microscopy images of TR-conA and LG21-rhodamine stained *P. aeruginosa* WFPA801 biofilms with and without treatment. Column 1 depicts the confocal images of the red fluorescence representing presence of TR-conA. Column 2 depicts the confocal images of the green fluorescence representing presence of LG21-rhodamine. The merged images were shown in Column 3. Bright field images of the biofilms are shown in column 4. Labels of the treatment conditions of *P. aeruginosa* WFPA801 biofilms are indicated in the left of each row. The experiments were performed in triplicate, and the representative image of each condition is shown as the result. Scale bar, 10 μm.

2.3. LG21 Interacts with Crude Extracted Psl Exopolysaccharide

Next, we investigated whether LG21 interacts with Psl exopolysaccharide by using the crude extracted Psl from the Pel deficient Δ*pelA* mutant (Pel-Psl+) and crude extracted Pel from the Psl deficient Δ*pslBCD* mutant (Pel+Psl-). Since all our *P. aeruginosa* strains are non-mucoid strains, the production of alginate exopolysaccharide is negligible. We then added LG21 to both crude extracted Psl and Pel, separately, to study the potential interaction between LG21 and exopolysaccharides via fluorescence and NMR methods.

Tryptophan is an excellent intrinsic fluorescent probe to monitor the interactions of peptides with polysaccharides because of its sensitivity towards polarity of the local environment. The fluorescence spectrophotometer data showed that there is a noticeable blue shift of the emission maxima ($\Delta\lambda = 8$ nm) and quenching of fluorescence when LG21 solution is titrated with Psl (Figure 3a). This indicates that LG21 is in the less polar environment and is interacting with Psl. However, addition of Pel to the LG21 solution does not yield any significant changes (Figure 3a). These results suggest that LG21 preferentially interacts with Psl.

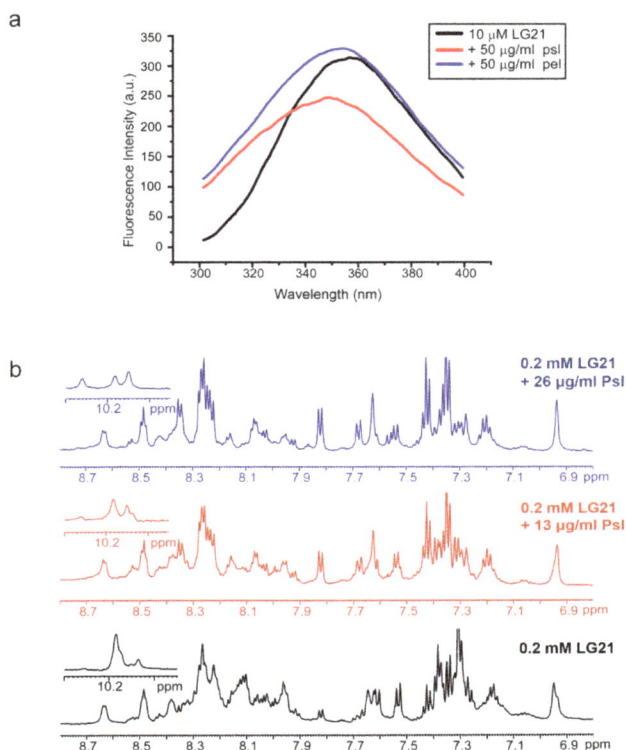

Figure 3. (**a**) Intrinsic tryptophan fluorescence of LG21 on titrating Pel and Psl recorded by fluorescence spectrophotometer; (**b**) 1-D NMR spectra of LG21 with and without addition of Psl. Multiple lines at 10.2 ppm suggested more than 1 conformation changes in LG21 upon binding to Psl.

Furthermore, addition of Psl to the LG21 solution showed conspicuous changes in its 1-D NMR spectra (Figure 3b, top) compared to the LG21 control (Figure 3b, bottom), which suggests Psl is able to interact with LG21. Moreover, the tryptophan NMR signal near 10.2 ppm showed multiple lines, indicating that there is more than one conformation of LG21 when in complex with Psl (Figure 3b, top).

2.4. Molecular Dynamics (MD) Simulation of Psl-LG21 Binding Mode

To monitor the binding of Psl and LG21, the minimum distance between these two molecules was calculated from the MD simulation trajectory. As shown in Figure S1b (see in the Supplementary Materials), the equilibrium distance between these two units is around 0.2 nm, which indicates the binding of the LG21 peptide to the Psl carbohydrate chain. To obtain more insights into the binding of LG21 to Psl, the minimum distances to each residue in the LG21 from Psl were calculated using the last 150 ns simulation trajectory. As shown in Figure 4a, the residues in LG21 numbered from 15 to 21 (WKRKRFG) consist of the main binding region to the Psl. The interaction energy between the LG21 and Psl polysaccharide were also calculated using the last 150 ns simulation frames. The contributions from polar and non-polar interactions (e.g., Van der Waals and hydrophobic interactions) in the interaction energy were calculated separately. In the total interaction energy of ≈ -401 kJ/mol, the contribution from polar interactions is slightly higher (≈ -223 kJ/mol) than that of the non-polar interactions (≈ -178 kJ/mol). The interaction energies of Psl with individual residues of the LG21 are plotted in Figure S3 along with the error bars. It is observed that the minimum interaction energy region in the plot is between the residue numbers 15–21 (WKRKRFG) and also with the residue numbers 7 (N) and 10 (K), which is consistent with the residues at minimum distances in Figure 4a. These regions are with either charged or polar residues which could initiate the hydrogen bonding with the hydrophilic groups of carbohydrate chain. Although, residues numbered from 15 to 21 (WKRKRFG) in LG21 are interacting more strongly with Psl than the residues 1–14, none of the specific residues has distinguishably high interaction energy. The major conformation of peptide is a random-coil in MD simulations, which could be one reason for having more than one residue interacting with Psl. Moreover, from the individual components of total interaction energy it is confirmed that both polar (hydrogen bonding) and non-polar (Van der Waals and hydrophobic) interactions play significant roles in the binding of LG21 with Psl.

a   **b**

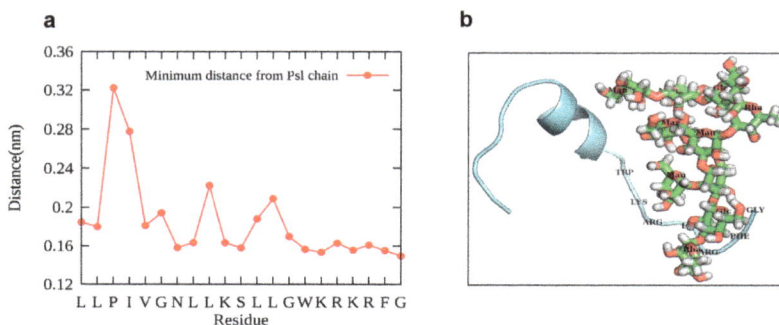

Figure 4. (**a**) The distribution of minimum distance between the Psl chain and each residue in LG21; (**b**) Snapshots of the Psl chain and LG21 peptide from the molecular dynamics (MD) simulation at *t* = 200 ns.

Figure 4b shows the simulation snapshot of the LG21 binding to the Psl at *t* = 200 ns, and the residues in LG21 numbered from 15 to 21 (WKRKRFG) which are at the minimum distances to the Psl are labeled. Based on the qualitative hydrogen-bond analysis and the observations from the trajectory, it is found that the LG21 peptide (residues 15–21) preferably interacts with the region near to the Rhamnose and Glucose monosaccharide groups at the terminal, compared to the Mannose rich region that is slightly hydrophobic in nature [28]. It is also worth noting that the LG21 peptide underwent a conformational change in its secondary structure by some of the residues forming a helical structure from a completely random-coil initial structure. Three independent MD simulations from different initial conformations resulted in the same binding behavior. Figure S4 shows the structural evolution of the LG21 peptide when it interacts with the Psl. It is observed that the middle residues (6–14) are the main contributors to the helical conformation. The reason could be that the presence of carbohydrate chain (Psl) in the solution decreases the polar nature of water to form hydrogen bonds with the middle residues and the increasing hydrophobic interactions within the peptide induces the helical structure, whereas the terminal residues still have a random-coil structure by being exposed to the aqueous solution. The percentage of secondary structure calculated using the total production run of simulation systems with LG21 and Psl in solution, and only LG21 in solution (control simulation) are shown in Table 1. It is observed that in the control simulation, the LG21 peptide is majorly found as a random coil structure. Previous experimental studies based on NMR and CD-spectra also observed that LG21 is a hybrid antimicrobial peptide that exists in a random-coil conformation in aqueous solution [29,30]. From Table 1, it is also observed that for the simulation system with Psl, the percentage of α-helix (≈0.26) is substantially higher than the one observed in the control simulation (≈0.05), which clearly demonstrates that interaction of LG21 with Psl induces the helicity. The forming of the helical conformation in the cationic antimicrobial peptides is observed earlier in the literature, in the presence of carbohydrate chains of biofilm [31].

Table 1. Percentage of secondary structures in the LG21 peptide calculated from systems containing Psl and LG21, and only LG21 (control simulation).

	Coil	Bend	Turn	α-Helix	3-Helix
LG21 + Psl	0.58	0.07	0.07	0.26	0.02
LG21 only	0.58	0.18	0.12	0.05	0.02

3. Conclusions

Exopolysacharides are abundant in bacterial biofilms as a key class of EPS component. Due to their structural complexity, exopolysacharides have distinct structural and functional roles in biofilm formation [17,18]. Psl exopolysacharide serves as the major EPS component for *P. aeruginosa* biofilms and confers resistance towards antibiotic treatment and immune clearance [22,32]. Our recent study showed that Psl attenuated the bactericidal effect of reactive oxidative species to *P. aeruginosa* [33]. The present work showed that Psl is able to interact with AMP LG21. Since AMPs are well known to function as signaling molecules in modulating the host's immunities [34], Psl over-producing clinical *P. aeruginosa* variants might thus be able to impair host immunity via AMP binding. Our work did not identify AMPs that bind to the Pel exopolysaccharide. However, it will be worth investigating AMPs that can target the Pel in the future as several *P. aeruginosa* linages only produce Pel while not Psl. Since AMPs and their mimetics are widely developed and used in different fields, our study suggests that engineering AMPs that target specific biofilm matrix components might facilitate development of strategies to detect and eradicate antibiotic resistant biofilms from both environmental and medical settings.

4. Materials and Methods

4.1. Bacterial Strains, Plasmids, Media and Growth Conditions

Batch cultivation of *P. aeruginosa* strains was carried out at 37 °C in ABT minimal medium [35] supplemented with 2 g L^{-1} glucose (ABTG) or 2 g L^{-1} glucose + 2 g L^{-1} casamino acids (ABTGC). When appropriate, the marker selection in *P. aeruginosa*, 30 µg mL^{-1} gentamicin (Gm), 50 µg mL^{-1} tetracycline (Tc), or 200 µg mL^{-1} carbenicillin (Cb) was used. *P. aeruginosa* slide biofilms were cultivated on the 24 × 50 mm Deckgläser microscope cover glass by inoculating 1:100 diluted overnight *P. aeruginosa* cultures in 50 mL BD falcon tubes (BD Biosciences, Singapore) that contained a cover glass, with 10 mL ABTGC medium. The cultures were then incubated at 37 °C for 24 h.

4.2. Screening of Psl-Binding AMPs

For identifying AMPs that bind to *P. aeruginosa* Psl exopolysaccharide, rhodamine-tagged peptides from a local peptide library were applied against both biofilms formed by *P. aeruginosa* wild-type PAO1 strain and its isogenic Psl defective Δ*pslBCD* mutant [17]. The Psl binding fluorescent stain Texas Red®-conjugated Concanavalin A (TR-ConA) (Molecular Probes, Eugene, OR, USA) was used as a positive control during the screening. The PAO1 and Δ*pslBCD* mutant slide biofilms were developed as described above. The biofilms were washed twice by dipping into a new falcon tube containing sterile 0.9% NaCl to remove the planktonic cells prior to staining with 100 µg mL^{-1} TR-ConA and 8 µM rhodamine-tagged LG21 for 15 min.

To monitor fluorescence of rhodamine (excitation 485 nm/emission 562 nm) [29], and TR-ConA (excitation 595 nm/emission 615 nm), the cells were imaged using an LSM780 confocal laser scanning microscope (CLSM; Carl Zeiss, Jena, Germany) with 100× objective lens and the images were processed using IMARIS software (Bitplane AG, Zurich, Switzerland). Three independent experiments were performed in triplicates and representative images were shown. Rhodamine-tagged AMPs that were able to bind to PAO1 biofilms but not Δ*pslBCD* mutant biofilms were identified and used for further characterization.

4.3. Binding of LG21 to Psl Overproducing Strain before and after PslG Treatment

The *P. aeruginosa* WFPA801 strain that contains an L-arabinose-inducible *psl* operon [36] was used to establish biofilms in ABTGC with 0, 0.02%, 0.5% and 2% L-arabinose at 37 °C for 24 h. To degrade Psl, 50 nM glycosyl hydrolase PslG was added to 24 hour-old WFPA801biofilms cultivated in ABTGC medium containing 2% L-arabinose for 0, 30, 45 min. Rhodamine-tagged LG21 and TR-ConA were applied to the treated WFPA801biofilms, followed by CLSM imaging as described above.

4.4. Crude Extraction of Psl and Pel Exopolysaccharides from P. Aeruginosa

Pel overproducing Δ*wspF*Δ*pslBCD* strain [13] and the Psl overproducing Δ*wspF*Δ*pelA* strain [13] were used for crude extraction of Psl and Pel exopolysaccharides from *P. aeruginosa*, respectively. Exopolysaccharides were extracted from *P. aeruginosa* cultures as previously described [17].

4.5. Tryptophan Fluorescence Spectroscopy

Intrinsic tryptophan fluorescence of 10 μM LG21 in 10 mM phosphate buffer (pH 7) was measured with a Cary Eclipse fluorescence spectrophotometer (Varian Inc., Palo Alto, CA, USA). The peptide samples were titrated with increasing concentrations of Pel or Psl and fluorescence spectra were recorded with excitation at 280 nm and emission at 300 nm–400 nm.

4.6. NMR Analysis

The interaction of LG21 with Psl was studied by performing a series of one-dimensional ^{1}H NMR experiments. LG21 (0.2 mM) in water at a pH 5.5 was titrated with 0, 13 and 26 μg/mL Psl, and ^{1}H NMR spectra were recorded at 298K on Bruker DRX 600 MHz NMR Spectrometer (Bruker Scientific Instruments, Billerica, MA, USA).

4.7. Molecular Dynamics (MD) Simulation of the Interaction between LG21 and Psl Chain

MD simulations were performed in explicit water to study the interaction of LG21 with Psl. The simulation system to study the interaction of LG21 with Psl was generated by placing one Psl chain containing at least two pentasaccharide repeat units [→3)-α-L-RhaP-(1→3)-β-D-Glcp-(1→3)-[α-D-Manp-(1→2)]-β-D-Manp-(1→3)-β-D-Manp-(1→]$_2$ and one LG21 peptide in a simulation box with charge balancing counterions. At the start of the simulation, the carbohydrate chain and peptide are separated by a distance of ≈2 nm as shown in Figure S1a. All simulations were carried out using GROMACS 4.6.5 simulation package [37]. A control simulation was also performed on a similar system as above but containing only LG21 peptide in the solution for comparison of peptide secondary structure evolution. Additional information on the description of the atomic models of Psl and LG21, force fields used and the details of MD simulations are provided in the supporting information.

Supplementary Materials: The following are available online at www.mdpi.com/2073-4441/9/9/681/s1, Figures S1–S4.

Acknowledgments: This research was supported by the National Research Foundation and the Ministry of Education of Singapore under its Research Centre of Excellence Program. YL is supported by the AcRF Tier 1 (2015-T1-002-100) and AcRF Tier 2 (MOE2014-T2-2-172) from Ministry of Education, Singapore.

Author Contributions: S.B. and L.Y. designed the research. J.S.F.C. and J.K.H.Y. performed the biofilm experiments. A.N. and L.L. performed MD simulation. S.S., Z.Q. and S.B. designed the peptides and performed fluorescence and NMR experiments. L.M. and Z.-X.L. analyzed the data. S.B., L.L. and L.Y. wrote the paper. All authors read and approved the final manuscript.

Conflicts of Interest: The authors declare no conflict of interest.

References

1. Pasupuleti, M.; Schmidtchen, A.; Malmsten, M. Antimicrobial peptides: Key components of the innate immune system. *Crit. Rev. Biotechnol.* **2012**, *32*, 143–171. [CrossRef] [PubMed]
2. Rosenfeld, Y.; Shai, Y. Lipopolysaccharide (endotoxin)-host defense antibacterial peptides interactions: Role in bacterial resistance and prevention of sepsis. *Biochim. Biophys. Acta* **2006**, *1758*, 1513–1522. [CrossRef] [PubMed]
3. Lan, Y.; Ye, Y.; Kozlowska, J.; Lam, J.K.; Drake, A.F.; Mason, A.J. Structural contributions to the intracellular targeting strategies of antimicrobial peptides. *Biochim. Biophys. Acta* **2010**, *1798*, 1934–1943. [CrossRef] [PubMed]
4. Alalwani, S.M.; Sierigk, J.; Herr, C.; Pinkenburg, O.; Gallo, R.; Vogelmeier, C.; Bals, R. The antimicrobial peptide ll-37 modulates the inflammatory and host defense response of human neutrophils. *Eur. J. Immunol.* **2010**, *40*, 1118–1126. [CrossRef] [PubMed]
5. Bader, M.W.; Sanowar, S.; Daley, M.E.; Schneider, A.R.; Cho, U.; Xu, W.; Klevit, R.E.; Le Moual, H.; Miller, S.I. Recognition of antimicrobial peptides by a bacterial sensor kinase. *Cell* **2005**, *122*, 461–472. [CrossRef] [PubMed]
6. Li, M.; Lai, Y.; Villaruz, A.E.; Cha, D.J.; Sturdevant, D.E.; Otto, M. Gram-positive three-component antimicrobial peptide-sensing system. *Proc. Natl. Acad. Sci. USA* **2007**, *104*, 9469–9474. [CrossRef] [PubMed]
7. Fernandez, L.; Jenssen, H.; Bains, M.; Wiegand, I.; Gooderham, W.J.; Hancock, R.E. The two-component system cprrs senses cationic peptides and triggers adaptive resistance in pseudomonas aeruginosa independently of parrs. *Antimicrob. Agents Chemother.* **2012**, *56*, 6212–6222. [CrossRef] [PubMed]
8. Thwaite, J.E.; Hibbs, S.; Titball, R.W.; Atkins, T.P. Proteolytic degradation of human antimicrobial peptide ll-37 by bacillus anthracis may contribute to virulence. *Antimicrob. Agents Chemother.* **2006**, *50*, 2316–2322. [CrossRef] [PubMed]
9. Ulvatne, H.; Haukland, H.H.; Samuelsen, O.; Kramer, M.; Vorland, L.H. Proteases in *Escherichia coli* and *Staphylococcus aureus* confer reduced susceptibility to lactoferricin b. *J. Antimicrob. Chemother.* **2002**, *50*, 461–467. [CrossRef] [PubMed]
10. Chua, S.L.; Yam, J.K.; Hao, P.; Adav, S.S.; Salido, M.M.; Liu, Y.; Givskov, M.; Sze, S.K.; Tolker-Nielsen, T.; Yang, L. Selective labelling and eradication of antibiotic-tolerant bacterial populations in pseudomonas aeruginosa biofilms. *Nat. Commun.* **2016**, *7*, 10750. [CrossRef] [PubMed]
11. Flemming, H.C.; Wingender, J. The biofilm matrix. *Nat. Rev. Microbiol.* **2010**, *8*, 623–633. [CrossRef] [PubMed]
12. Seviour, T.; Hansen, S.H.; Yang, L.; Yau, Y.H.; Wang, V.B.; Stenvang, M.R.; Christiansen, G.; Marsili, E.; Givskov, M.; Chen, Y.; et al. Functional amyloids keep quorum-sensing molecules in check. *J. Biol. Chem.* **2015**, *290*, 6457–6469. [CrossRef] [PubMed]
13. Chen, Y.; Yuan, M.; Mohanty, A.; Yam, J.K.; Liu, Y.; Chua, S.L.; Nielsen, T.E.; Tolker-Nielsen, T.; Givskov, M.; Cao, B.; et al. Multiple diguanylate cyclase-coordinated regulation of pyoverdine synthesis in pseudomonas aeruginosa. *Environ. Microbiol. Rep.* **2015**, *7*, 498–507. [CrossRef] [PubMed]
14. Driscoll, J.A.; Brody, S.L.; Kollef, M.H. The epidemiology, pathogenesis and treatment of pseudomonas aeruginosa infections. *Drugs* **2007**, *67*, 351–368. [CrossRef] [PubMed]
15. Bjarnsholt, T. The role of bacterial biofilms in chronic infections. *APMIS* **2013**, *121*, 1–58. [CrossRef] [PubMed]
16. Tolker-Nielsen, T. Pseudomonas aeruginosa biofilm infections: From molecular biofilm biology to new treatment possibilities. *APMIS* **2014**, *122*, 1–51. [CrossRef] [PubMed]
17. Yang, L.; Hu, Y.; Liu, Y.; Zhang, J.; Ulstrup, J.; Molin, S. Distinct roles of extracellular polymeric substances in pseudomonas aeruginosa biofilm development. *Environ. Microbiol.* **2011**, *13*, 1705–1717. [CrossRef] [PubMed]
18. Chew, S.C.; Kundukad, B.; Seviour, T.; van der Maarel, J.R.; Yang, L.; Rice, S.A.; Doyle, P.; Kjelleberg, S. Dynamic remodeling of microbial biofilms by functionally distinct exopolysaccharides. *MBio* **2014**, *5*. [CrossRef] [PubMed]
19. Ma, L.; Conover, M.; Lu, H.; Parsek, M.R.; Bayles, K.; Wozniak, D.J. Assembly and development of the pseudomonas aeruginosa biofilm matrix. *PLoS Pathog.* **2009**, *5*, e1000354. [CrossRef] [PubMed]
20. Haussler, S.; Tummler, B.; Weissbrodt, H.; Rohde, M.; Steinmetz, I. Small-colony variants of pseudomonas aeruginosa in cystic fibrosis. *Clin. Infect. Dis.* **1999**, *29*, 621–625. [PubMed]

21. Haussler, S.; Ziegler, I.; Lottel, A.; von Gotz, F.; Rohde, M.; Wehmhohner, D.; Saravanamuthu, S.; Tummler, B.; Steinmetz, I. Highly adherent small-colony variants of pseudomonas aeruginosa in cystic fibrosis lung infection. *J. Med. Microbiol.* **2003**, *52*, 295–301. [CrossRef] [PubMed]

22. Mishra, M.; Byrd, M.S.; Sergeant, S.; Azad, A.K.; Parsek, M.R.; McPhail, L.; Schlesinger, L.S.; Wozniak, D.J. Pseudomonas aeruginosa psl polysaccharide reduces neutrophil phagocytosis and the oxidative response by limiting complement-mediated opsonization. *Cell. Microbiol.* **2012**, *14*, 95–106. [CrossRef] [PubMed]

23. Byrd, M.S.; Sadovskaya, I.; Vinogradov, E.; Lu, H.; Sprinkle, A.B.; Richardson, S.H.; Ma, L.; Ralston, B.; Parsek, M.R.; Anderson, E.M.; et al. Genetic and biochemical analyses of the pseudomonas aeruginosa psl exopolysaccharide reveal overlapping roles for polysaccharide synthesis enzymes in psl and lps production. *Mol. Microbiol.* **2009**, *73*, 622–638. [CrossRef] [PubMed]

24. Kocharova, N.A.; Knirel, Y.A.; Shashkov, A.S.; Kochetkov, N.K.; Pier, G.B. Structure of an extracellular cross-reactive polysaccharide from pseudomonas aeruginosa immunotype 4. *J. Biol. Chem.* **1988**, *263*, 11291–11295. [PubMed]

25. Tran, C.S.; Rangel, S.M.; Almblad, H.; Kierbel, A.; Givskov, M.; Tolker-Nielsen, T.; Hauser, A.R.; Engel, J.N. The pseudomonas aeruginosa type iii translocon is required for biofilm formation at the epithelial barrier. *PLoS Pathog.* **2014**, *10*, e1004479. [CrossRef] [PubMed]

26. Borlee, B.R.; Goldman, A.D.; Murakami, K.; Samudrala, R.; Wozniak, D.J.; Parsek, M.R. Pseudomonas aeruginosa uses a cyclic-di-gmp-regulated adhesin to reinforce the biofilm extracellular matrix. *Mol. Microbiol.* **2010**, *75*, 827–842. [CrossRef] [PubMed]

27. Yu, S.; Su, T.; Wu, H.; Liu, S.; Wang, D.; Zhao, T.; Jin, Z.; Du, W.; Zhu, M.J.; Chua, S.L.; et al. Pslg, a self-produced glycosyl hydrolase, triggers biofilm disassembly by disrupting exopolysaccharide matrix. *Cell Res.* **2015**, *25*, 1352–1367. [CrossRef] [PubMed]

28. Janado, M.; Yano, Y. Hydrophobic nature of sugars as evidenced by their differential affinity for polystyrene gel in aqueous-media. *J. Solut. Chem.* **1985**, *14*, 891–902. [CrossRef]

29. Mohanram, H.; Bhattacharjya, S. Resurrecting inactive antimicrobial peptides from the lipopolysaccharide trap. *Antimicrob. Agents Chemother.* **2014**, *58*, 1987–1996. [CrossRef] [PubMed]

30. Mohanram, H.; Bhattacharjya, S. 'Lollipop'-shaped helical structure of a hybrid antimicrobial peptide of temporin b-lipopolysaccharide binding motif and mapping cationic residues in antibacterial activity. *Biochim. Biophys. Acta* **2016**, *1860*, 1362–1372. [CrossRef] [PubMed]

31. Chan, C.; Burrows, L.L.; Deber, C.M. Helix induction in antimicrobial peptides by alginate in biofilms. *J. Biol. Chem.* **2004**, *279*, 38749–38754. [CrossRef] [PubMed]

32. Billings, N.; Millan, M.; Caldara, M.; Rusconi, R.; Tarasova, Y.; Stocker, R.; Ribbeck, K. The extracellular matrix component psl provides fast-acting antibiotic defense in pseudomonas aeruginosa biofilms. *PLoS Pathog.* **2013**, *9*, e1003526. [CrossRef] [PubMed]

33. Chua, S.L.; Ding, Y.; Liu, Y.; Cai, Z.; Zhou, J.; Swarup, S.; Drautz-Moses, D.I.; Schuster, S.C.; Kjelleberg, S.; Givskov, M.; et al. Reactive oxygen species drive evolution of pro-biofilm variants in pathogens by modulating cyclic-di-gmp levels. *Open Biol.* **2016**, *6*, 160162. [CrossRef] [PubMed]

34. Brown, K.L.; Hancock, R.E. Cationic host defense (antimicrobial) peptides. *Curr. Opin. Immunol.* **2006**, *18*, 24–30. [CrossRef] [PubMed]

35. Chua, S.L.; Liu, Y.; Yam, J.K.; Chen, Y.; Vejborg, R.M.; Tan, B.G.; Kjelleberg, S.; Tolker-Nielsen, T.; Givskov, M.; Yang, L. Dispersed cells represent a distinct stage in the transition from bacterial biofilm to planktonic lifestyles. *Nat. Commun.* **2014**, *5*, 4462. [CrossRef] [PubMed]

36. Ma, L.; Jackson, K.D.; Landry, R.M.; Parsek, M.R.; Wozniak, D.J. Analysis of pseudomonas aeruginosa conditional psl variants reveals roles for the psl polysaccharide in adhesion and maintaining biofilm structure postattachment. *J. Bacteriol.* **2006**, *188*, 8213–8221. [CrossRef] [PubMed]

37. Pronk, S.; Pall, S.; Schulz, R.; Larsson, P.; Bjelkmar, P.; Apostolov, R.; Shirts, M.R.; Smith, J.C.; Kasson, P.M.; van der Spoel, D.; et al. Gromacs 4.5: A high-throughput and highly parallel open source molecular simulation toolkit. *Bioinformatics* **2013**, *29*, 845–854. [CrossRef] [PubMed]

MDPI

St. Alban-Anlage 66

4052 Basel

Switzerland

Tel. +41 61 683 77 34

Fax +41 61 302 89 18

www.mdpi.com

Water Editorial Office

E-mail: water@mdpi.com

www.mdpi.com/journal/water

www.ingramcontent.com/pod-product-compliance
Lightning Source LLC
Chambersburg PA
CBHW051855210326
41597CB00033B/5902